Shaping Space

A Polyhedral Approach

Design Science Collection

Series Editor
Arthur L. Loeb
Department of Visual and Environmental Studies
Harvard University

Amy C. Edmondson	*A Fuller Explanation: The Synergetic Geometry of R. Buckminster Fuller*, 1987
Marjorie Senechal and George Fleck (Eds.)	*Shaping Space: A Polyhedral Approach*, 1988
Judith Wechsler (Ed.)	*On Aesthetics in Science*, 1988
Arthur L. Loeb	*Concepts and Images*, in preparation

Marjorie Senechal and George Fleck
Editors

Shaping Space

A Polyhedral Approach

With 188 Line and 174 Halftone Illustrations

A Pro Scientia Viva Title

Birkhäuser
Boston · Basel

Shaping Space: A Polyhedral Approach
Edited by Marjorie Senechal and George Fleck

CODEN: DSCOED
First printing, 1988

AMS Subject Classifications (1980): *Primary:* 51-01, 51-02, 51A25, 51M20; *Secondary:* 00A06, 00A11, 00A25, 00A69, 01-01, 82A60, 92A40

Library of Congress Cataloging-in-Publication Data
Shaping space.
(Design science collection)
"A pro scientia viva title."
Includes material from the Shaping Space Conference held at Smith College, Apr. 6--8, 1984.
Includes index.
1. Visual perception. 2. Space (Art) 3. Form (Aesthetics) 4. Polygons. 5. Polyhedra. 6. Polytopes.
I. Senechal, Marjorie. II. Fleck, George M. III. Shaping Space Conference (1984 : Smith College) IV. Series.
N7430.5.S52 1988 701 87-10378

CIP-Kurztitelaufnahme der Deutschen Bibliothek
Shaping space : a polyhedral approach / ed. by Marjorie Senechal ; George Fleck.—Boston ; Basel : Birkhäuser, 1988.
(A pro scientia viva title) (Design science collection)
ISBN 3-7643-3351-0 (Basel)
ISBN 0-8176-3351-0 (Boston)
NE: Senechal, Marjorie [Hrsg.]
-1. print.

Figures 1-38, 1-39, and 1-40 are from David Plowden, *Bridges: The Spans of North America* (New York: Viking Press, 1974; W. W. Norton and Co., 1984) and are reprinted by permission of Curtis Brown, Ltd. Copyright © David Plowden 1974. Other acknowledgments are given in the text.

ISBN 0-8176-3351-0
3-7643-3351-0

Typeset by Bi-Comp, Inc., York, Pennsylvania.
Printed and bound by Arcata Graphics/Halliday, West Hanover, Massachusetts.
Manufactured in the United States of America.

Dedicated to the Memory of

Elizabeth Angela McBeath
1960–1981

A.B., Smith College, conferred posthumously, 1982

Contents

Part III. Roles of Polyhedra in Science

Part IV. Theory of Polyhedra

Series Editor's Foreword

In a broad sense design science is the grammar of a language of images rather than of words. Modern communication techniques enable us to transmit and reconstitute images without needing to know a specific verbal sequential language such as the Morse code or Hungarian. International traffic signs use international image symbols which are not specific to any particular verbal language. An image language differs from a verbal one in that the latter uses a linear string of symbols, whereas the former is multidimensional.

Architectural renderings commonly show projections onto three mutually perpendicular planes, or consist of cross sections at different altitudes capable of being stacked and representing different floor plans. Such renderings make it difficult to imagine buildings comprising ramps and other features which disguise the separation between floors, and consequently limit the creative process of the architect. Analogously, we tend to analyze natural structures as if nature had used similar stacked renderings, rather than, for instance, a system of packed spheres, with the result that we fail to perceive the system of organization determining the form of such structures.

Perception is a complex process. Our senses record; they are analogous to audio or video devices. We cannot, however, claim that such devices perceive. Perception involves more than meets the eye: it involves processing and organization of recorded data. When we name an object, we actually name a concept: such words as *octahedron, collage, tessellation, dome,* each designate a wide variety of objects sharing certain characteristics. When we devise ways of transforming an octahedron, or determine whether a given shape will tessellate the plane, we make use of these characteristics, which constitute the grammar of structure.

The Design Science Collection concerns itself with various aspects of this grammar. The basic parameters of structure such as symmetry, connectivity, stability, shape, color, size, recur throughout these volumes. Their interactions are complex; together they generate such concepts as Fuller's and Snelson's tensegrity, Lois Swirnoff's modulation of surface through color, self-reference in the work of M. C. Escher, or the synergetic stability of ganged unstable polyhedra. All of these occupy some of the professionals concerned with the complexity of the space in which we live, and which we shape. The Design Science Collection is intended to inform a reasonably well educated but not highly specialized audience of these professional activities, and particularly to illustrate and to stimulate the interaction between the various disciplines involved in the exploration of our own three-dimensional, and, in some instances, more-dimensional spaces.

Shaping Space is a polyhedral anthology. Like the conference of the same name which inspired it, it is polyglot and polydisciplinary. It is unlikely that as many scholars and artists actively involved with polyhedra will be together again in the near future; it was therefore deemed important to leave a physical imprint of the event by means of publication of this book, and to share its concerns with a broader audience. The volume reflects the exuberance of the conveners of that conference, George Fleck and Marjorie Senechal. Its high picture-to-word ratio characterizes the visual character of the gathering. Senechal and Fleck strived to preserve much of the spirit of the event without producing a mere transcript of the proceedings (which, at best, would have been a pale reflection). To this purpose the editors, with the cooperation of the contributors, adapted and modified the original contributions. It is hoped that the reader will not remain passive, but will be stirred into action by the recipes, and will be challenged by some of the as yet unsolved problems. The conference was a participatory one; the book should be the same.

Shaping Space addresses itself to designers, artists, architects, engineers, chemists, mathematicians, bioscientists, crystallographers, earth scientists—in short, to all scholars and educators interested in, and working with, two- and three-dimensional structures and patterns. There is a broad range of abstraction; some readers may find the more mathematical chapters challenging, but the editors have endeavored to entice the readers to rise to that challenge. Conversely, we know that at least one of the contributing mathematicians took off time from his purely mathematical research to gain "hands-on" experience in building polyhedral models.

Shaping Space celebrates the coming of age of polyhedrics; with it we must note with a certain regret the passing of youth. A discipline experiences sigmoidal growth: a slow initial phase followed by a period of rapid growth, and finally saturation characterized by an asymptotically decreasing growth rate. During the initial slow-growth phase workers tend to operate individually and in isolation. As pioneers they must fashion their own tools, and their products may in retrospect appear crude and dilettantish. We should not forget, however, that asking the right significant questions is usually more difficult than finding the answers to these questions, and that the pioneers generate the questions which characterize the discipline. Rapid growth usually results when the isolated groups find each other and communicate; the conference at Smith College catalyzed such communication. In this phase the "professionals" from older disciplines enter to help in finding answers. Thus we find, in *Shaping Space*, some mathematicians referring to "folklore." The mathematicians help codify the discipline, identify (and sometimes fill) the gaps, and ask new questions generated by the solutions to the old ones. On the other hand, every discipline needs cross-fertilization for long-term survival. When the folklorists cease to come up with significant questions, the discipline will settle into its terminal saturation phase. It is evident from this volume that polyhedrics is not yet prepared to enter old age.

Cambridge, Massachusetts ARTHUR L. LOEB

Preface

Shaping Space: A Polyhedral Approach is inspired by the Shaping Space Conference, which was held at Smith College on April 6–8, 1984. The conference was very successful, attesting to the aesthetic and intellectual appeal of its subject matter. It attracted a broad audience of all ages (see Figs. P-1–P-3) including elementary school, middle school, high school, college, and graduate students, teachers, artists, scientists, mathematicians, engineers, architects, model-building enthusiasts, mystics, and townspeople. The three days of exhibits, workshops, and lectures were designed for a wide variety of interests and levels of expertise.

The exhibits ranged from a display of polyhedra made by children at the Smith College Campus School (see Fig. P-4) through the beautiful multicolored creations of Morton Bradley, Jr. (see Figs. P-5, P-6). They included several rooms of models brought by conference participants, and a very impressive exhibit of fake artifacts from the history of polyhedra, "discovered" by a group of Smith College mathematics majors. Photographs of many of the models exhibited at the conference are included in this volume. Five of the workshops are recreated in Part I. The main lectures of the conference are presented (in slightly adapted form) in Part II. Parts III and IV deal with research material

Fig. P-1. Brian Julin of Holyoke, Massachusetts. Photograph by Stan Sherer.

Fig. P-2. Audience at a general session of the Shaping Space Conference. Photograph by Stan Sherer.

Fig. P-3. R. M. Langer of Arlington, Massachusetts. Photograph by Stan Sherer.

presented in the conference sessions titled "Applications of Polyhedra" and "Theory of Polyhedra."

This book, however, is much more than the proceedings of the Shaping Space Conference. In addition to trying to translate the spirit and substance of the conference onto the printed page, we have added material not presented at the conference itself to ensure that the book will be accessible to a broad audience. The book also reflects the continuing evolution of our ideas of the

Fig. P-4. A decorated icosahedron in the Campus School exhibit. Photograph by Stan Sherer.

Fig. P-5. Morton Bradley sculptures, exhibited in the Clark Science Center at the Shaping Space Conference, courtesy of the artist. Photograph by Stan Sherer.

roles that three-dimensional geometry can play in the curriculum at all levels (see Part V).

We hope you will join us on this exploration of the world of polyhedra, beginning with an introductory "Visit to the Polyhedron Kingdom" (Chapter 1) and concluding with an examination of the significance of polyhedral models in contemporary science and a survey of some recent advances and unsolved problems in mathematics.

Fig. P-6. Morton Bradley, Jr. Photograph by Stan Sherer.

The conference was the first stage of a three-part project. The publication of this book completes the second stage, and we will soon embark on the third stage, the development of an interdisciplinary high school geometry course which emphasizes the study of three-dimensional forms and their roles in science, technology, mathematics, and art.

We are pleased to have this opportunity to acknowledge the advice and assistance which we have received from our colleagues, students, and other friends at every stage of this project, including A. Lee Burns of the Smith College Art Department, who co-organized the Shaping Space Conference with us (Robert Whorf, the creator of *Symmetrics*, was also a co-organizer; his untimely death has saddened all who knew him); the conference speakers and other participants; and the many people who have sent us suggestions for this book. H. S. M. Coxeter, A. C. Laan, R. O. Erickson, Joseph Malkevitch, Godfried Toussaint, Gerry Segal, and Timothy Brown read portions of early drafts of the manuscript and offered helpful criticisms. Special thanks are due to Wendy Klemyk, Smith College Class of 1987, without whose dedicated and efficient assistance this book would never have been completed; to Stan Sherer for conveying the breadth and spirit of the conference through his expert photography; and to Arthur Loeb for his encouragement and advice over many years. The financial support of the National Science Foundation (Grant DPE 84-00339) is gratefully acknowledged.

Northampton, Massachusetts

MARJORIE SENECHAL
GEORGE FLECK

Part I
The First Steps

1

A Visit to the Polyhedron Kingdom

Marjorie Senechal

What is a polyhedron? If you would like to know, we invite you to join us on a fanciful visit to the Polyhedron Kingdom. Although you may not have heard of it before, you will find that this Kingdom is nearly as vast, and as varied, as the animal, mineral, and vegetable kingdoms (and that it overlaps all three of them). You will meet aristocrats and workers, families and individuals, old polyhedra with long and interesting histories and young polyhedra who were born yesterday or the day before. You will even catch a glimpse of some polyhedral ghosts who live in four-dimensional space. You will take a brief walking tour of polyhedral architecture, visit a nature preserve and an art gallery, and end the visit by browsing at an artisans' polyhedra fair.

The boundaries of the Polyhedron Kingdom are in dispute (as are those of most kingdoms) but it is safe to visit the border areas. You need not worry about the nature of the disputes until later in this book.

The language of the Polyhedron Kingdom is mathematics, but for this brief first visit you

Fig. 1-1. Cube with face, by a fifth-grade student at the Smith College Campus School, exhibited at the Shaping Space Conference. Photograph by Stan Sherer.

Fig. 1-2. A polyhedral monster, also in the Campus School exhibit. Photograph by Stan Sherer.

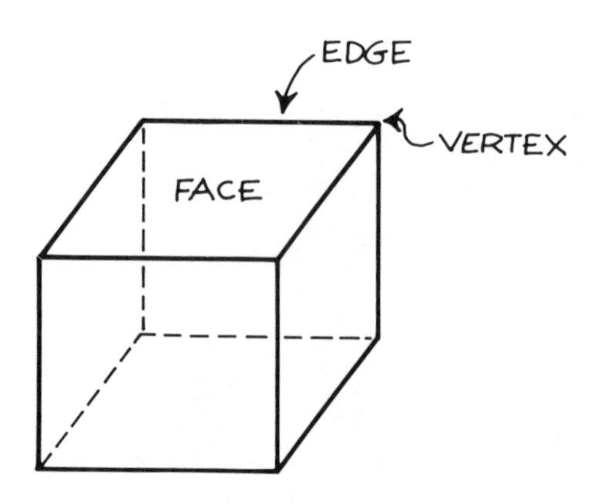

Fig. 1-3. The cube has six faces, twelve edges, and eight vertices.

can get by if you learn three important words: face, edge, and vertex. The word *polyhedron* comes from the Greek and means an object with many faces. In Figs. 1-1 and 1-2 we see polyhedra with faces. But this is not what we mean when we speak of the faces of a polyhedron. For our purposes, the *faces* of a polyhedron are the polygons from which it is constructed. The *edges* of a polyhedron are the lines bounding its faces; its *vertices* are the corners where three or more faces (and thus three or more edges) meet (Fig. 1-3). You will see as we go along that these terms can have somewhat more general meanings, but the definitions just given are adequate for the moment. As you tour the Polyhedron Kingdom, you will become more comfortable with an increasing vocabulary and with a wider range of common usages.

We begin our tour with a visit to the rulers of the Kingdom.

The Regular "Solids"

At the gates of the Kingdom live its rulers, the famous and venerable regular "solids" pictured in Fig. 1-4. Each of these polyhedra is called *regular* because of certain very special properties: its faces are identical regular polygons,[1] and the same number of polygons meet at each vertex. So the faces of each polyhedron are all alike and their vertices (or, more precisely, the arrangements of polygons at their vertices) are all alike.

There is a simple argument, given by Euclid (300 B.C.), which shows that there are only five regular polyhedra. Let's try to build polyhedra with the regularity property just described; we will quickly find that there are only five possibilities. We start by constructing polyhedra whose faces are equilateral triangles. First, we can put three triangles together to form one vertex of a polyhedron. If we continue this pattern at all the other corners we obtain a pyramid that has four triangular faces, four vertices, and six edges; this is the regular *tetrahedron* (Fig. 1-4a). If we put four triangles at each vertex, we can build an *octahedron* (Fig. 1-4b); if we put five together then we get the *icosahedron* (Fig. 1-4c). Six equilateral triangles fit together around a point to form a plane surface, not a closed polyhedron. And if we try to fit seven or more together—well, try it and see what happens! So these three polyhedra are the only regular ones that can be built out of equilateral triangles. Now let us try to build a regular polyhedron out of squares. We see that there is just one possibility, the *cube* (Fig. 1-4d), in which three faces meet at each vertex, because four squares in a plane fit together around a point. (What happens if we try to fit five?) If we use regular pentagons, we can again build just one solid, the *pentagonal dodecahedron* (Fig. 1-4e). We cannot continue this procedure with regular polygons with a greater number of sides because three regular hexagons lie flat, three or more heptagons or octagons buckle, and so forth. We conclude that there are no other regular polyhedra.

The regular polyhedra are also known as the "Platonic solids" because the Greek philosopher Plato (427–347 B.C.) immortalized them in his dialogue *Timaeus*. In this dialogue Plato discussed his ideas about the "elements" of which he believed the universe to be composed: earth, air, fire, and water. Today when we think of "element," we usually think of the chemical elements in the Periodic Table. (We recognize the solid, gas, plasma, and liquid *states* of matter.) But notice that we still speak of needing protection from the "elements," and when we say this we mean snow, wind, lightning, and rain. In *Timaeus,* Plato argued that the geometric forms of the smallest particles of the elements are the cube, the octahedron, the tetrahedron, and the icosahedron, respectively. (The fifth regular solid, the dodecahedron, was assigned to the Great All, the cosmos.) This association of the regular solids with the elements has captured the imagination of many people from Plato's time to this. We see the interpretation of the astronomer Johannes Kepler (1571–1630) in Fig. 10.32. In our own time, the artist M. C. Escher has presented it in various ways. Figure 1-5 shows an icosahedral candy box decorated by Escher. Figure 1-6 might be subtitled "Platonic Puzzle," because all of the five Platonic solids appear in it in one form or another!

Plato aside, do the regular polyhedra have any special significance outside the Polyhedron Kingdom? Maybe not. About four hundred years ago, the astronomer Johannes Kepler believed that he had at last discovered their true meaning. He wrote that the spheres in which they can be inscribed, nested one inside another, were the divine model for the orbits of the six planets. This was the reason why there were only six planets! (Kepler's ideas are discussed in detail by H. S. M. Coxeter in Chapter 3.) The beauty of the regular

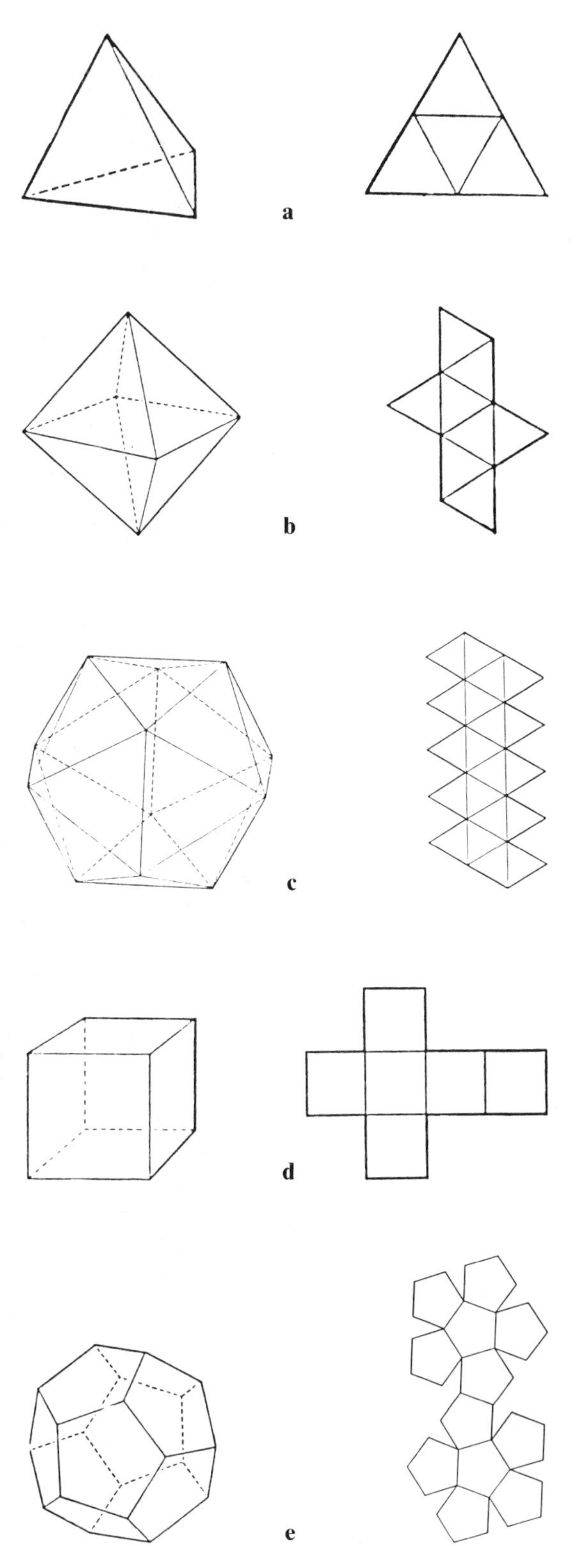

Fig. 1-4. The regular polyhedra and their plane nets. From *Mathematical Models* by Henry Martyn Cundy and A. P. Rollett, 2nd edition, 1981, published by Oxford University Press.

Fig. 1-5. *Icosahedron with Starfish and Shells,* a candy box by M. C. Escher. From Bruno Ernst, *The Magic Mirror of M. C. Escher* (New York: Ballantine Books, 1976). © M. C. Escher Heirs c/o Cordon Art–Baarn–Holland.

Fig. 1-6. *Reptiles.* Woodcut by M. C. Escher. © M. C. Escher Heirs c/o Cordon Art–Baarn–Holland.

polyhedra has led scientists astray in our own time as well. A patterned octahedron was the first model proposed for the molecular structure of proteins, by Dorothy Wrinch in 1934 (Fig. 1-14); unfortunately the structures of proteins have turned out to be much less elegant (see p. 10).

Still, once you have become acquainted with them, you will find that you meet the regular polyhedra in the most unexpected places:

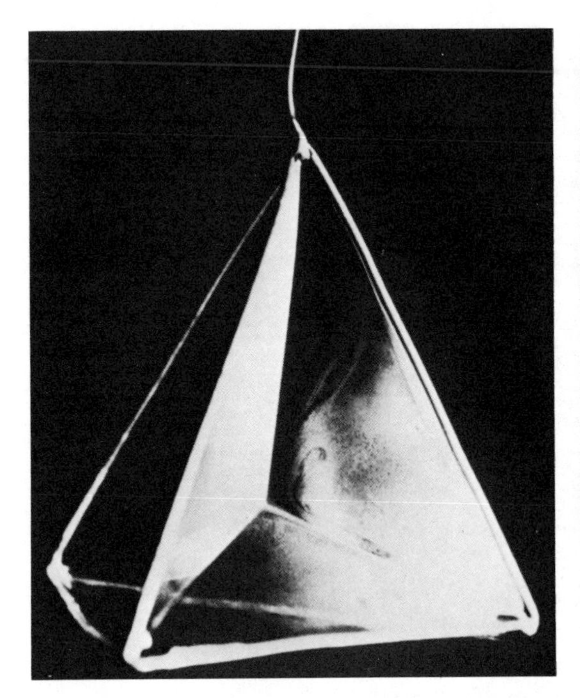

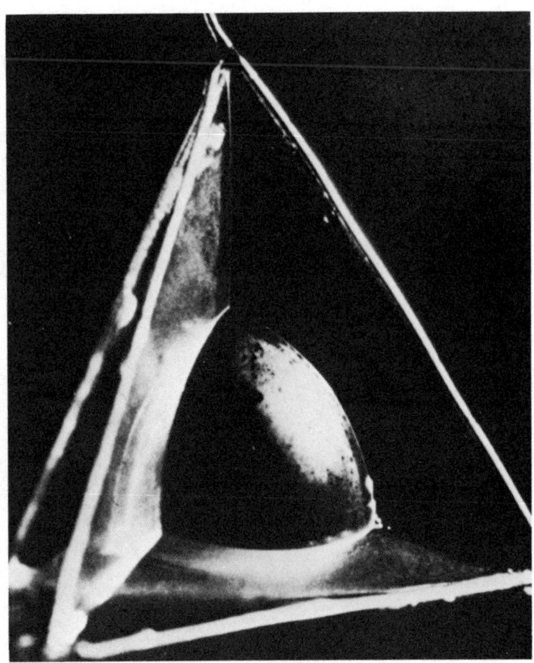

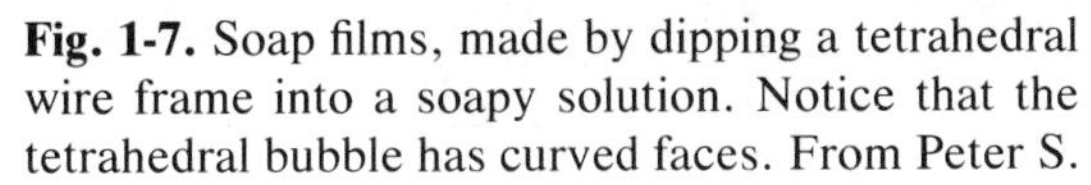

Fig. 1-7. Soap films, made by dipping a tetrahedral wire frame into a soapy solution. Notice that the tetrahedral bubble has curved faces. From Peter S. Stevens, *Patterns in Nature*. (Boston: Little, Brown and Company, 1974). Reprinted by permission.

for example, in the soap films shown in Fig. 1-7 (if we agree that a polyhedron can have curved faces and edges), in decorative ornament (Fig. 1-8), in ordinary viruses (Fig. 1-9) and, if we agree that edges and vertices alone can constitute a polyhedron, perhaps even in outer space (Fig. 1-10). The shapes of many molecules are thought to be closely related to the regular polyhedra (Fig. 1-12). Many crystals have cubic, octahedral, or dodecahedral forms; others are tetrahedral or icosahedral. But most dodecahedral (and icosahedral) crystals, like the pyrite crystals in Fig. 1-11, are not regular. (Indeed, until November 1984, it was believed that regular dodecahedral and icosahedral crystals could not exist, because their symmetry is theoretically impossible for a crystal. Then some crystals with this symmetry were discovered, posing some challenging problems for symmetry theory!) Perhaps to make up for its limited role in the mineral kingdom, the regular dodecahedron with its twelve faces has been used by people in imaginative ways, such as streetcorner recycling bins in France (Fig. 1-13).

Today we believe that it is not the classical form of the regular polyhedra that is significant: instead it is the high degree of order which they represent. Indeed, by now you have noticed that the regular "solids" are not always found in solid form. In some contexts, they have hollow interiors; in others, they have perforated surfaces; in yet others they have no faces, but appear as skeletons made of edges and vertices. Still, they are usually recognizable because of their high degree of *symmetry*. For example, all of the regular polyhedra have *mirror symmetry*: they can be divided into mirror-image halves in many different ways. They also have *rotational symmetry*: there are many ways in which they can be rotated without changing their apparent position. Both the mirror symmetry and rotational symmetry are due to the fact that, for each of these polyhedra, every face, every vertex, and every edge is like every other. In other words, they are highly organized; this is one of the reasons that they are found so often in nature. This organization is also aesthetically pleasing, and it is largely because of their

Fig. 1-8. The icosahedron and other polyhedra often appear as decorative elements in Baroque architecture; here, the church of Santissimi Apostoli by Borromini. From Paolo Portoghesi, *The Rome of Borromini: Architecture as Language* (New York: George Braziller, 1968). Photograph by courtesy of Electa, Milano.

TUESDAY, FEBRUARY 12, 1985

Science Times

With Education, Style, Arts

The New York Times

C1

The Many Guises of Papilloma Virus

Only recently has it been possible to differentiate chemically various types of human papilloma viruses. Following are some types that have been identified and the parts of the body that they infect. A few have been closely associated with human cancers, although they have not been proved to cause cancer.

TYPE 1: Plantar warts (sole of foot)
TYPE 2: Common warts of the hands
TYPE 3: Flat warts of the face, particularly in young people
TYPE 4: Planar (hands) and plantar (feet) warts
TYPE 6: Condyloma, a kind of wart of the female genital tract
TYPE 7: Common warts of the hands of butchers, meat handlers
TYPE 11: Laryngeal papillomas, wartlike growths in the larynx that can be life threatening because they grow rapidly and can shut off the flow of air
TYPE 13: Flat wartlike lesions of the oral cavity (mouth)
TYPE 16 and TYPE 18: Have been associated with cervical cancers.

Several other types, 5, 8, 9, 10, 12, 14, 15, 17, 19 and 26, have been associated with a rare skin disease called epidermodysplasia verruciformis, which has also been linked with the later development of cancer. At present there are 31 known human papilloma virus types.

Common Virus Linked to Cancer

New findings are expected to affect efforts to fight genital afflictions.

By HAROLD M. SCHMECK Jr.

VIRUSES related to those that cause common warts are coming under increasing suspicion as factors in causing human cancer of the cervix and other types of malignancy.

Scientists in the United States and abroad are finding strong evidence linking some of the viruses in a family called papilloma with genital cancers, notably cancers of the cervix and vulva, and with an uncommon kind of skin cancer.

The recent findings seem largely to eclipse earlier evidence linking genital cancers to herpes viruses. The virus called herpes simplex or herpes hominus type 2 has been under study for decades as a possible cause of genital cancers, but scientists say the evidence linking papilloma viruses with these cancers is now far stronger than the evidence relating to herpes.

The new findings are expected to have an important impact on efforts to prevent and treat genital cancers, particularly in women. Many specialists believe that in time it should be possible to develop a vaccine against the types of papilloma virus that are under the greatest suspicion.

Scientists do not yet claim they have proved papilloma viruses cause human genital cancers. In recent interviews, however, several of the leaders in the research said the evidence available today comes as close to proving a cause and effect relationship as is possible short of developing a vaccine against the virus and proving that it prevents the cancers. Efforts to develop such a vaccine are expected in the near future, but are likely to take several years.

It used to be thought that papilloma viruses in humans cause only noncancerous growths, mainly various kinds of warts. This benign nature still holds true of many of the known types of papilloma virus, but in a few types the links to cancer are described as strong. The new evidence has come as a surprise to many. Indeed one research worker said several of his past proposals for study of links between papilloma viruses and cancer were turned down by funding agencies as unpromising.

The National Cancer Institute has set up a special laboratory, headed by Dr. Peter M. Howley, to study the links between some of these viruses and cancer. "The institute has made a major commitment to the papilloma viruses and their role in cervical carcinoma and human cancers in general," Dr. Howley said.

Some of the most compelling recent evidence on the relationships of papilloma viruses to cancer in women was found recently by medical scientists of Georgetown University. They detected traces of papilloma viruses of several types now under suspicion in 146 patients, according to a report the group gave recently at a meeting in Miami of the Society of Gynecologic Oncologists. Leaders of the research team are Dr. A. Bennett Jenson, Dr. Wayne D. Lancaster, Dr. Robert J. Kurman and Dr.

Continued on Page C[illegible]

Fig. 1-9. The icosahedron is a common form of viruses. From *The New York Times*, Tuesday, February 12, 1985. © 1985 by The New York Times Company. Reprinted by permission.

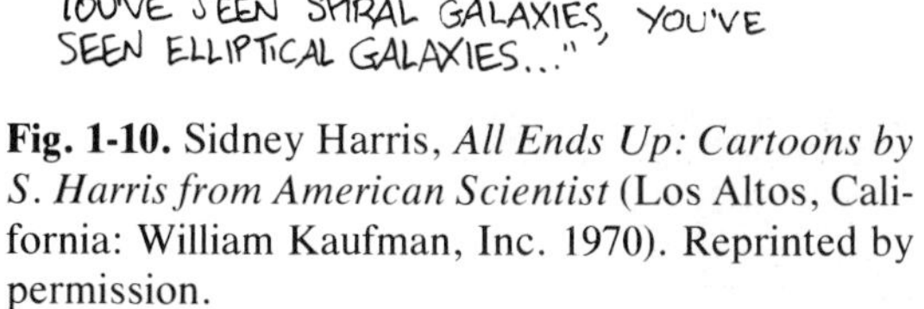

Fig. 1-10. Sidney Harris, *All Ends Up: Cartoons by S. Harris from American Scientist* (Los Altos, California: William Kaufman, Inc. 1970). Reprinted by permission.

Fig. 1-11. Pyrite crystals. From Cedric Rogers, *Rocks and Minerals*. London, Triune Books, 1973. Used by permission.

Fig. 1-12. An artist's conception of a methane molecule. From Linus Pauling and Roger Hayward, *The Architecture of Molecules,* San Francisco, W. H. Freeman, 1964. Used by permission.

Fig. 1-13. Dodecahedral recycling bin for glass, on a street corner in Paris, France. Photograph by Marjorie Senechal.

Fig. 1-14. The model for protein structure proposed by Dorothy Wrinch in 1934. Sophia Smith Collection, Smith College. Used by permission of Schenkman Books.

symmetry that they are considered to be beautiful. The regular solids have the highest possible symmetry among polyhedra that are finite in extent. This is one reason why we can justly say that the regular solids are the rulers of the Polyhedron Kingdom. As you read through this book you will learn a great deal about symmetry.

Direct Descendants

There are many variations on the theme of the regular polyhedra. First let us meet the eleven (in Fig. 1-15) which can be made by cutting off (*truncating*) the corners, and in some cases the edges, of the regular polyhedra so that all the faces of the faceted polyhedra obtained in this way are regular polygons. These polyhedra were first discovered by Archimedes (287–212 B.C.) and so they are often called Archimedean solids. Notice that vertices of the Archimedean polyhedra are all alike, but their faces, which are regular polygons, are of two or more different kinds. For this reason they are often called *semiregular*. (Archimedes also showed that in addition to the eleven obtained by truncation, there are two more semiregular polyhedra: the snub cube and the snub dodecahedron (see Fig. 1-15).)

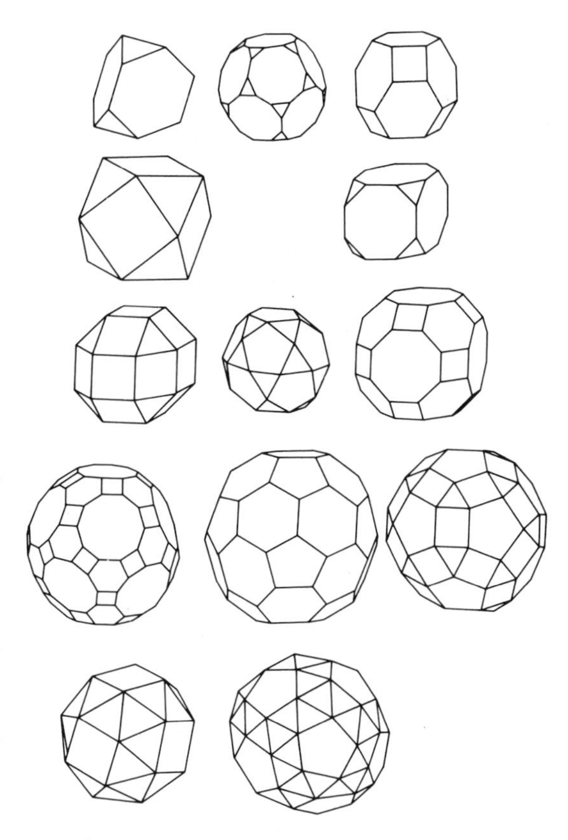

Fig. 1-15. The Archimedean or semiregular polyhedra; The first eleven can be obtained from the regular polyhedra by truncation. Redrawn from *Mathematical Models* by Henry Martyn Cundy and A. P. Rollett, published by Oxford University Press.

According to our definition, *prisms* (see Fig. 1-16) with regular polygonal bases and square sides are semiregular solids too. Prisms are quite common in nature and in architecture, as we will see later on in the tour. *Antiprisms* also have two identical polygonal faces, but the "top" face is rotated relative to the "bottom" one, so that the two polygons are joined by triangles (see Fig. 1-17); when its faces are regular polygons, an antiprism is a semiregular polyhedron.

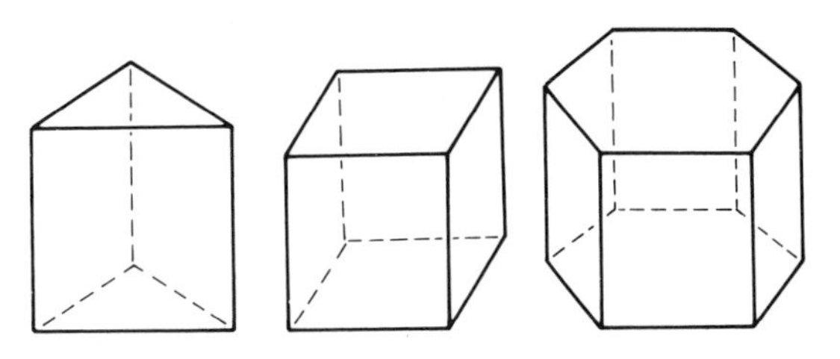

Fig. 1-16. Three semiregular prisms.

Perhaps the most elaborate variations on the theme of the regular polyhedra are those of the sixteenth-century Nuremberg goldsmith Wenzel Jamnitzer, who engraved a fascinating and extensive series of polyhedra in honor of Plato's theory of matter. In his book *Perspectiva Corporum Regularium,* published in 1568, each of the five regular solids is presented in exquisite variation. Can you tell which solid is being varied in Fig. 1-18? Jamnitzer's figures show us that polyhedra need not be *convex*;

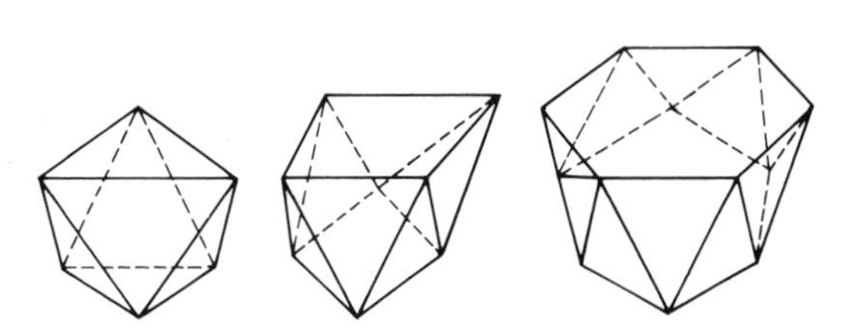

Fig. 1-17. Three semiregular antiprisms.

Fig. 1-18. Plate D.II. from Wenzel Jamnitzer, *Perspectiva Corporum Regularium,* 1568; facsimile reproduction, Akademische Druck- u. Verlagsanstalt, 1973, Graz, Austria.

that is, they can have indentations. Regular *polygons* that are not convex, such as the famous pentagram (Fig. 1-19), are familiar to

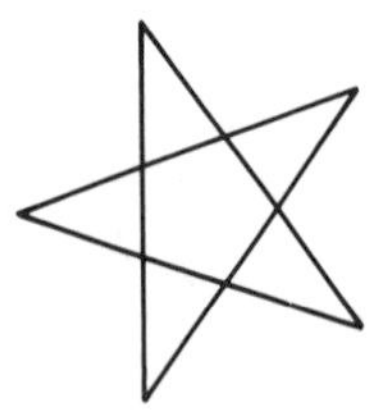

Fig. 1-19. The pentagram has equal sides and equal angles.

most of us. Such "star polygons" can be used to build regular "star polyhedra." There are exactly four regular star polyhedra (see Fig. 1-20). Notice that all their faces are regular polygons and the same number of faces meet at each vertex. In this case, however, either the faces or the vertex arrangements are pentagrams. The lineage of these polyhedra can be traced to fourteenth-century Venice (see Fig. 1-21), but no general theory seems to have been developed at that time. Later Kepler investigated regular star polyhedra and found two of them; after that star-shaped poly-

Fig. 1-20. The four regular star polyhedra. From *Mathematical Models* by Henry Martyn Cundy and A. P. Rollett, 2nd edition, 1961, published by Oxford University Press.

Fig. 1-21. Marble tarsia (1425–1427) in the Basilica of San Marco, Venice, attributed to Paolo Uccello. By courtesy of Scienza e Tecnica 76, Mondadori.

Fig. 1-22. Courtyard of Borromini church. From Paolo Portoghesi, *The Rome of Borromini: Architecture as Language* (New York: George Braziller, 1968). Photograph by courtesy of Electa, Milano.

hedra (not necessarily regular) became ubiquitous (see for instance Fig. 1-22). But it was not until the early nineteenth century that two more regular star polyhedra were found and the French mathematician Augustin-Louis Cauchy (1789–1857) showed that there are no others.

The *uniform* polyhedra are polyhedra, star or otherwise, whose vertices are all symmetrically equivalent. (They are generalizations of the Archimedean polyhedra.) Perhaps the most spectacular uniform polyhedron is the Yog-Sothoth, shown in Fig. 1-23. Although its existence had been predicted (on theoretical grounds) for many years, no one had ever seen it until one was built by Bruce Chilton several years ago. Chilton's Yog-Sothoth was presented to society for the first time at the Shaping Space Conference, and it was a spectacular success. It has 112 faces: 12 are pentagrams, 40 are triangles of one type, and 60 are triangles of another. Despite its complexity, the Yog-Sothoth has the symmetry of the icosahedron and dodecahedron, no more no less!

There are many other interesting lines of descent from the regular solids. For example, there are polyhedra whose faces are all alike but whose vertices are not. Closely related to the semiregular solids, these polyhedra are especially important in the study of crystal forms. But it is time to move on to other parts of the Kingdom.

An Architectural Walking Tour

The first polyhedral buildings we see on our tour are perhaps the most famous of all: the pyramids of Egypt, built about 2500 B.C. (Fig. 1-24). Yes, a pyramid is a polyhedron; one of its faces is a polygonal base (of 3, 4, 5, . . ., n sides) and the others are congruent isosceles triangles joined to the base along its edges, meeting above it in a single point. The bases of the Egyptian pyramids are squares.

As we walk along, we see several buildings based on prisms. Shown in Fig. 1-25 is a Hungarian hut, a triangular prism resting on one of its rectangular sides (like our modern A-frames). Nearby we see a notorious pentagonal prism located near Washington, D.C. (Fig. 1-26), and a much older building in the form of

Fig. 1-23. Three plan views of the Yog-Sothoth, along five-, three-, and twofold axes, drawn by Bruce L. Chilton. From Bruce L. Chilton and George Olshevsky, *How to Build a Yog-Sothoth* (George Olshevsky, P.O. Box 11021, San Diego, Calif, 92111-0010, 1986). Used by permission.

Fig. 1-24. The pyramids of Mycerinus, Chefren, and Cheops at Giza. Photograph by Hirmer Verlag, Munich.

Fig. 1-25. This hut in Hungary is a triangular prism resting on one of its rectangular faces. Photograph by Werner Bischof. Magnum Photos, New York.

Fig. 1-27. The Baptistry of S. Giovanni, Florence. Photograph by D. Anderson. Alinari/Art Resource, New York.

Fig. 1-26. The Pentagon is an enormous pentagonal prism. U.S. Air Force Photo by Eddie McCrossan.

an octagonal prism (Fig. 1-27). It is rare to see a prismatic building with more than eight flat sides, but if we allow the meaning of "prism" to include polyhedra with curved sides, then we find that they are quite common. Actually, most buildings are prisms, since the rectangular "boxes" that constitute much familiar architecture are prisms with rectangular bases.

Even boxes can become interesting polyhedral structures when juxtaposed in imaginative ways, as in the "Habitat" housing project in Montreal shown in Fig. 1-28.

Interesting polyhedral structures are often designed for world fairs (later to be moved to their permanent sites in the Kingdom). The Coca-Cola Building (Fig. 1-29), designed by Erwin Hauer for the 1964 New York World Fair, is a prism with many curved sides. The detail of the outer grill shown in Fig. 1-30 shows that the design can be considered a slice though a packing of truncated octahedra. Other interesting polyhedral buildings are almost spherical in form. In Fig. 1-31 we see one of Buckminster Fuller's geodesic domes, the United States Pavilion at Expo '67 in Montreal. Its faces are triangles, grouped into hexagons and pentagons (see Chapter 3). The geodesic dome has been the inspiration for countless buildings, large and small. The

Fig. 1-30. Detail of the Coca-Cola Building at the 1964 World Fair, showing a slice through a tight packing of opaque acrylic truncated octahedra. By Irwin Hauer.

Fig. 1-28. Polyhedral "Habitat" housing project by Moshe Safdie in Montreal, Quebec. Photograph by Carol Moore-Ede, from C. Moore-Ede, *Canadian Architecture 1960/70* (Toronto: Burns and MacEachern, Ltd., 1971).

Fig. 1-29. Coca-Cola Building at the 1964 World Fair. By Irwin Hauer.

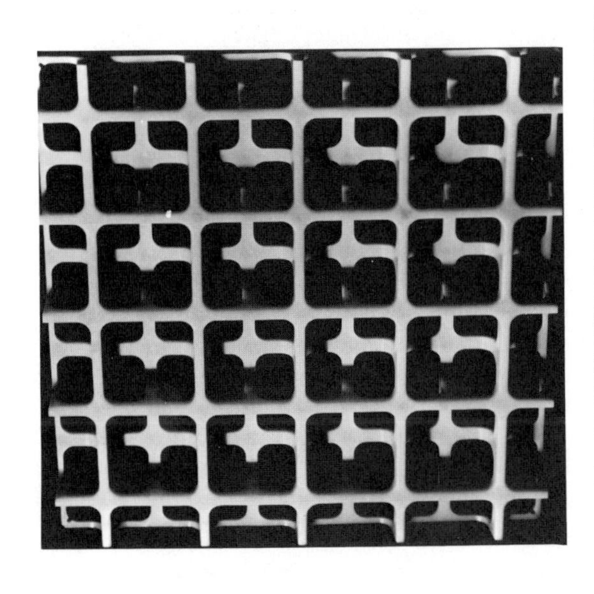

Fig. 1-31. The U.S. Pavilion at the Expo '67 World Fair Montreal. Courtesy of the Buckminster Fuller Institute, Los Angeles.

Fig. 1-32. Geodesic dome house under construction in Hadley, Massachusetts. Photographs by Wendy Klemyk.

Fig. 1-33. Baer's fused triple rhombicosidodecahedra at Drop City, Colorado. *Domebook 2* (Bolinas, Calif.: Shelter Publications, 1971). Reprinted by permission of Steve Baer.

Fig. 1-34. The Five-College Radio Astronomy Observatory is the largest millimeter-wavelength radio telescope in the United States. Photograph by Steve Long. University of Massachusetts Photocenter.

house shown in Fig, 1-32 was recently constructed by a family in Hadley, Massachusetts. Many other interesting domes are described in do-it-yourself publications (see, for instance, Fig. 1-33). The faces of some dome structures are deliberately arranged in asymmetric ways. For example, the dome grid (shown in Fig. 1-34) of the Five-College Radio Astronomy Observatory at Quabbin Reservoir, Massachusetts, is deliberately random to prevent interference patterns with the incoming signal.

Once your eyes are opened, you will find many interesting examples of polyhedral architecture in your own neighborhood. Shown in Fig. 1-35, for example, are some "geometric" student residences at Hampshire College in Amherst, Massachusetts.

The rulers of the Polyhedron Kingdom have recently instituted a Polyhedral Hall of Fame to honor human beings who use polyhedra in especially unexpected and delightful ways. The first person to be elected to the Hall was the Israeli architect Zvi Hecker, cited for his

Fig. 1-35. Modular residences, Hampshire College, Amherst, Mass. Photographs by Wendy Klemyk.

a

b

Fig. 1-36. Synagogue in the Negev Desert, Israel, 1969–1970. (a) Exterior view. (b) Interior view. Zvi Hecker, architect.

Fig. 1-37. Housing complex in Ramot, Israel, 1972–1980. Designed by Zvi Hecker.

Fig. 1-38. Former highway bridge, St. Louis Bay, Duluth, Minnesota–Superior, Wisconsin. David Plowden, *Bridges: the Spans of North America.* (New York: The Viking Press, 1974; W. W. Norton and Co., 1984). Reprinted by permission of Curtis Brown, Ltd. Copyright © David Plowden 1974.

multipolyhedral synagogue in the Negev desert (Fig. 1-36) and his dodecahedral housing complex in Ramot, Israel (Fig. 1-37).

Architecture reminds us that the most important part of some polyhedral structures is the network of edges and vertices. If we agree that such networks themselves constitute polyhedra, then we see (Figs. 1-38–1-40) that many bridges are polyhedra, and so are common (and uncommon) jungle gyms (such as those shown in Figs. 1-41 and 1-42). In fact there is no end to the polyhedral structures around us.

The Nature Preserve

The nature preserve is a vast region of the Polyhedron Kingdom, whose known extent keeps growing larger as it becomes possible to study structures on increasingly smaller scales. On this brief visit we will only have time to glance casually at some natural poly-

Fig. 1-40. Bridge with polygonal entrances, Carabasset River, North New Portland, Maine. David Plowden, *Bridges: the Spans of North America* (New York: The Viking Press, 1974; W. W. Norton and Co., 1984). Reprinted by permission of Curtis Brown, Ltd. Copyright © David Plowden 1974.

Fig. 1-39. Boston and Maine Railroad Bridge, Connecticut River, Northampton, Mass. David Plowden, *Bridges: the Spans of North America* (New York: The Viking Press, 1974; W. W. Norton and Co., 1984). Reprinted by permission of Curtis Brown, Ltd. Copyright © David Plowden 1974.

Fig. 1-41. Jungle gym at the Nonotuck Community Child-Care Center adjacent to the Smith College campus. Photograph by Wendy Klemyk.

Fig. 1-42. Children surrounded by polyhedra at the Bleecker Street Playground in New York. Marilynn K. Yee/The New York Times.

Fig. 1-43. Quartz crystals. From Wolf Strache, *Forms and Patterns in Nature* (New York: Pantheon Books, a Division of Random House, Inc., copyright 1956).

Fig. 1-44. Icositetrahedral leucite crystal. From Vincenzo de Michele, *Minerali* (Milan: Istituto Geografico de Agostini-Novara, 1971).

Fig. 1-45. Benitoite crystal. From Earl H. Pemberton, *Minerals of California* (New York: Van Nostrand Reinhold Company Inc., 1983).

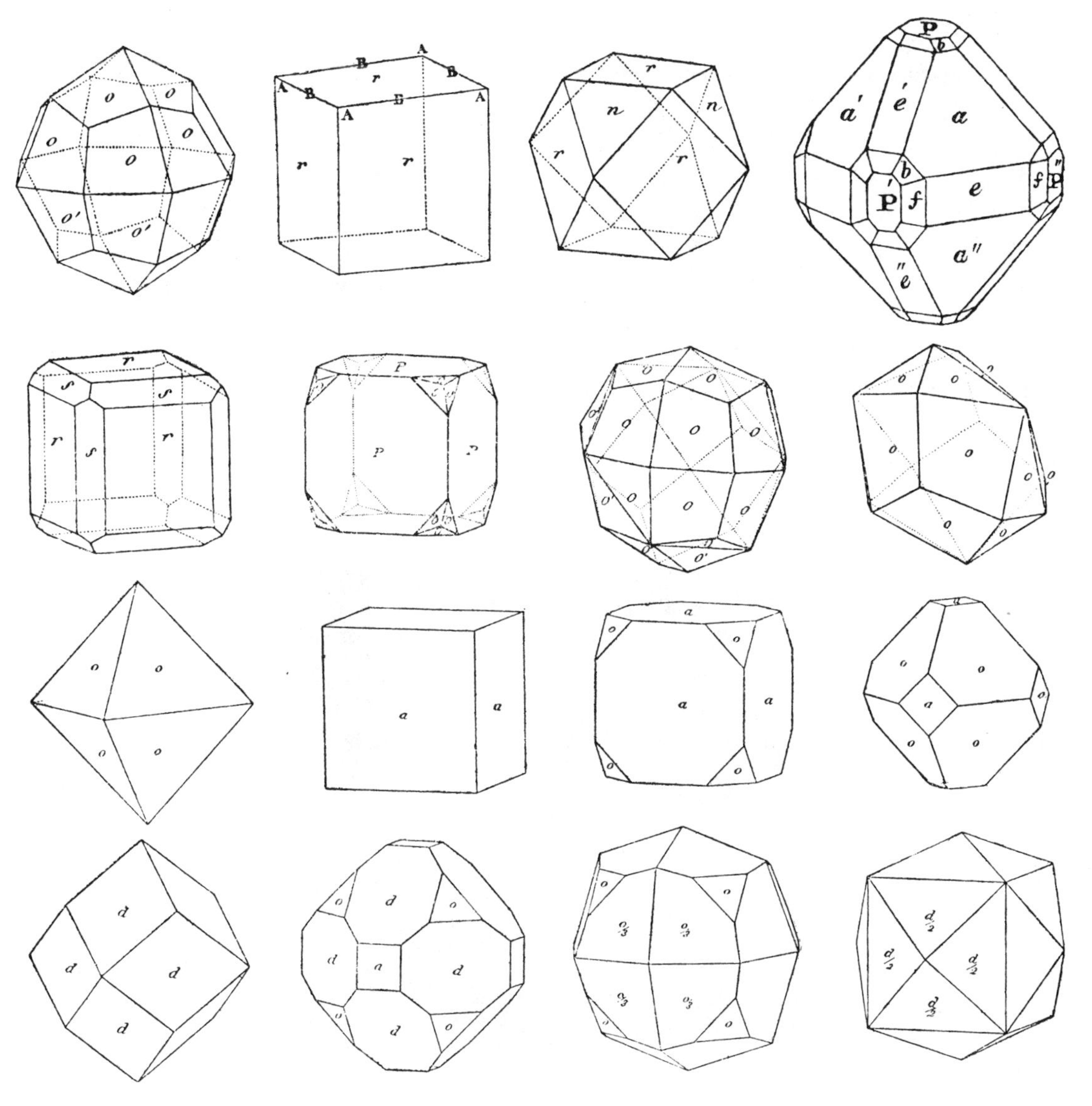

Fig. 1-46. Drawings of crystals of gold. From Viktor Goldschmidt, *Atlas der Krystallformen* (Heidelberg: Carl Winters, 1912).

hedra which can be seen with the naked eye or with a simple microscope.

Our first stop is at a mine, where polyhedral crystals of many different kinds can be found. Look carefully in Fig. 1-43 at crystals of the familiar mineral quartz. Quartz crystals are essentially prisms with terminating facets, which are arranged in interesting ways. In Fig. 1-44 we see a leucite crystal in the shape of an *icositetrahedron;* it has 24 trapezoidal faces. In other crystals, the faces are truncated, as in the crystal of benitoite shown in Fig. 1-45. Some kinds of crystals come in many forms. Sixteen drawings of gold from the famous twenty-volume *Atlas der Krystallformen* are shown in Fig. 1-46.

The polyhedra that occur as plants and animals are usually less standard in form than polyhedral crystals, but are no less intriguing. The purple sea urchin of Peru (shown in Fig. 1-47) combines features reminiscent of both star polyhedra and geodesic domes. The pufferfish is a polyhedron of uncommon charm (see Fig. 1-48). Radiolaria are single-celled sea creatures whose skeletons have very interesting polyhedral forms. The sketches in Fig.

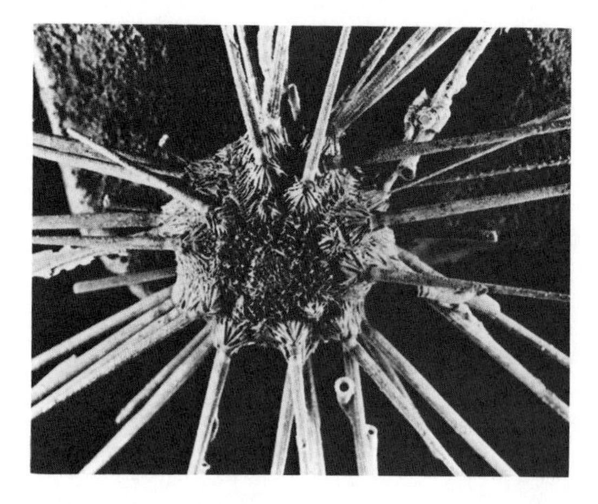

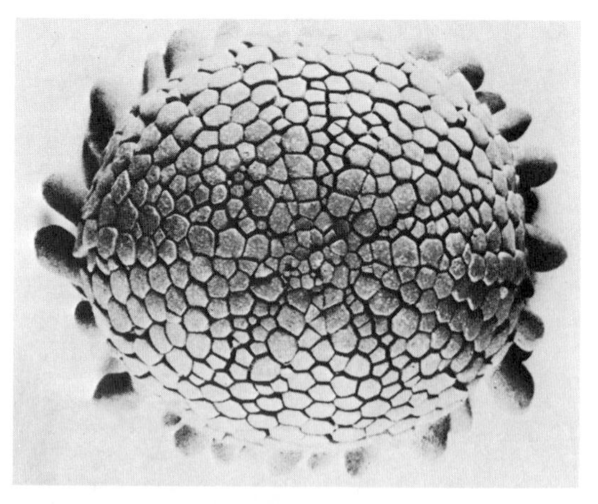

Fig. 1-47. Purple sea urchin, "front" and "back." Photograph by Dr. V. J. Staněk. From V. J. Staněk, *Krásy přírody/The Beauty of Nature* (Prague: Artia, 1955).

Fig. 1-48. Inflated spiny pufferfish. Photograph by Carl Roessler.

1-49 were made by Ernst Haeckel on his trip aboard H.M.S. *Challenger* with Charles Darwin. Some insects—for example, the bees—build polyhedra for their own purposes. Among the culinary delights of the Nature Preserve are its many honeycombs. As shown in Fig. 1-50, a comb is an aggregate of half-open polyhedra.

Aggregated polyhedra are also found in plants. How would you describe the examples shown in Fig. 1-51? Aggregates of polyhedra will be discussed in more detail later on this tour and in this book (see Chapter 5).

This concludes our tour of the Nature Preserve. A much deeper discussion of polyhedra in nature is found in Chapters 9 and 10.

At the edge of the Nature Preserve, we come to the Gallery of Polyhedral Art.

The Gallery of Polyhedral Art

Polyhedral art can be found throughout the world. In honor of your visit to the Kingdom, a small but exquisite collection of sculpture, paintings, and graphics in which polyhedra are an important theme has been assembled. The Renaissance and Modern exhibits are especially strong. We see in Fig. 1-52 the famous

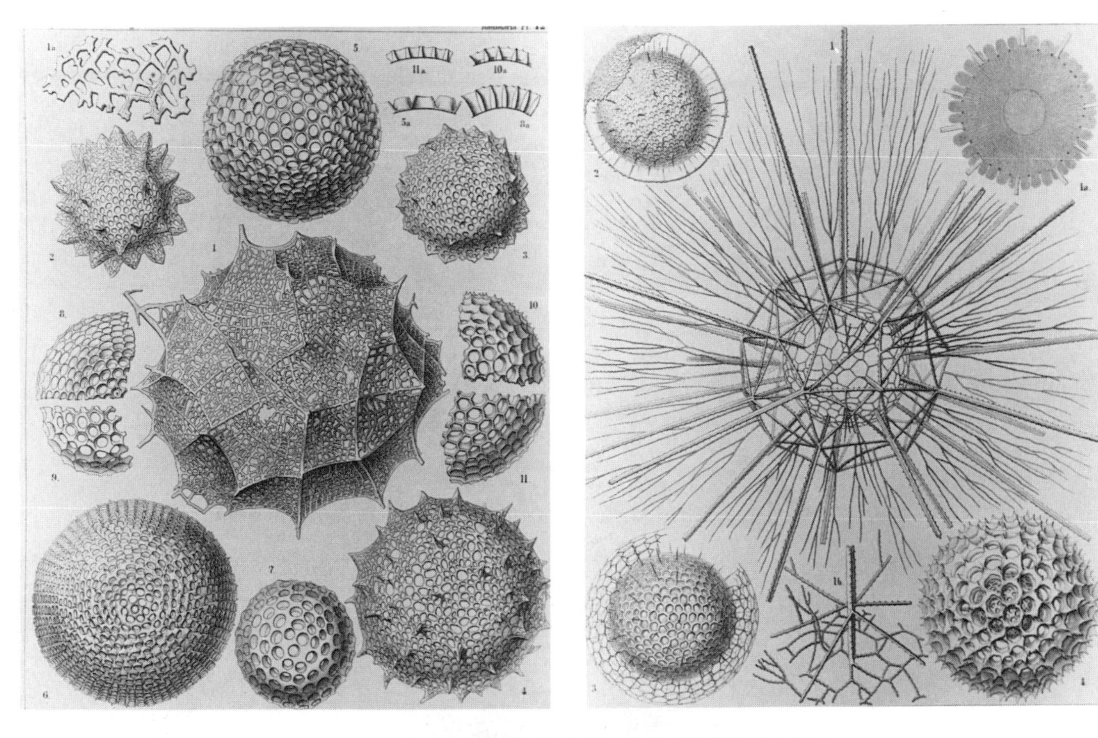

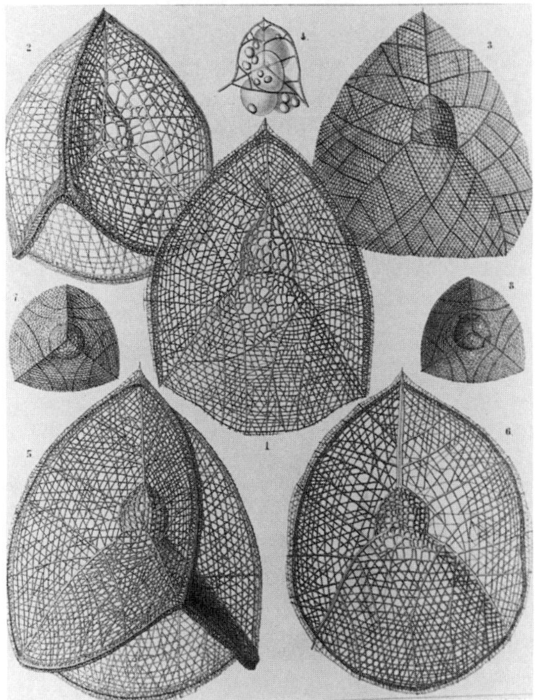

Fig. 1-49. Radiolaria. E. Haeckel, *The Voyage of H.M.S. Challenger* (Berlin: Georg Reimer, 1887), plates 12, 20, and 63.

Fig. 1-50. A honeybee comb. Photograph by Lawrence Conner, Ph.D., entomologist.

a

b

c

Fig. 1-51. (a) *Carex grayi*. (b) *Adonis pernalis*. (c) *Cornus kousa*. Karl Blossfeldt, *Wundergarten der Natur* (Berlin: Verlag für Kunstwissenschaft, 1932.)

engraving *Melencolia I* by Albrecht Dürer (1471–1528). This work simply groans with symbolism, not all of it understood. In particular, the meaning of the enormous polyhedron continues to be disputed.

Renaissance artists seem to have been very fond of the regular polyhedra, both because of their association with Plato and because they offered opportunities for the study of perspective (see Chapter 4). We saw one of Jamnitzer's engravings in Fig. 1-18; Leonardo da Vinci (1452–1519) also drew many polyhedra. Polyhedra often appear in Renaissance paintings; the gallery is proud to display Jacopo de Barbari's portrait of Fra Luca Pacioli (author of *Divina Proportione*), shown in Fig. 1-53. The mazzocchio, a doughnut-shaped polyhedral hat popular in fourteenth-century Florence, appears in many paintings by Paolo Uccello; details of two of his paintings are reproduced in Fig. 1-54. The mazzocchio was revived for the Shaping Space Conference, and you will see it at the Polyhedral Artisan Fair (see Figs. 1-55 and 1-89).

In the exhibition of modern polyhedral art, we find three striking paintings by Salvador

Fig. 1-52. *Melencolia I.* Dated 1514. Albrecht Dürer, German, 1471–1528. Engraving, 243 × 187 mm. Centennial gift of Landon T. Clay. 68.188. Courtesy, Museum of Fine Arts, Boston.

Fig. 1-53. Jacopo de Barbari, Portrait of Fra Luca Pacioli and His Student Guidobaldo, Duke of Urbino. Museo e Gallerie Nazionali di Capodimonte, Naples. Illustration by permission of Soprintendenza ai B.A.S. di Napoli.

Fig. 1-54. *Top:* Paolo Uccello, *The Rout of San Romano* (1456–60, tempera on panel. Louvre, Paris.), detail. Museé de Louvre, Paris. *Bottom:* Paolo Uccello, *After the Flood* (Frescoed lunette in terraverde. Green Cloister of Santa Maria Novella, Florence), detail. Alinari/Art Resource, New York.

Dali: *The Sacrament of the Last Supper* (Fig. 1-56), *Cosmic Contemplation* (Fig. 1-57), and *Corpus Hypercubicus* (Fig. 1-58). At first glance the polyhedron in the second of these appears to be a pentagonal dodecahedron (cf. Plato), but then we notice that it has one hexagonal face. Such a structure is impossible (see Chapter 3)! How do you think Dali envi-

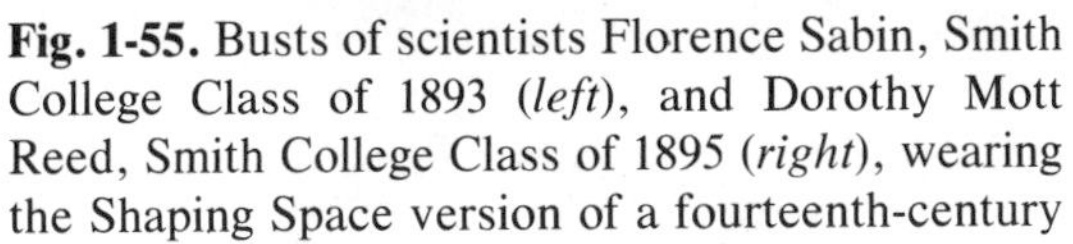

Fig. 1-55. Busts of scientists Florence Sabin, Smith College Class of 1893 (*left*), and Dorothy Mott Reed, Smith College Class of 1895 (*right*), wearing the Shaping Space version of a fourteenth-century mazzocchio (see Fig. 1-89). The busts, by Joy Buba, are in Sabin-Reed Hall of Smith College. Photographs by Stan Sherer. Used by permission of the Trustees of the Smith College.

Fig. 1-56. 1963.10.115 *The Sacrament of the Last Supper*. Salvador Dali. National Gallery of Art, Washington. Chester Dale Collection.

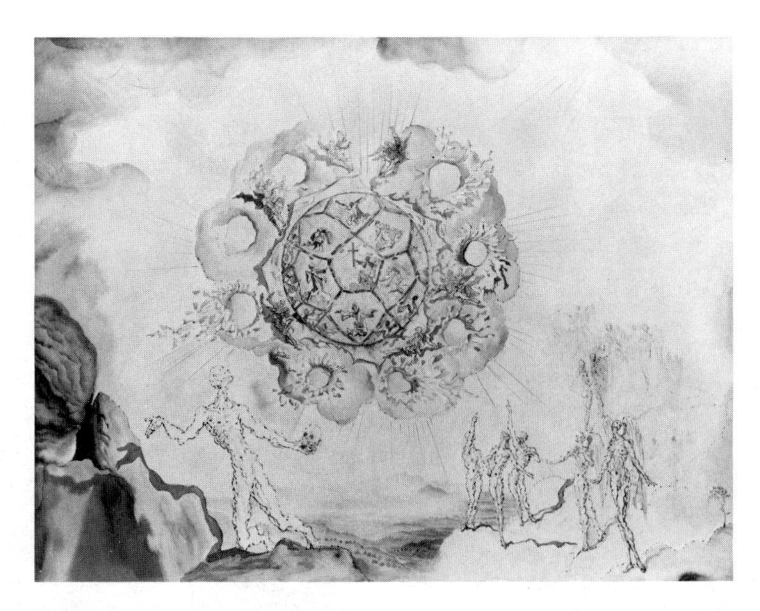

Fig. 1-57. Salvador Dali, *Cosmic Contemplation*. Watercolor and ink, 1951. The Salvador Dali Foundation, Inc., St. Petersburg, Fla.

Fig. 1-58. Salvador Dali, *Corpus Hypercubicus*, 1954. Oil on canvas. $76\frac{1}{2} \times 48\frac{3}{4}$ in. #55.5. Metropolitan Museum of Art, New York. Gift of the Chester Dale Collection. © S.P.A.D.E.M., Paris/ V.A.G.A., New York.

Fig. 1-59. *Fall-Out* from the series *Unsculptable* by Mary Bauermeister. Photograph supplied by artist.

Fig. 1-60. Pablo Picasso, *Girl with a Mandolin* (Fanny Tellier). Paris, early 1910. Oil on canvas, $39\frac{1}{2} \times 29''$. Collection, The Museum of Modern Art, New York. Nelson A. Rockefeller Bequest.

Fig. 1-61. Georges Braque, *Maisons à l'Estaque*, 1918. Kunstmuseum Bern/Hermann and Margrit Rupf-Stiftung. Copyright © 1985, 1986 by Cosmopress, Geneva; A.D.A.G.P., Paris; V.A.G.A., New York.

Fig. 1-62. Roger de la Fresnaye, *The Conquest of the Air,* 1913. Oil on canvas. $7'8\frac{7}{8}'' \times 6'5''$. Collection, The Museum of Modern Art, New York. Mrs. Simon Guggenheim Fund.

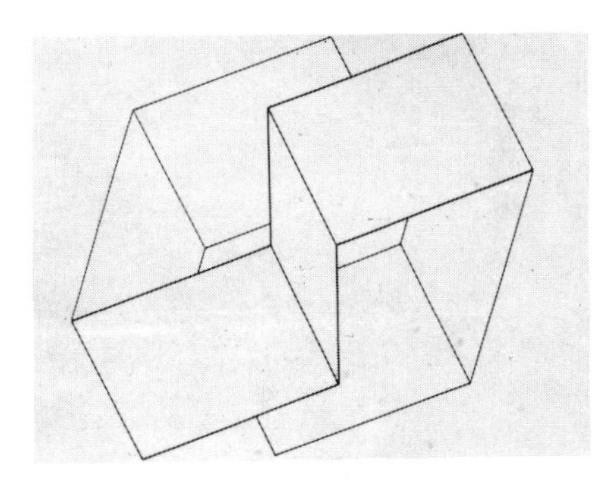

Fig. 1-63. Josef Albers, *Structural Constellation*. Reproduced by permission of Mrs. Anni Albers and the Josef Albers Foundation, Inc.

sioned its other side? A well-known impossible structure appears in the painting by Mary Bauermeister reproduced in Fig. 1-59.

The gallery's exhibit of cubist painting is very good; it includes works by Pablo Picasso (Fig. 1-60), Georges Braque (Fig. 1-61), and Roger de la Fresnaye (Fig. 1-62). There are also important works by Josef Albers (Fig. 1-63) and M. C. Escher (Figs. 1-64 and 1-65).

Fig. 1-64. M. C. Escher, *Waterfall.* © M. C. Escher Heirs c/o Cordon Art–Baarn–Holland.

Fig. 1-65. M. C. Escher, *Order and Chaos.* © M. C. Escher Heirs c/o Cordon Art–Baarn–Holland.

Fig. 1-66. Isamu Noguchi, *Red Rhombohedron,* in the plaza of the Marine Midland Bank Building, New York. Photograph by permission of the Marine Midland Bank, Corporate Communications Group.

Fig. 1-67. Charles O. Perry, *Eclipse.* The helical explosion of every face rotating from a dodecahedron through the icosidodecahedron to the small rhombicosidodecahedron. Hyatt Regency Hotel, San Francisco. Photograph by Jeremiah O. Bragstad.

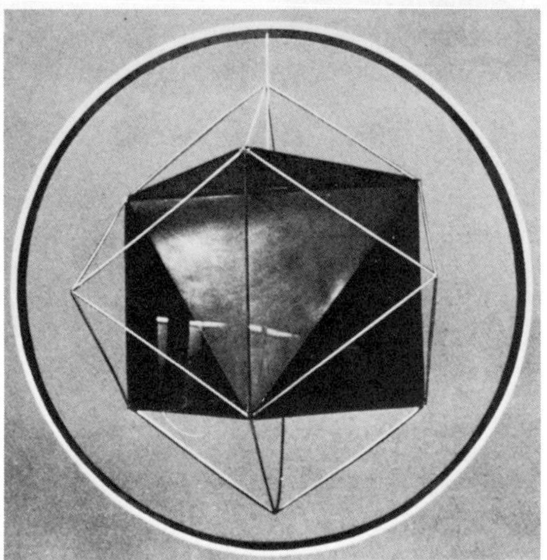

Fig. 1-68. *Top:* Arthur L. Loeb standing next to his sculpture *Polyhedral Fancy* in the lobby of Burton Hall, Smith College. Photograph by Stan Sherer. *Bottom: Polyhedral Fancy,* a part of the permanent collection of Smith College, is a copper tetrahedron within a Plexiglas® cube within an octahedral framework within a brass cross-section of a sphere.[2] Photograph reproduced from Arthur L. Loeb, *Space Structures: Their Harmony and Counterpoint* (Reading, Mass.: Addison-Wesley, Advanced Book Program, 1976).

The crowded sculpture court contains a wide variety of noted works, several of which were exhibited at the Shaping Space Conference. The largest sculptures in the court are Isamu Noguchi's *Red Rhombohedron* (Fig. 1-66) and Charles Perry's *Eclipse* (Fig. 1-67). There is also *Polyhedral Fancy* by Arthur Loeb (Fig. 1-68) and *Tetrahedron* by Lee Burns (Fig. 1-69). The gallery is also proud to display Hugo Verheyen's sculpture with movable parts (Fig. 1-70) and Max Bill's *Construction with 30 Equal Elements* (Fig. 1-71).

It is perhaps here in the sculpture court that we first become acquainted with the boundaries of the Polyhedron Kingdom. As we move away from the center of the Kingdom, the population variation becomes greater and greater, until we cannot really say what is a polyhedron and what is not. Is *Eclipse* a polyhedron? If not, why not? Figures 1-72 and 1-73 are two sculptures by Erwin Hauer. Is either of them a polyhedron? What about Alan Holden's *Ten Tangled Triangles* (Fig. 1-74)?

Before leaving the gallery, take a close look

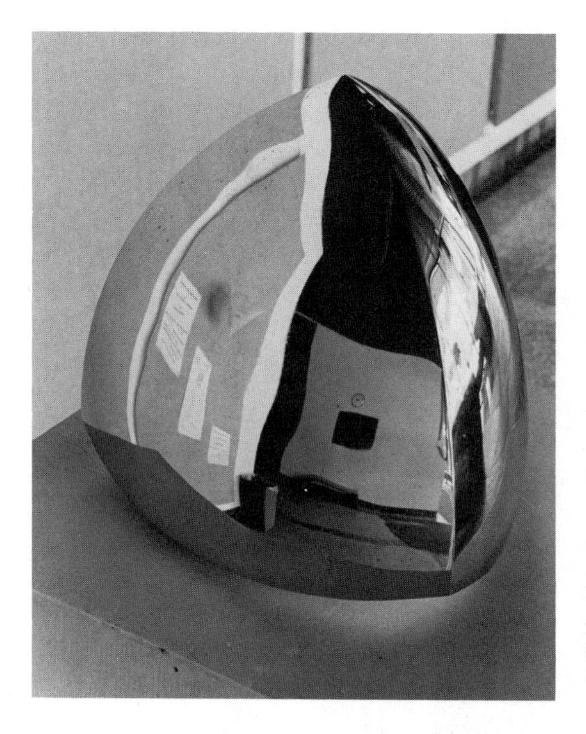

Fig. 1-69. A. Lee Burns, *Tetrahedron*. A polished brass tetrahedral "soap bubble," inspired by a soap bubble in a tetrahedral frame (recall Fig. 1-7). Photograph by Stan Sherer.

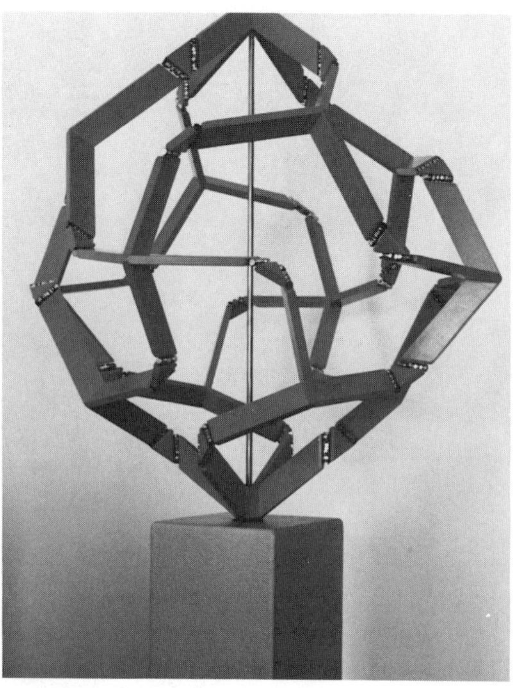

Fig. 1-70. *IRODO*, an expandable polyhedral sculpture based on the impandable rhombic dodecahedron. Hugo F. Verheyen.

Fig. 1-71. Max Bill, *Construction with 30 Equal Elements*, from Gyorgy Kepes, *The New Landscape in Art and Science* (Chicago: Paul Theobald and Company, 1956).

Fig. 1-72. Erwin Hauer, *Rhombidodeca.* An excerpt from an infinite, continuous and periodic surface, WPI. The inner labyrinth is expressed as a solid volume. Produced of organic composite materials, 28 × 26 × 26 inches. Photograph by Erwin Hauer.

Fig. 1-74. Alan Holden, *Ten Tangled Triangles,* Smith College, sent by the artist as his surrogate representative to the Shaping Space Conference. Photograph by Stan Sherer.

Fig. 1-73. Erwin Hauer, *Obelisk,* also an excerpt from WPI, but along the diagonal bisectors of the constituent cubes. The outer labyrinth is maximized in volume and appears as the large perforations through the sculpture. The shallow exterior spaces are what remains of the inner labyrinth. Produced in cast stone, 1 inch thick, the sculpture measures 9 × 2 × 2 ft. Photograph by Erwin Hauer.

Fig. 1-75. Harriet E. Brisson sitting in the *Truncated 600-Cell,* a four-dimensional form made by Harriet E. Brisson and Curtis LaFollete, 1984. Photograph by Bob Thayer. © 1984 by the Providence Journal Company.

(in Fig. 1-75) at the *Truncated 600-Cell* by Harriet Brisson. The 600-cell is the name of a four-dimensional polyhedron whose 600 "faces" are three-dimensional regular tetrahedra! The viewer entering the large tetrahedron is surrounded by mirror images approximating the experience of the fourth dimension extending to infinity. It is fascinating to think about the ways in which four-dimensional polyhedra can be represented in our three-dimensional world (see Chapter 17).

Fig. 1-76. *Ginger and Fred* by Robinson Fredenthal. Photograph supplied by the sculptor.

A Note on Polyhedral Society

Polyhedra communicate with one another in a variety of subtle ways (see Fig. 1-76). Indeed, the sociology of polyhedra is extremely complicated, as polyhedra tend to be related to one another through many different kinship structures. Some of them are related by geometry; for example, some can be inscribed inside one another, as in Fig. 1-68.

Others can be grouped into families whose members are related by truncation, that is, by successively slicing off larger and larger corners and edges (see Fig. 1-77). (As we have seen, this is the way that the Archimedean polyhedra are related to their regular forebears.)

Crystals of the same kind are often related by truncation, and the discovery of this fact by J. B. L. Romé de Lisle in 1783 was a milestone in our understanding of crystal structure. Some such relationships are recorded in Fig. 1-78.

A major eighteenth-century discovery was that of Leonhard Euler (1707–1783), who found a simple equation that has great theoretical importance. Euler discovered that for

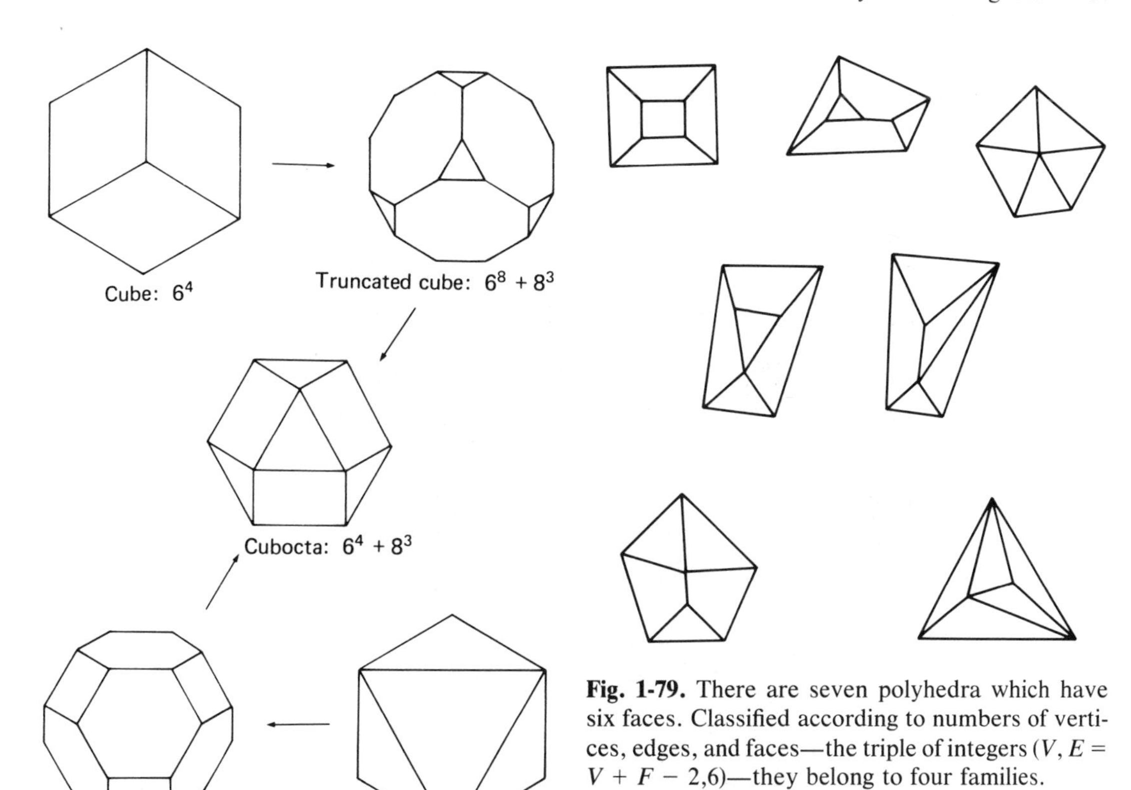

Fig. 1-79. There are seven polyhedra which have six faces. Classified according to numbers of vertices, edges, and faces—the triple of integers (V, $E = V + F - 2$,6)—they belong to four families.

Fig. 1-77. From cube to octahedron. From William Blackwell, *Geometry in Architecture,* p. 155. Copyright © 1984 by John Wiley & Sons, Inc. Reprinted by permission of John Wiley & Sons, Inc.

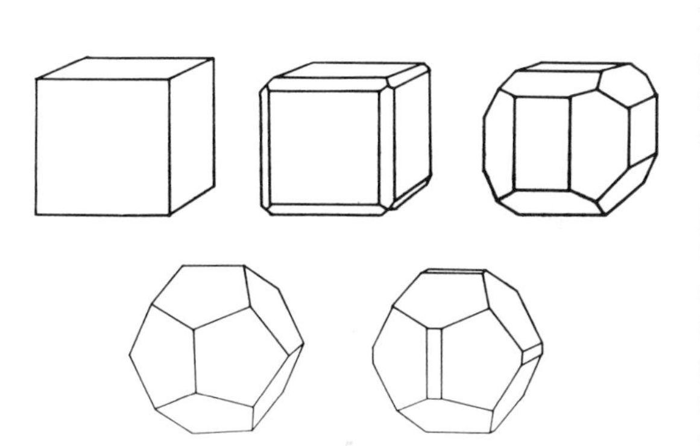

Fig. 1-78. Drawings of pyrite. From Viktor Goldschmidt, *Atlas der Krystallformen* (Heidelberg: Carl Winters, 1912).

every *convex* polyhedron, the numbers of faces (F), edges (E), and vertices (V) are related by the equation $V - E + F = 2$. This suggests another way of classifying polyhedra into families: classification according to the triple of numbers (V, E, F) (see Fig. 1-79 and Part IV). Yet another important relationship among polyhedra is duality, which is rather intimate, and about which there is still a great deal to be learned (see Chapter 13).

In addition to belonging to such families, polyhedra often form voluntary associations to provide important services to nature and to society. Like human and animal associations, these associations require a great deal of conformity but can be very effective in achieving their goals. Bubbles in a froth have polyhedral forms which, although appearing to be quite varied, have a very restrictive property: in each bubble, exactly three faces must meet at every vertex. You can see some explorations into the nature of soap films in Figs. 1-80 and 1-81. Froths are important models for many biological structures (see Chapter 9). Crystal architecture is another cooperative polyhedra endeavor (see Chapter 5). The atoms in a crys-

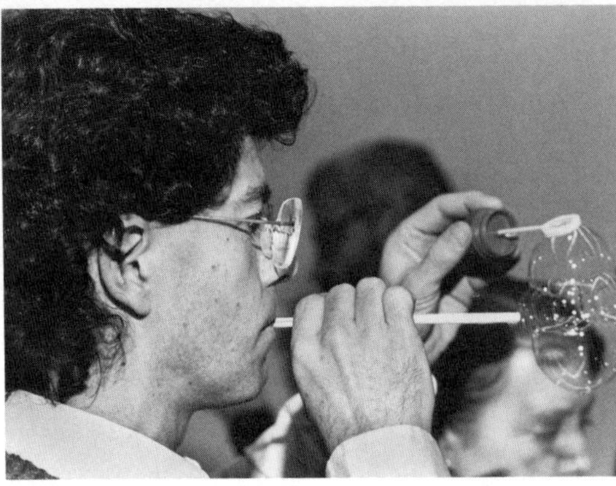

Fig. 1-80. *Top:* A. Lee Burns leading the Gala Soap Bubble Workshop at the Shaping Space Conference. *Bottom:* Godfried Toussaint creating a polyhedral bubble. Photographs by Stan Sherer.

Fig. 1-81. Soap bubbles in a froth. From Peter S. Stevens, *Patterns in Nature* (Boston: Little, Brown and Company, 1974).

tal come together in more or less regular arrays, like building blocks, to form the crystals that we see with our eyes. With an electron microscope, we can "see" the arrays themselves (Fig. 1-82). The Russian crystallographer E. S. Federov showed 100 years ago that there are exactly five polyhedral building blocks. That is, there are five types of polyhedra whose copies fill space completely when they are stacked face to face in parallel position; they are shown in Fig. 1-83. Many other examples of polyhedral cooperation are found in human-constructed architecture. We have already seen some examples on our walking tour. And of course the bees' cells stack together to make the honeycomb in Fig. 1-50.

Fig. 1-82. Electron micrographs of crystals, showing arrays of individual molecules. Protein from southern bean mosaic virus (magnification 30,000) (*left*). Protein from tobacco necrosis virus (magnification 73,000) (*right*). Photographs by R. W. G. Wyckoff.

New relations among polyhedra are being found all the time (for example, how are the polyhedra of the family in Fig. 1-84 related to one another?) This will continue as the theory of polyhedra expands to include the study of function as well as form (see Chapter 9).

The Polyhedral Fair

Our tour concludes with a stop at one of the polyhedral artisan fairs that are held from time to time in the Kingdom. Here you can browse among the many delightful items that the more

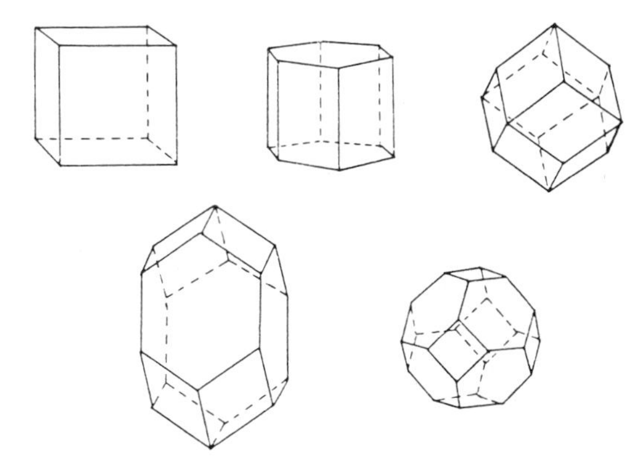

Fig. 1-83. The five kinds of polyhedra which fill space in parallel position.

Fig. 1-84. Cardboard models of twelve deltahedra-regular polyhedra by Lucio Saffaro.

Fig. 1-85. Wooden Puzzles: "Rhombics" display at Shaping Space Conference. Photograph by Stan Sherer.

Fig. 1-86. M. C. Escher contemplating a homemade polyhedron. Reproduced from Bruno Ernst, *The Magic Mirror of M. C. Escher* (New York: Ballantine Books, 1976). © M. C. Escher Heirs c/o Cordon Art–Baarn–Holland.

Fig. 1-87. Uttara Coorlawala and Matthew Solit at the Shaping Space Conference, building a tetrahedron with Rhombics parts. Photograph by Stan Sherer.

Fig. 1-88. A plywood dome, from *Domebook 2*, © 1971 by Shelter Publications, Inc., Bolinas, California. Reprinted by permission.

artistic natives of the Polyhedron Kingdom have created for your enjoyment. As you wander among the many displays, you will find such things as:

- Wooden puzzles (see Fig. 1-85)
- Polyhedra kits. Kits for building star polyhedra (and other polyhedra, too) are sold in many stores. In Fig. 1-86 we see M. C.

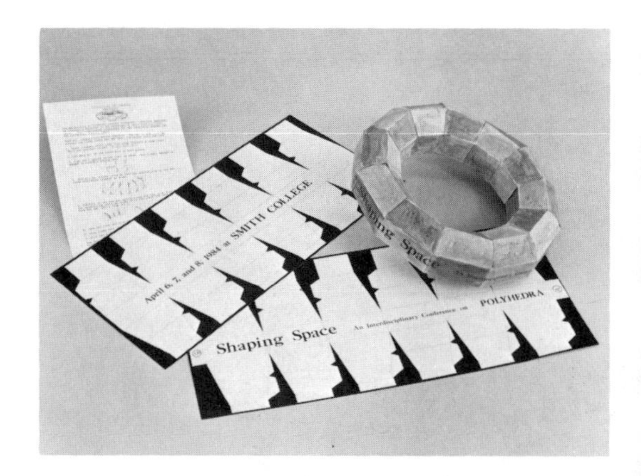

Fig. 1-89. The fourteenth-century mazzocchio was adapted for the Shaping Space Conference by Helen Connolly, who prepared a do-it-yourself kit. Photograph by Stan Sherer.

Fig. 1-90. Lampshade inspired by the hexagonal structure of a poem by Rumi (1207–1273), teacher of Islamic Sufiism. By Bahman Negahban, architect, and Ezat O. Negahban, calligrapher.

Escher contemplating a polyhedron constructed from parts provided in such a kit. Uttara Coorlawala and Matthew Solit are shown (in Fig. 1-87) building a Rhombics structure.

- Dome kits. Figure 1-88 shows a dome built from instructions in *Domebook 2*.
- Minerals
- Polyhedral jewelry

Fig. 1-91. Corner-posed cubical vase by unknown Kyoto potter, 1964: collection of A. Taeko Brooks. Photograph by Wendy Klemyk.

- The mazzocchio. A modern adaptation is shown in Figs. 1-55 and 1-89.
- An unusual lampshade is shown in Fig. 1-90, inspired by the rhythm and repetition of a Persian mystical poem.
- Vases and other pottery (see Fig. 1-91).
- "Total photos." A total photo is reproduced in Fig. 1-92.
- Unusual toys. Deltahedra are polyhedra whose faces are equilateral triangles but which are not regular because the numbers of triangles at the vertices can vary. One of them seems to have been the inspiration for the crawl-through toys shown in Fig. 1-93. You can build this deltahedron and all the other convex ones by following instructions in Chapter 2.

But if you do not have time to linger at the fair, do not be disappointed; you will find many delightful polyhedra for sale in shops everywhere.

Fig. 1-92. Total photo. Unfolded dodecahedron total photo of the Chicago Art Center. Reprinted by permission of Dick A. Termes.

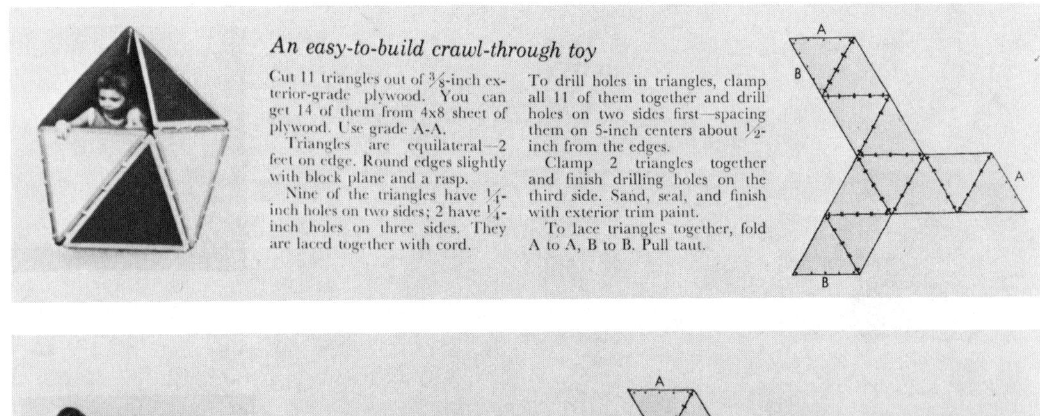

An easy-to-build crawl-through toy

Cut 11 triangles out of 3/8-inch exterior-grade plywood. You can get 14 of them from 4x8 sheet of plywood. Use grade A-A.

Triangles are equilateral—2 feet on edge. Round edges slightly with block plane and a rasp.

Nine of the triangles have 1/4-inch holes on two sides; 2 have 1/4-inch holes on three sides. They are laced together with cord.

To drill holes in triangles, clamp all 11 of them together and drill holes on two sides first—spacing them on 5-inch centers about 1/2-inch from the edges.

Clamp 2 triangles together and finish drilling holes on the third side. Sand, seal, and finish with exterior trim paint.

To lace triangles together, fold A to A, B to B. Pull taut.

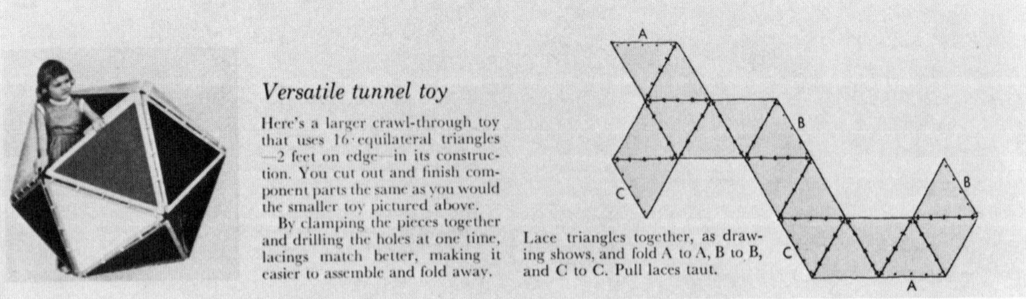

Versatile tunnel toy

Here's a larger crawl-through toy that uses 16 equilateral triangles—2 feet on edge—in its construction. You cut out and finish component parts the same as you would the smaller toy pictured above.

By clamping the pieces together and drilling the holes at one time, lacings match better, making it easier to assemble and fold away.

Lace triangles together, as drawing shows, and fold A to A, B to B, and C to C. Pull laces taut.

Fig. 1-93. Plywood crawl-through toys, with construction plans. Reprinted from *Better Homes and Gardens Christmas Ideas*. Copyright Meredith Corporation, 1957. All rights reserved.

Where to Go from Here

There is much more to see in the Polyhedron Kingdom, and much more to learn about it; indeed, as you go through this book you will be surprised to learn how much of it is still unexplored. But first, before reading further, we urge you to build a few of the polyhedra you have met, with your own hands; a famous Chinese proverb says: "I hear, I forget; I see, I remember; I do, and I understand." In Chapter 2 you will find "recipes" for making polyhedra; some are for beginners, other recipes are intermediate or advanced. You can use these recipes, or devise your own, or consult any of a number of excellent books (see Chapter 20). We are sure you will enjoy building the models, and that doing so will give you a much deeper understanding of the chapters which follow.

Notes

[1] Regular polygons are polygons whose edges have equal lengths and whose angles have equal measure. Thus a regular polygon of three edges is an equilateral triangle, of four edges a square, and so on.

[2] Plexiglas® is a registered trademark of Rohm and Haas, Philadelphia, for acrylic safety glazing.

2

Five Recipes for Making Polyhedra

In this section five of the world's greatest polyhedra chefs explain some of their remarkable creations. The creations range in difficulty, from beginning to advanced. Take the time now to make one—or more—yourself! Sources of additional recipes can be found in the list of Resources at the end of this book.

A. Constructing Polyhedra without Being Told How to!

MARION WALTER

Getting Started: How to Attach Polygons

Put some cut-out regular polygons on a table. Put a little glue on a flat tile, a plastic lid, or a piece of plastic, and spread out the glue a little so that you can dip a whole edge of a polygon into the glue.

Choose two polygons that you want to glue together along an edge, and dip one of these edges in the glue. Dip lightly; if polygons don't stick well it is usually because there is too much glue (Fig. 2-A1).

Hold the two edges together firmly. The joint will remain flexible but the polygons will stick together (Fig. 2-A2).

If you find later that you need extra glue on an edge of a polygon that you have already attached, you can (lightly) dip a toothpick or applicator stick in the glue to smear some along an edge.

What Shape Are You Going to Make?

It is most fun and most rewarding to make a shape you yourself create, rather than follow-

Fig. 2-A1.

Fig. 2-A2.

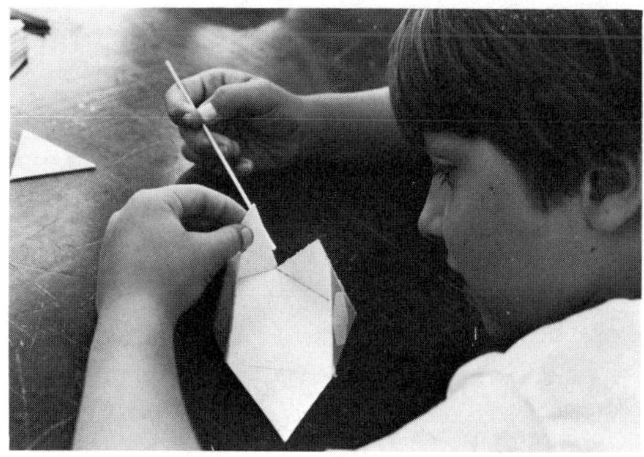

Fig. 2-A3.

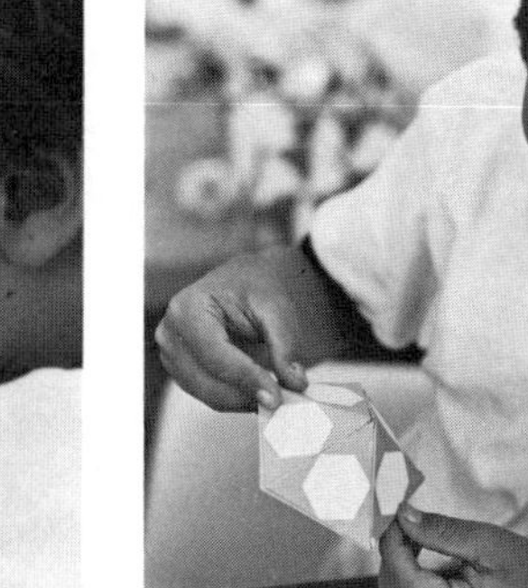

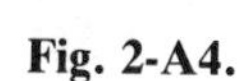

Fig. 2-A4.

Fig. 2-A5.

ing someone else's plans. How can you do this?

There are many ways to start. One way is to limit yourself to using only one or two different shapes—say triangles, or triangles and pentagons, or triangles and squares.[1] What shapes can you make using triangles and only one pentagon? (See Fig. 2-A3.)

In Fig. 2-A4 is the first shape the boy pictured made; the base is a pentagon and all the sides are triangles. It is called a pentagonal pyramid. Now make up another question of your own. What will your first shape look like? When you experiment freely, you may get a few surprises and you will learn a lot. For example, six triangles lie flat.

What a surprise: the shape in Fig. 2-A5 lies flat too! Notice that the twelve triangles that surround the hexagon help to make a bigger hexagon. The student shown in the photograph also had a surprise after she attached only six triangles to the hexagon. Do you think it will make a pyramid with a hexagonal base?

What shapes can you make with hexagons and squares? (See Fig. 2-A6 and 2-A7.)

Making shapes requires thinking ahead. Try to make a shape using only pentagons. What a relief: the two edges in Fig. 2-A8 really do seem to meet! How will the boy shown go on? Do the girls in Fig. 2-A9 and 2-A10 seem to be making the same shape?

The shape in Fig. 2-A11 is made entirely of pentagons: how many of them were used? Turn it around and look at it. How many edges does it have? How many corners? How many edges meet at one corner? How many faces

Fig. 2-A6.

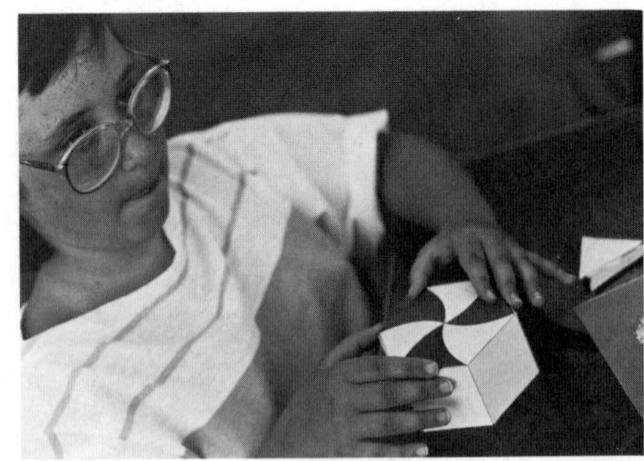

Fig. 2-A7.

Fig. 2-A8.

Fig. 2-A9.

Fig. 2-A10.

Fig. 2-A11.

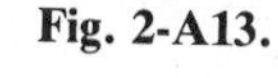

Fig. 2-A13.

Fig. 2-A12.

Fig. 2-A14.

meet at a corner? This shape is a *dodecahedron*.

When you are experimenting, don't expect that your shape will always close! (Figure 2-A12.) Some shapes may have holes that you cannot fill with the shapes that we have; remember that we are using only regular polygons.

Shapes You Can Make with Triangles

The shape in Fig. 2-A13 is only one of the many you can make using just triangles. It is an *icosahedron*. Look at it from many sides. How many faces, edges, and corners does it have? Compare these numbers to the corresponding numbers you found for the dodecahedron.

In Fig. 2-A14 two identical shapes are being glued together. Each is made of four triangles without a base. The finished shape will be an *octahedron*. What other shapes can you make with triangles?

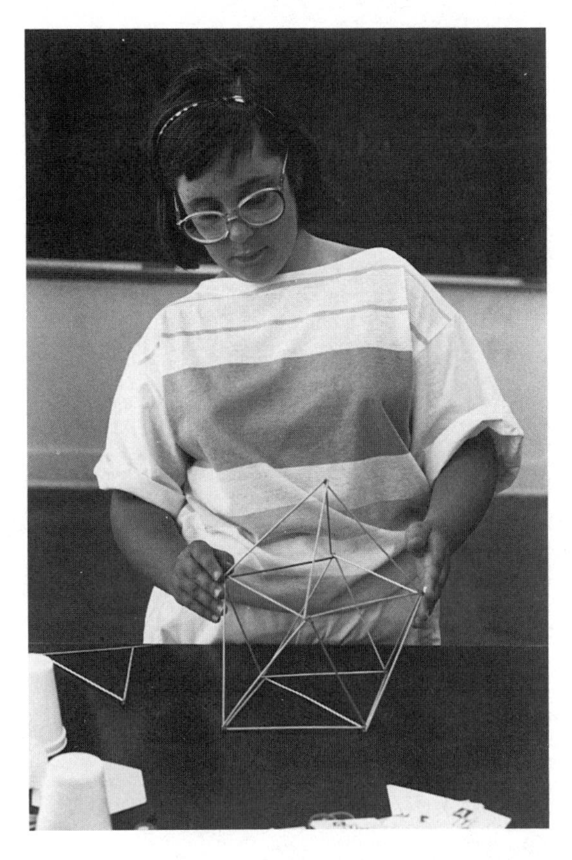

Fig. 2-A15.

A Note to the Teacher

Every problem leads to new observations and questions. For example, even the simple problem "Make all possible convex shapes using only equilateral triangles" is very rich in possibilities. These shapes are called *deltahedra,* after the triangular Greek letter Δ. Usually, after some experimentation, students will discover the tetrahedron, the octahedron, the triangular and pentagonal bipyramids, and the icosahedron. Later the search also yields the 12-, 14-, and 16-sided deltahedra. Figure 2-A15 shows a 14-sided deltahedron.

The observation that each deltahedron has an even number of faces leads to the question of why this should be so. The reason is straightforward once one sees it! Each triangle has three edges. If the shape has F faces, then there are $3F$ edges altogether. These $3F$ edges are glued in pairs, so there must be an even number of edges. Hence $3F$ and therefore F must be even. Noticing that there exist 4-, 6-, 8-, 10-, 12-, 14-, 16-, and 20-sided deltahedra immediately sets off a search for an 18-sided one. Can an 18-sided deltahedron be made? It was not until 1947 that the answer was proved to be *no*.

Looking at deltahedra is one thing; visualizing them without models is quite another. I found it difficult to close my eyes and visualize the 12-, 14-, and 16-sided deltahedra. One day, while I was looking at a cube made from applicator sticks and glue,[2] I decided to pose problems by using the "What-If-Not Strategy."[3] The idea is that one starts with a situation, a theorem, a diagram, or in our case an object, lists as many of its attributes as one can, and then asks, "What if not?" For example, among the many attributes (not necessarily independent) of a cube that I had listed were the following:

1. All edges are equal.
2. All faces are squares.
3. The object is not rigid.
4. The top vertices are directly above the bottom ones.
5. Opposite faces are parallel.

While working on attribute 4, I asked myself: "What if the top vertices were *not* directly above the bottom ones?" And because the contact glue gives movable joints, it was easy to give the top square a twist. As my twist approached 45°, I began to see an antiprism emerge. I attached sticks to complete the antiprism, but the shape wasn't rigid. The obvious thing to do to make it rigid was to add diagonals to the top and bottom squares. Since all the applicator sticks are of the same length, I had to squeeze the squares into "diamonds." The resulting shape was rigid—and was built of 12 equilateral triangles! (See Fig. 2-A16).

How else could I have made the antiprism rigid? I hastily removed the top diagonal, and added four sticks which meet above the square to form a square pyramid (Fig. 2-A17a). Lo and behold, I had made a 14-sided deltahedron! From there it was a quick step to remove the bottom diagonal also, build another four-sided pyramid, and thus obtain the 16-sided deltahedron (Fig. 2-A17b).

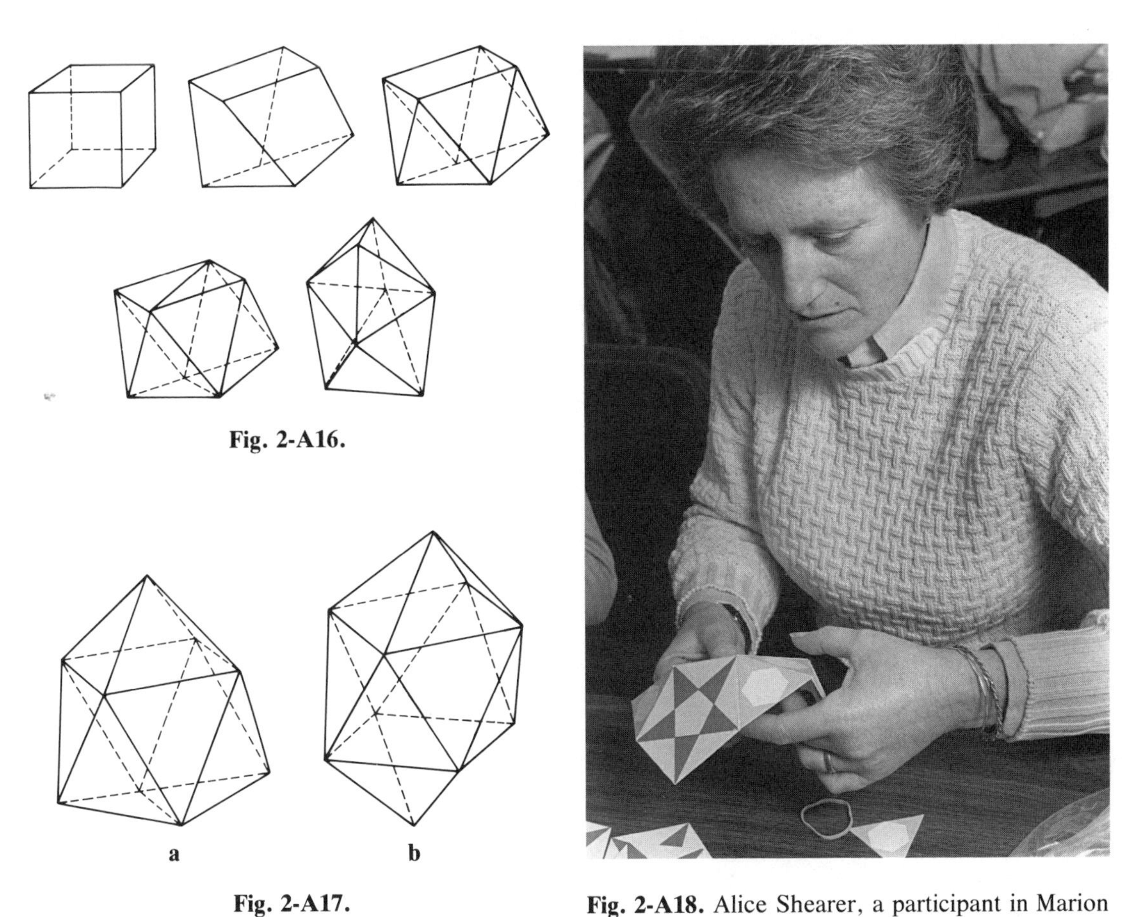

Fig. 2-A16.

Fig. 2-A17.

Fig. 2-A18. Alice Shearer, a participant in Marion Walter's workshop at the Shaping Space Conference, beginning construction of a model. Photograph by Stan Sherer.

Fig. 2-A19. By the end of the workshop, many lovely models had been created. Jane B. Phipps contemplating a polyhedron constructed from MATs. Photograph by Stan Sherer.

Not only have these deltahedral "villains" now become friends, I see now that they are closely related to one another. One can also place the icosahedron in this family, since it is a pentagonal antiprism capped with two pentagonal pyramids. (Indeed the octahedron itself is an antiprism, and the tetrahedron can be viewed as an antiprism in which the two bases have shrunk to an edge. As Arthur Loeb has pointed out, two opposite edges may be considered degenerate polygons, which are here in antiprism orientation.) That leaves us only with the 6- and 10-sided deltahedra as "odd ones out," but they are both bipyramids and are easy to visualize.[4]

A Word about Materials

Cardboard always works well; you should experiment with different weights. I prefer MATs, described in the next paragraph. All the polygons shown in these photographs are MATs. A glue used for carpets, such as Flexible Mold Compound—Mold It® is excellent, as is the English Copydex.[5]

Adrien Pinel found that hexagonal cardboard beer mats (used in English pubs) were excellent for making polyhedra with holes and, when augmented by triangles and squares cut from the hexagons, became even more useful. It was not long before the Association of Teachers of Mathematics of Great Britain had regular polygons of three, four, five, six, and eight sides produced from the same easy-to-glue material as the beer mats. They call them Mathematics Activity Tiles (MATs for short). They also produce rectangles and isosceles triangles. The polygons may be ordered separately or in two different kits: Kit A has 100 each of equilateral triangles, squares, pentagons, and hexagons, and Kit B has 200 each of triangles and squares and 50 each of pentagons, hexagons, and octagons.[6]

Acknowledgments: Photographs in Figs. 2-A1, 2-A3, and 2-A15 are by Ken R. O'Connell. Photographs in Fig. 2-A2 and Figs. 2-A4–2-A14 are by Marion Walter.

B. Constructing Pop-Up Polyhedra*

Jean Pedersen

Required Materials

- One 22 × 28 inch piece of brightly colored posterboard
- Six rubber bands (of the type that come around the morning newspaper in some localities)
- One yardstick or meter stick
- One ballpoint pen
- One pair of scissors

General Instructions for Preparing the Pattern Pieces

Begin by drawing the pattern pieces on the posterboard as shown in Fig. 2-B1. Press hard with the ballpoint pen so that the posterboard will fold easily and accurately in the final assembly. Label the points indicated. Be certain to put the labels on what will become the cube (or octahedron) when the model is finished—not on the paper that surrounds it. Cut out the pattern pieces and snip the notches at A and B (but not the notches at C and D).

Constructing the Cube

1. Crease the pattern piece with square faces on all of the indicated fold lines, remembering that the unmarked side of the paper should be on the outside of the finished cube. Thus each individual fold along a marked line should hide that marked line from view.

2. Position the pattern piece so that it forms a cube with flaps opening from the top and the bottom as shown in Fig. 2-B2.

3. Temporarily attach the two rectangles together inside the cube with paper clips. Then, with the cube still in its "up" position, cut through both thicknesses of paper at once to produce the notches at the positions which you already labeled C and D.

4. Connect three rubber bands together as shown in Fig. 2-B3.

5. Slide one end-loop of this chain of rubber bands through the slot which you labeled A, and the other end-loop through the slot labeled B, leaving the knots on the outside of the cube.

6. Stretch the end loops of the rubber bands so that they hook into slots C and D as shown in Fig. 2-B4. The bands must produce the right amount of tension in order for the model to work. If they are too tight the model will not

* I should like to thank Les Lange, Editor of *California Mathematics,* for giving permission to use in this article some of the ideas that were originally part of "Pop-up Polyhedra," *California Mathematics* (April 1983):37–41.

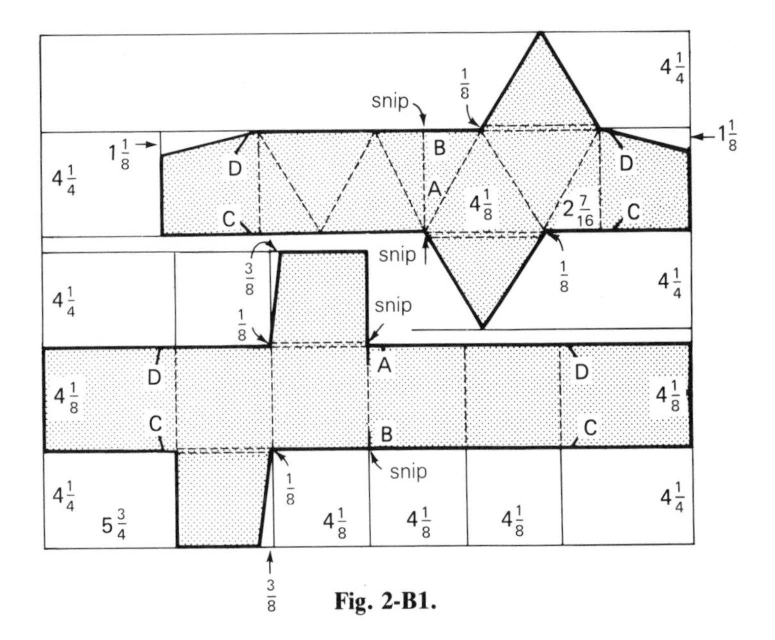

Fig. 2-B1.

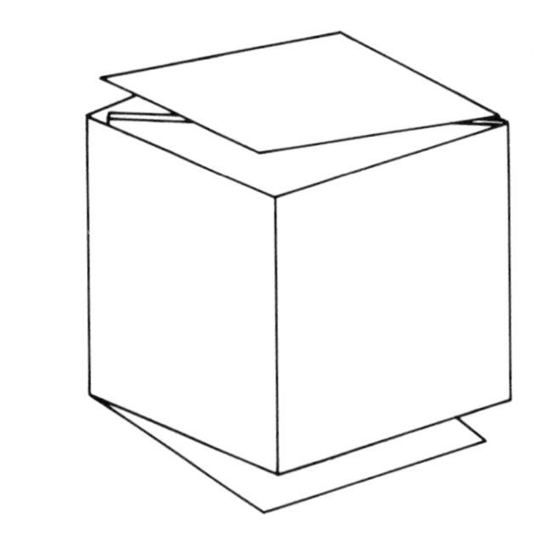

Fig. 2-B2.

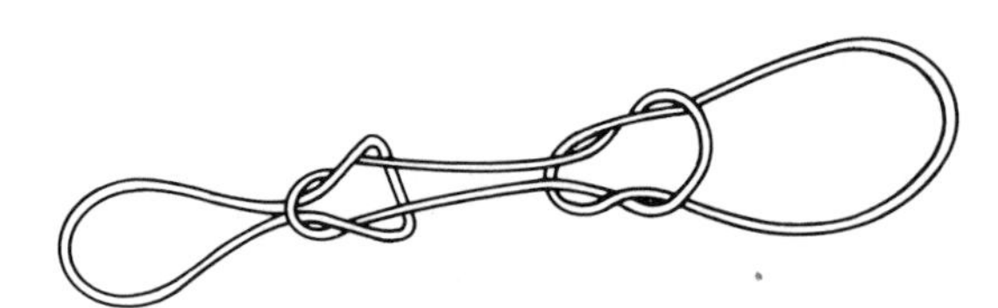

Fig. 2-B3.

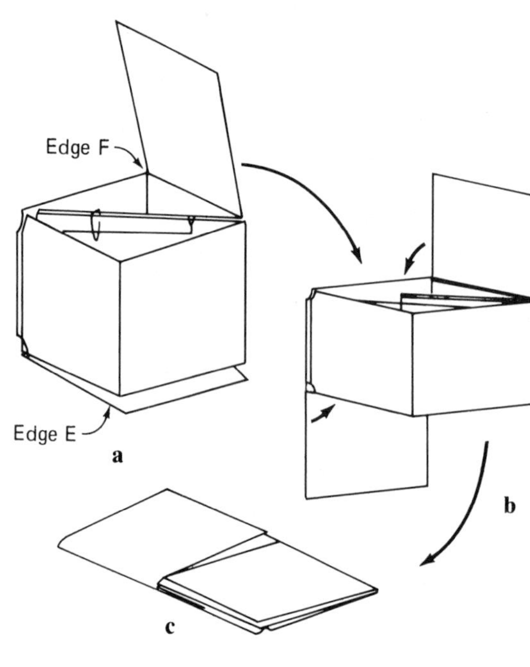

Fig. 2-B4.

go flat and if they are too loose the model won't pop up. You may need to do some experimenting to obtain the best arrangement.

7. Remove the paper clips when you are satisfied that the rubber bands are performing their function.

8. To flatten the model push the edges labeled E and F toward each other as shown in Figure 2-B4b and wrap the flaps over the flattened portion as in Figure 2-B4c.

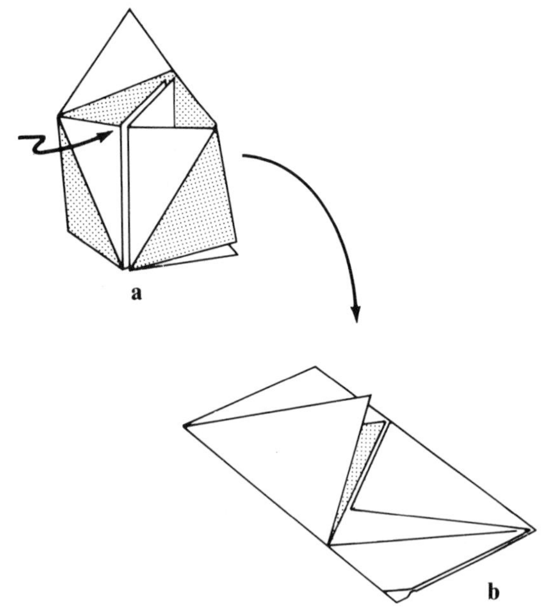

Fig. 2-B5.

9. Holding the flaps flat, toss the model into the air and watch it *pop up*. If you want it to make a louder noise when it snaps into position, glue an additional square onto each visible face of the cube in its "up" position. This also allows you to make the finished model very colorful.

Constructing the Octahedron

1. Crease on all the indicated fold lines so that the marked lines will be on the inside of the finished model.

2. Position the pattern piece so that it forms an octahedron with triangular flaps opening on the top and bottom, as shown in Fig. 2-B5a. Don't be discouraged by the complicated look of the illustration; the construction is so similar to the cube that once you have the pattern piece in hand, it becomes clear how to proceed.

3. Secure the quadrilaterals inside the octahedron with paper clips and cut through both thicknesses of paper to make the notches at C and D. Angle these cuts toward the center of the octahedron (so that the rubber bands will hook more securely). Gluing the quadrilaterals

inside the model to each other in their proper position produces a sturdier model.

4. Connect three rubber bands together as shown in Fig. 2-B3.

5. Slide one loop-end of the rubber band arrangement through the slot A and the other loop-end through the slot B, leaving both knots on the outside of the octahedron.

6. Stretch the end loops of the rubber bands so that they hook into the slots at C and D. Some adjustment in the size of the rubber bands may be necessary, so experiment to find the best arrangement.

7. Remove the paper clips when you have a satisfactory arrangement of rubber bands.

8. To flatten the model put your fingers inside and pull at the points A and D so that you are pulling those opposite faces away from each other until each one is folded along an altitude of that triangular face. Then wrap the triangular flaps over the flattened portion so that it looks like Fig. 2-B5b.

9. Holding the triangular flaps flat, toss the model and watch it *pop up*. Just as with the cube, this model will make more noise if you glue an extra triangle on the exposed faces. Of course, if you use colored pieces the resulting model is more interesting.

Note

If you store either the cube or the octahedron in its flattened position for several hours, or days, it may fail to pop up when tossed in the air. This is because the rubber bands lose their elasticity when stretched continuously for long periods of time. If the rubber bands have not begun to deteriorate, the model will behave normally as soon as you let the rubber bands contract for a short while.

C. The Great Stellated Dodecahedron

MAGNUS WENNINGER

The great stellated dodecahedron (Fig. 2-C1) makes a lovely Christmas decoration or indeed an interesting ornament for any time or place. It is very attractive, and when made as suggested here it is also very sturdy and rigid, even though it is entirely hollow inside. The pattern to use for making this model is simply an isosceles triangle with base angles of 72° and a vertex angle of 36°. The length of the base should be between 1 and 2 inches (between 2.5 and 5.0 cm). The angular measures just given will automatically make the equal sides of the triangle τ times longer than the base, where τ is 1.618034 (the golden section number). You will need 60 such triangles to complete one model. Very attractive results can be obtained by using different colors of index card, namely ten triangles of each of six different colors. Astonishingly beautiful results can be obtained by using glitter film with pressure sensitive adhesive backing to cover the index card.[7]

Getting Started

Begin the work by first cutting all the glitter film triangles to exactly the same size. You can lay out a tessellated network of such triangles by marking the back or waxy side of a sheet of glitter film with a scoring instrument and then cutting out the triangles with scissors. Next peel off one corner of the waxy backing from the film and attach this to a piece of index card. Finally, remove the entire backing while you smooth out the film on the card. Now trim the card with scissors, leaving a border of card all around the film. A quarter inch or so is suitable (about 7 or 8 mm). Next trim

Fig. 2-C1. Smith College student Katherine Kirkpatrick studying models made in Magnus Wenninger's workshop at the Shaping Space Conference. Photograph by Stan Sherer.

the vertices of the triangle as suggested in Fig. 2-C2. You will now find it easy to bend or fold the card down along the edges of the film even without scoring the card. This edging of card serves as a tab for joining the triangles together. Use ordinary white paper glue (such as Elmer's Glue-All®[8]) for this purpose.

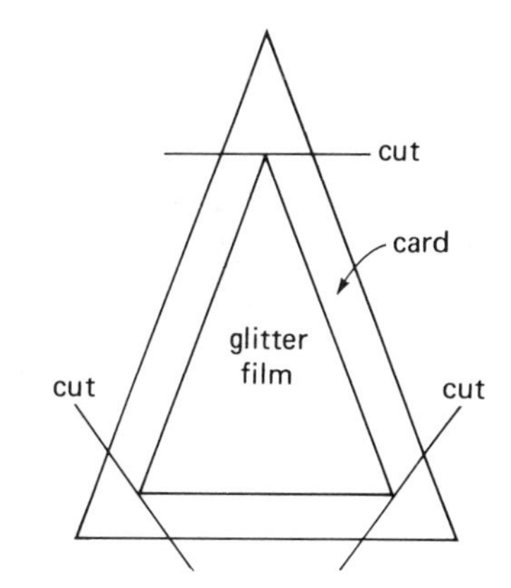

Fig. 2-C2.

Assembling the Model

Glue three triangles together as shown in Fig. 2-C3. Shape this part into a triangular pyramid without a base. This will then form one trihedral vertex of the great stellated dodecahedron. The color arrangement for ten vertices is as follows:

(1) B Y G	(6) B W G	Y = yellow or gold
(2) O B Y	(7) O W Y	B = blue
(3) R O B	(8) R W B	O = orange
(4) G R O	(9) G W O	R = red
(5) Y G R	(10) Y W R	G = green
		W = white or silver

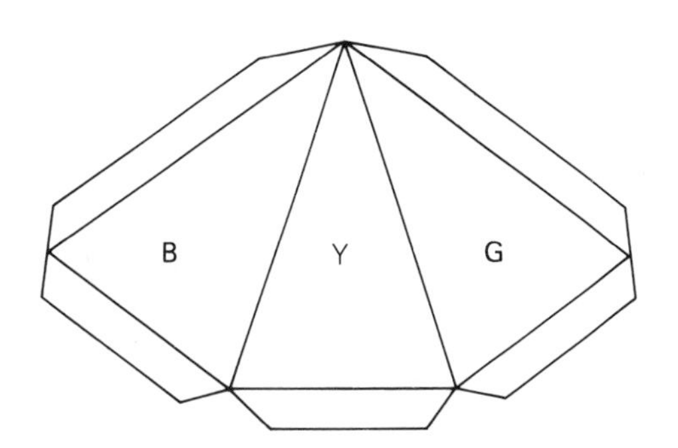

Fig. 2-C3.

The first five vertices or triangular pyramids are joined in a ring with the bottom edges of the middle Y, B, O, R, G of (1), (2), (3), (4), (5), forming an open pentagon. Then the next five parts are added to each edge of this pentagon, so that the W of (6) is glued to the Y of (1) and so on around. This completes half the model. You may find it a bit tricky to get the colors right at first, but the arrangement suggested here makes each star plane the same color. The triangles are star arms, so once you get started right it is not hard to continue.

The remaining ten vertices or parts have their colors in reverse order. They are the mirror image arrangement of the first ten. To make these just read the color table in reverse order and from right to left. For example, vertex (11) will be R W Y, the reverse of (10) which is Y W R. And this is glued in place diametrically opposite to its counterpart on the model. Watching the colors of the star arms will help you get all the remaining parts in their proper places. As the model closes up it is helpful to use tweezers to get the tabs to adhere. The secret is to do only one pair of tabs at a time. On the last part glue one pair of tabs first. Then, when this has set up firmly,

Fig. 2-C4.

Fig. 2-C5. Magnus Wenninger leading a workshop at the Shaping Space Conference. Photograph by Stan Sherer.

put glue on the remaining two sets of tabs and close the triangular opening. The model now has sufficient rigidity so that the tabs will adhere by applying gentle pressure from the outside with your hands. An extra drop of glue at the base of each pyramid corner will provide extra strength where you may perceive a small opening remaining.

You should now see, if you have not already noticed this, that parallel star planes are the same color. Hence twelve star planes complete this model, two of each of the six colors. The twelve stars give this model its name: stellated dodecahedron (Fig. 2-C4). It is called ''great'' because it is the final stellation of the dodecahedron, truly a beautiful thing to behold!

''A thing of beauty is a joy forever.''

D. Creating Kaleidocycles and More

DORIS SCHATTSCHNEIDER

The transition from a flat pattern to a three-dimensional form can be fascinating to explore. Even the youngest child can shape a simple basket by cutting squares from the corners of a rectangular piece of construction paper and folding it up. But except for this well-known pattern learned as a preschool exercise, the two-dimensional pattern of an unfamiliar three-dimensional object often seems to yield little information about the object. Perhaps part of the reason for this is that we are rarely asked to imagine what shape will result from folding up a flat pattern. The following exercises provide hands-on exploration of some of the relations between flat nets and three-dimensional forms, and provide an extra surprise in the creation of kinetic forms.

Folding Strips of Triangles

Begin by constructing strips of four connected congruent triangles like the one sketched in Fig. 2-D1. You should construct several different kinds of strips—those whose triangles are (1) equilateral triangles, (2) isosceles acute triangles, (3) isosceles right triangles, (4) isosceles obtuse triangles, (5) scalene triangles (a strip of acute triangles, or of right, or obtuse). *Note:* If you are in a hurry, use graph paper for rapid layout of the strips of congruent triangles. If more time is available, carry out rule-and-compass construction of the strips (brush up on the congruence theorems!).

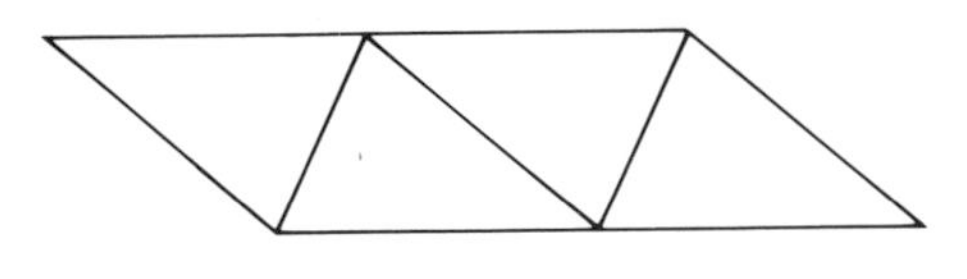

Fig. 2-D1.

Question: For each of the constructed nets, what three-dimensional shape will be formed when the net is folded along the common edges of its triangles?

First guess at the answers to this question. Then score the connecting edges of the triangles (use a medium ballpoint pen held against a straightedge), cut out the strips of triangles, and fold each of them to see what happens. (All folds should be of the same type, folding the pattern back-to-back.)

After this, other strips of four triangles can be explored: for instance, a strip of four triangles, all acute, but not all congruent; a strip of triangles with some triangles right, others acute, and so forth. Exploring what happens when these nets are folded up leads to some natural questions:

1. When will four congruent triangles form a tetrahedron?
2. What must be true of four triangles if they are to form a tetrahedron?
3. Are there different flat nets (other than the strips of four triangles) that will fold up to make the same shapes as those formed by the strips of four triangles?

Kaleidocycles

Next, we will create and explore nets of connected strips of triangles. For ease and accuracy of construction, large paper and long (18 inch) rulers should be used. Graph paper can be purchased in size 17 × 22 inch, just right for two constructions. Drawing paper can easily be purchased in large sizes. Lay out each of the two grids shown in Figs. 2-D2a and 2-D3. The grid in Fig. 2-D2a is made up of six connected vertical strips of congruent isosceles triangles which are characterized by the property that base equals altitude. The grid is easily laid out using graph paper; it is also easily constructed with ruler and compass because of the simple defining property of the triangles. (There is a grid of squares which underlies the triangular grid; this is shown in Fig. 2-D2b).

The grid in Fig. 2-D3 is made up of twelve connected vertical strips of isosceles right triangles, where the top and bottom triangles have been cut in half. This grid is obviously based on a grid of squares, and so is easily laid out on graph paper or constructed with ruler and compass.

Before the grids are turned into three-dimensional objects, ask yourself the question that was asked earlier for the single strips.

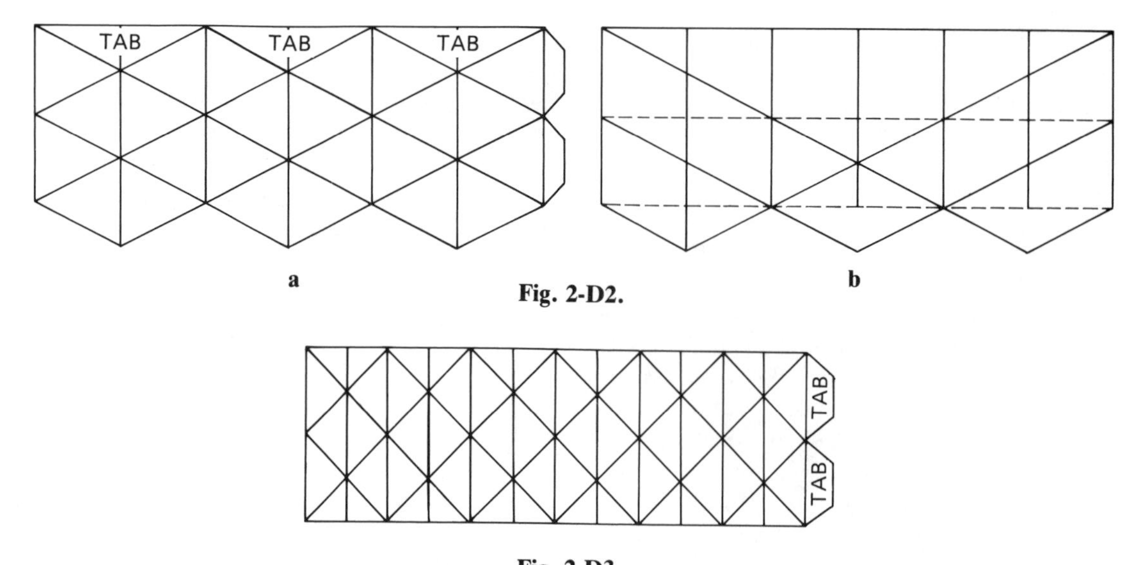

Fig. 2-D2.

TAB
TAB

Fig. 2-D3.

Question: From each of the constructed nets (as in Figs. 2-D2a and 2-D3), what three-dimensional shape will be formed when they are folded along the common edges of the triangles?

Of course, you will need to use the earlier answers to the question in attempting to answer the question for the more complex nets. An auxiliary question that is worth asking is: Will all of the lines in the grid (which are common edges of pairs of triangles) play the same role in the three-dimensional form?

Now score all the lines in each grid (use a medium ballpoint pen), and cut out the nets around the outline (be sure to cut around the tabs). Fold the nets as follows:

1. Fold the net face-to-face (valley fold) on all vertical lines, including those to which the tabs are attached.
2. And fold the net back-to-back (mountain fold) on all diagonal lines.

Then cup the folded net in both hands, and gently squeeze it to encourage the top and the bottom to come together.

The net in Fig. 2-D2a should come together easily, with the half-triangles labeled as tabs completely covered. A chain of linked tetrahedra is formed. (Glue or tape the edges of the tetrahedra fitted over the tabs.) Holding the ends of the chain, bring the ends of the chain together, fitting the tabs at one end of the chain into the open edge at the other end of the chain. (If the chain does not come together easily, turn it until it does.) Glue or tape these last two edges to the tabs, completing the model.

The ring of six linked tetrahedra is a (carefully) crinkled torus (doughnut), and has the property that it can be *endlessly turned through its center hole*. Simply grasp the model in both hands and turn the tetrahedra inward, pushing the points through the center hole!

The Isoaxis

The net in Fig. 2-D3 will not come together to form a closed three-dimensional form, but rather it will form a (carefully) crinkled cylinder that will also turn through its center hole, changing its shape and appearing to "bloom" as it is turned. This form was discovered by graphic designer Wallace Walker, and is called Isoaxis®.[9] Assemble Isoaxis as follows. Gently squeeze the scored and folded net, so that it begins to curl and collapse along the fold lines. When fully collapsed, it will look like an accordion-folded paper with square cross section. (One method of achieving this state of the model is to begin at one end of the net, collapsing the net along the folds to form a square cross section, and holding the collapsed part between thumbs and forefingers, "gathering" the rest of the net into the col-

Fig. 2-D4. Corraine Alves and Diana Weimer making kaleidocycles in Doris Schattschneider's workshop at the Shaping Space Conference.

lapsed state with the middle fingers.) The accordion-folded net should be pressed firmly; it is best if it can be pressed under a heavy object for 12 hours or more to set the folds fully. The two ends of the folded net are then joined (use tape or glue), matching tabs to the inside of opposite triangles. Join one tab at a time; the model will be tight, and so turn it through its center to join the second tab. To rotate this model, hold it in both hands and bring points to the center; push on the points. The crinkled cylinder will turn continuously through its center hole!

Further Exploration

There are many avenues for follow-up. A few are suggested below.

1. Explore the symmetry properties of the three-dimensional models. This can be enhanced by decorating the faces of the models to display various symmetries. One question that will need to be answered is: What faces in the fat net are adjacent (or become adjacent during rotation) in the three-dimensional forms?
2. Create other similar nets, varying the kinds of triangles chosen, and the number of triangles in the net. Fold in the same manner to see what three-dimensional forms result. A good challenge that can be met using only a knowledge of elementary geometry is: create other rings of tetrahedra having more tetrahedra, but such that the center hole in the ring is (in theory) a point, as is the case for the model in Fig. 2-D2. (The model in Fig. 2-D2 has been called a "hexagonal kaleidocycle" by Walker and Schattschneider, because the center cross section of the assembled form is a regular hexagon.[10]

Information on Construction Materials

The basic necessities for the above constructions are

- Paper
- 18-inch ruler
- Medium ballpoint pen
- Scissors
- White glue or tape

If the models are to be decorated, then coloring materials that will not weaken or warp the paper should be used. Since the models rotate, the paper chosen must not easily tear or break when bent repeatedly. Ordinary construction paper is not suitable. In addition, the paper should be heavy enough so that the three-dimensional models have suitable firmness. Medium-weight drawing paper, 100% rag, is excellent, and takes decoration well. Ordinary graph paper is too thin, but there is an excellent gridded layout bristol (made by Tara) that comes in large sizes. The nets should not be made too small, or they become very difficult to put together and to manipulate. A good size for the nets is 2.5 to 3 inches width for each "panel" of linked triangles for Fig. 2-D2, and 1.25 to 1.75 inches for each "panel" of linked triangles for Fig. 2-D3. The overall width of the nets should be in the range of 15 to 22 inches.

White glue seems to be best for assembling the models; in any case, the glue chosen should not warp the paper, nor should it be the "instant hold" variety since tabs need to be manipulated into place before the glue sets. If tape is used, then it must be of the type used for hinging (such as Mylar®[11]); ordinary clear plastic tape will break after a few turns of the model.

Giftwrap paper which has a pattern based on a square grid can be laminated (use spray glue) to drawing paper to create a nice all-over decorated Isoaxis. (The square grid of the pattern must be carefully followed for the lines of the net of Isoaxis.)

E. The Rhombic Dodecahedron: Its Relation to the Cube and the Octahedron

ARTHUR L. LOEB

This is a recipe for constructing modules that generate the rhombic dodecahedron in two fundamentally different ways. The first construction stellates a cube with six square pyramids; the second stellates a regular octahedron with eight triangular pyramids.

The Pyramids

The first step is the construction of the sides of the pyramids. The square pyramids have an apex angle whose cosine equals 1/3, while the triangular pyramids have an apex angle whose cosine equals −1/3. In order to produce mutually congruent dodecahedra by both methods, we construct the template shown in Fig. 2-E1 by the following steps:

1. Draw two mutually perpendicular lines. Call their intersection O.
2. Choose a point A, different from O, on one of the mutually perpendicular lines.
3. Draw a circle having radius equal to three time the distance OA, whose center is located on A.
4. Call the intersections of this circle with the extension of the line OA C and B, as shown. Call an intersection of the circle with the line perpendicular to OA D, as shown.
5. Connect C and D, as well as B and D.

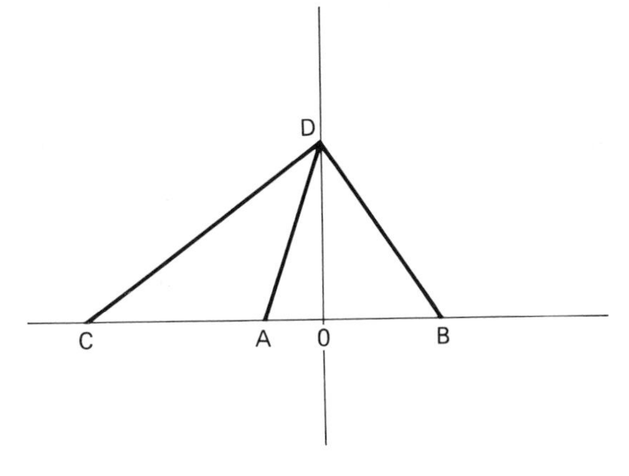

Fig. 2-E1.

The resulting template furnishes the following linear and angular dimensions:

- The length of the line segment CD is the edge length of the octahedron to be built.
- The length of the line segment BD is the edge length of the cube to be built.
- The triangle CAD is the shape of the sides of the triangular pyramids to be built.
- The triangle BAD is the shape of the sides of the square pyramids to be built.

Construction of the Polyhedra

1. Construct a regular octahedron whose edge length equals the length of line segment CD.
2. Construct a cube whose edge length equals the length of line segment BD.
3. Construct eight triangular pyramids whose bases are equilateral triangles having edge length equal to the length of line segment CD, and whose sides have the shape of triangle CAD.
4. Construct six square pyramids whose bases are square having edge length equal to the length of line segment BD, and whose sides have the shape of triangle BAD.

Juxtaposition of Polyhedra

Arrange the six square pyramids so that their square bases are in the configuration shown in Fig. 2-E2. Hinge them together so that they can rotate with respect to each other around their shared edges. When the pyramids are folded inward until their six apices touch, they will form a cube congruent with the cube also constructed.

Arrange four of the triangular pyramids with their triangular bases in the configuration shown in Fig. 2-E3, and hinge them together as above. When folded in till their apices touch, they will form a regular tetrahedron. Repeat for the remaining four tetrahedra.

Place the six square pyramids around the cube, square faces joined to square faces. Place the eight triangular pyramids around the octahedron, with the equilateral triangles joined. The result should be two mutually congruent rhombic dodecahedra. Note that the cube edges constitute the shorter, the octahe-

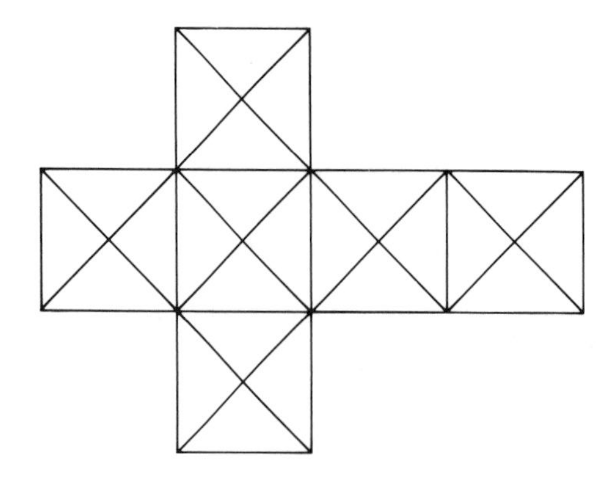

Fig. 2-E2.

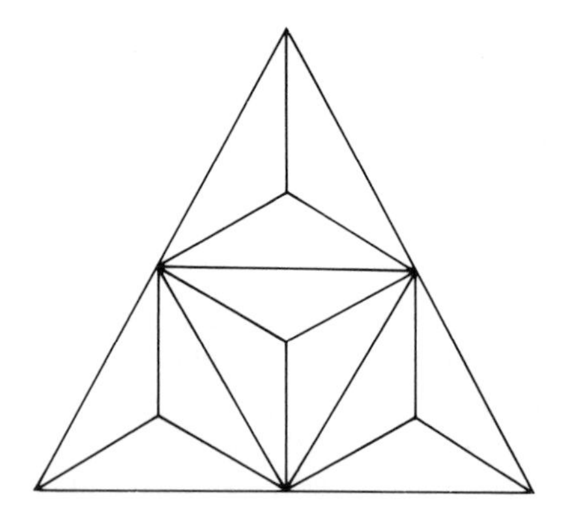

Fig. 2-E3.

dron edges the longer, diagonals of the rhombic faces.

Place two square pyramids with their square bases joined. The result is an octahedron which is not regular, because its faces are not equilateral. Six of these irregular octahedra can be put together to form a rhombic dodecahedron. (*Note:* this would require *twelve* square pyramids rather than the six already constructed.)

Space-Fillers

Of the polyhedra constructed, the following will fill space without interstitial spaces.

- Cube
- Rhombic dodecahedron
- Square pyramid
- Irregular octahedron
- Regular octahedron combined with eight triangular pyramids

Combinations of these (say cube in combination with square pyramids) are, of course, also possible.

A Note on Materials

Contributed by JACK GRAY

Any useful polyhedral model is formed on the spectrum between "a rough sketch" and "a long-lasting work of art." The position on the spectrum is determined by the choices of materials, tools, and techniques as well as by the time and care used in the construction process. A rough sketch is always a valid precursor to a work of art. Expect to make a few mistakes on the sketch, and then try to conquer those in a second model.

Transparent tape is a good hinging material, while paper tape is thicker and more cumbersome. Use permanent tape, taking care to position it as follows. Place a strip of tape, sticky side up, on a flat surface. The strip should be longer by a good amount than the edge to be hinged. Weight down the ends of the tape so that it cannot move while you are connecting the polyhedron to it.

Carefully lower the edge of the first polyhedron to the tape. Before letting it contact, make sure that it is in the center of the width and the length of the tape. Make contact along the whole length of the edge.

Orient the second polyhedron to the first. Slide the second polyhedron down the face of the first until its edge touches the tape; then rotate it about that edge, so that contact with the tape is made along the entire edge. Trim off the excess tape with an X-Acto®[12] knife or a single-edge razor blade.

Flip over the pair of joined polyhedra and inspect the tap hinge. Burnish it with your finger to complete contact along the full surface of the tape.

Place another piece of tape of the same length on the flat surface. This will be used to tape the other side of the joined edges, creating a hinge that is equally strong on both sides. The tape should be weighted down as before, and care should be taken in centering the already-joined edges on the strip of tape before making contact. Let the joined faces lie flat against the tape to make contact along the full

Fig. 2-E4. Arthur L. Loeb demonstrating his models in his workshop at the Shaping Space Conference. Photograph by Stan Sherer.

width of tape surface. Remove the excess tape and burnish as before.

If your sketch looks like a work of art, plan a finished model. Visit a local art store or hobby shop to examine the sheet materials that are available.

Various colored art papers, colored and transparent acetate, mirrored Mylar®, oak tag, and construction paper can be found. Fine rice papers are good for finishes, though they are too flexible for the body of such models. (In adding a surface finish of thin sheet material to your model, cut out each polygonal face so that it will not go across the hinge. Otherwise, the finish material will buckle when flexed.) Your experience making a sketch model will prepare you to pick materials "by feel."

Thicker sheet materials like mat board and Plexiglas®[13] need to have their edges mitered to half the dihedral angle between faces to prevent the thickness of the material from creating inaccuracies. Great care should be used in gluing such joints, so that glue does not spill onto the surfaces.

On a finished model, the hinging should be done with transparent polyester hinging tape. Cloth tape can be used on larger models.

Notes (for Chapter Two)

[1] These types of starting points were suggested by Adrian Pinel. For many more suggestions for polyhedra-making activities, see Pinel's 24-page booklet *Mathematical Activity Tiles Handbook* (Association of Teachers of Mathematics, Kings Chambers, Queen Street, Derby DE1 3DA, United Kingdom), ISBN 0900095-59-8, $2.50.

[2] Use sticks all of the same length. Some drugstores sell applicator sticks which are ideal; be sure to get the kind without cotton at each end. Hobby and craft stores often sell small-diameter wooden dowel rods which work well. Put a small amount of contact glue on the ends of the sticks and let it dry for about 15 minutes, until the glue is tacky. Then the sticks will join well and yet stay flexible. Don't be surprised if a cube or dodecahedron made of applicator sticks won't stand up, however. Unlike structures built entirely of triangles, these structures are nonrigid.

[3] S. I. Brown and M. I. Walter, *The Art of Problem Posing* (Hillsdale, N.J.: Lawrence (Erlbaum, 1983).

[4] Marion Walter, "On Constructing Deltahedra," *Wiskobas Bulletin, Jaargang 5/6* [I.O.W.O., Utrecht] (Aug. 1976).

[5] Mold It® is a registered trademark of Joli Plastics and Chemical Corporation, 14922 Garfield Avenue, Paramount, Calif. 90723.

[6] MATs may be ordered from the Association of Teachers of Mathematics, Kings Chambers, Queen Street, Derby DE1 3DA, United Kingdom. Kit A costs £5.75 plus postage. Kit B costs £6.90 plus postage.

[7] This is obtainable from Coburn Corporation, 1650 Corporate Road West, Lakewood, N.J. 08701.

[8] Elmer's® is a registered trademark of Borden, Inc., Columbus, Ohio, for a white liquid adhesive.

[9] Isoaxis® can be obtained from Wallace Walker, 41 West 46 Street, New York, N.Y. 10036.

[10] D. Schattschneider and W. Walker, *M. C. Escher Kaleidocycles* (New York: Ballantine Books, 1977); Tarquin Publications, Stradbroke, Diss, Norfolk, IP21 5JP, United Kingdom, 1982; Pomegranate Artbooks, Corte Madera, Calif., 1987.

[11] Mylar® is a registered trademark of E. I. du Pont de Nemours and Company, Wilmington, Del.

[12] X-Acto® is a registered trademark of the X-Acto Company, Long Island City, N.Y.

[13] Plexiglas® is a registered trademark of Rohm and Haas, Philadelphia, for acrylic safety glazing.

Part II
Lectures from the Shaping Space Conference

3

Regular and Semiregular Polyhedra

H. S. M. Coxeter

The cube, the octahedron, and the tetrahedron obviously have been admired for thousands of years. It is impossible to say who first described them. Certainly the Pythagoreans knew all about them. I understand that a dodecahedron was found in Italy which was apparently made in 500 B.C. or perhaps even earlier, and that icosahedral dice were used by the ancient Egyptians. They can be seen in the British Museum, although there is some doubt about their exact date. All the five so-called Platonic solids are described in the later books of Euclid. Subsequent writers have made it much easier to see that the number of Platonic solids is just five.

First of all, perhaps, one should define what one means by a regular solid. It is rather strange that not many people realize how very simple the definition can be. If one starts in the plane defining a regular polygon, one can say that a polygon is regular if it has a circumcircle and an incircle which are concentric. All the vertices lie on a circle and all the sides touch a circle and those two circles have the same center. That is a very obvious way of defining a regular polygon. The same thing works in the analogous situation for a polyhedron in three dimensions. A polyhedron is regular if it has three spheres, all with the same center: one through all the vertices, one touching all the edges and one touching all the faces. And that is all one needs. It is very easy to show from this that the faces are regular polygons and that they are all alike.

Of course, if you are dealing with a honeycomb (that is, a tessellation of the plane with regular polygons), you have to make the definition a little bit different and say that the polygons are regular and all alike. And then you know that the ways of filling the plane with regular polygons are just three: triangles, six at a vertex, which I call $\{3,6\}$; squares, four at vertex, which I call $\{4,4\}$; and hexagons, three at a vertex, which I call $\{6,3\}$. Those are what we call Schläfli symbols. The first entry is the number of sides of a face and the second is the number of faces at a corner. So a cube, for instance is called $\{4,3\}$. Figure 3-1 gives such symbols for all five Platonic solids.

There is a very nice book on the history of these things by van der Waerden called *Science Awakening*.[1] He has a fine chapter about Pythagoras, and he says that an Etruscan dodecahedron made of soapstone was found near Padua, dated from before 500 B.C. The faces of a dodecahedron are pentagons. You know that if you draw the diagonals of a pentagon you get a star pentagon inside. The star pentagon is the ancient symbol of the Pythagoreans. The story is told in van der Waerden's book that one of the Pythagoreans was lying on his deathbed in a foreign country, unable to pay the man who had taken care of him. And he advised this man to paint a star pentagon on the door of the house so that any Pythagorean who might enter would make inquiries. And many years later, a Pythagorean did come, and the man was richly rewarded. A rather nice little story.

Coming to much more recent times, René Descartes (1596–1650) wrote a book called *De Solidorum Elementis*.[2] Although the manu-

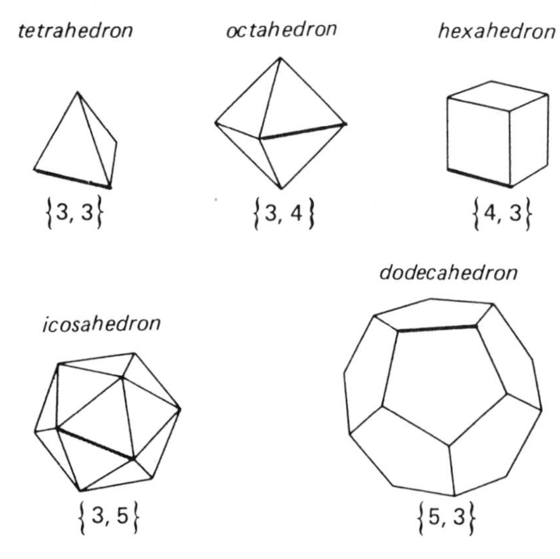

Fig. 3-1. Schläfli symbols for the Platonic solids.

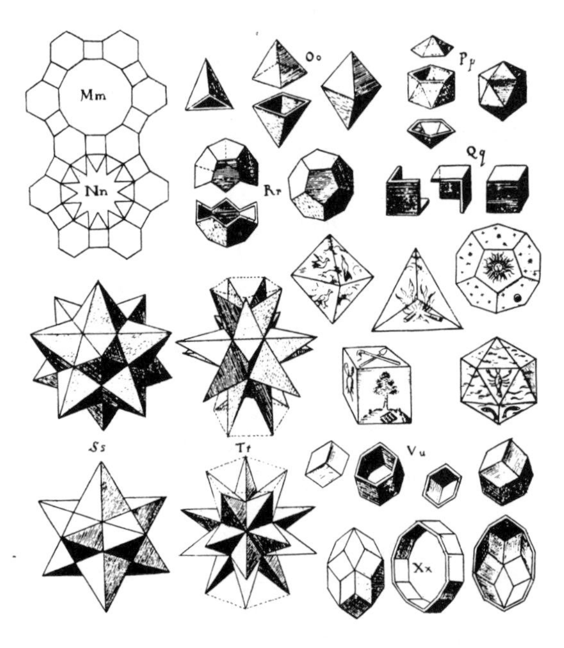

Fig. 3-2. The regular solids (and others) in Kepler's *Harmonices Mundi* (Linz, 1619).

script was soaked for three days after a shipwreck on the river Seine, it was copied in 1676 by Leibniz before it was lost forever. His copy was lost too, but that loss was only temporary; two hundred years later Leibniz's copy was found in Hannover, Germany. Shortly before Descartes came Leonardo da Vinci and Luca Pacioli. Pacioli wrote a book called *Divina Proportione* in which he had pictures of regular and Archimedean solids based on models made by da Vinci.

Johannes Kepler was very interested in these things, and his *Harmonices Mundi* of 1619 contains one of his famous illustrations. In Fig. 3-2 you see at the top (Mm) his attempt to fill the plane with polygons of various kinds such as a dodecagon with hexagons and squares around it. Next you see a tetrahedron, and then two halves of an octahedron (Oo) showing that the octahedron is just two square pyramids put base to base. Rr is the dodecahedron and here he has divided it into two parts by cutting along a set of ten edges which form a Petrie polygon. See how they fit together. And in Qq the cube is divided by a skew hexagon, which is *its* Petrie polygon. Those two halves fit together to make the cube. Pp is the icosahedron with little caps taken off the top and bottom, leaving a pentagonal antiprism between the two pentagonal pyramids. Alternatively, just take the antiprism and stick the two pyramids on top and bottom and there is the icosahedron.

One of the ideas that Kepler got from the ancient Greeks was making four of the five Platonic solids correspond to the four elements: earth, air, fire, and water. In Fig. 3-2 you see the tetrahedron with a bonfire drawn on it because the tetrahedron represents *fire;* the octahedron, representing *air*, has birds; the icosahedron has a lobster and fishes because the icosahedron represents *water;* and you see a hoe and spade on one face of the cube, a carrot on another and a tree on a third, because the cube represents *earth*. Well of course there was a fifth solid and no fifth element, so the ancients just said that the dodecahedron should correspond to the whole universe. That was curiously echoed by the Japanese; I have a model in which, if you look closely, you find that on the twelve faces are drawn the twelve Japanese signs of the Zodiac. It's a little bit different from the Greek Zodiac: one sign is a dragon, one is a doe, one is a dog, one is a chicken, and so on.

In Fig. 3-2 you see also two *stellated* dodecahedra, each derived from the convex dodecahedron by extending the planes of the faces. Kepler's drawings Tt of the *great* stellated dodecahedron are a little bit inaccurate,

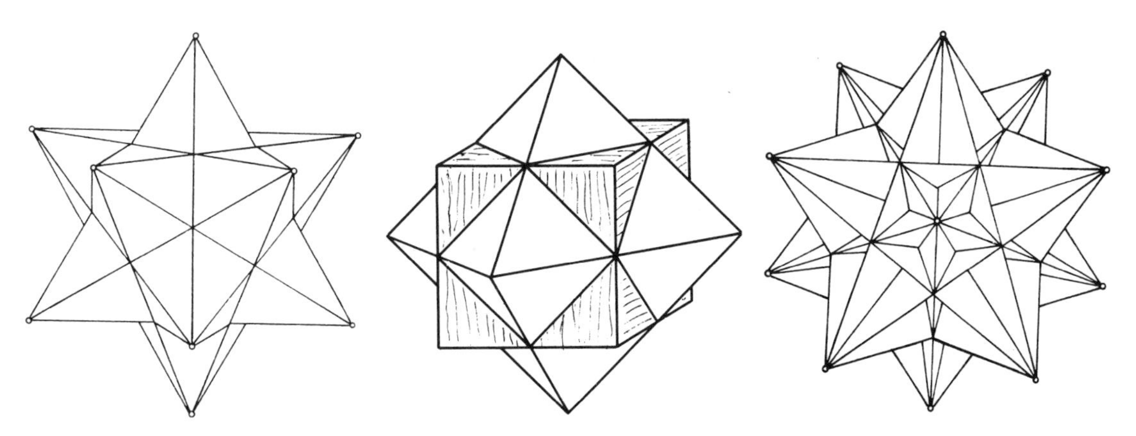

Fig. 3-3. The small stellated dodecahedron.

Fig. 3-4. The cube and the octahedron.

Fig. 3-5. The great icosahedron.

but they give us the right idea. In Ss we see two views of the simpler *small* stellated dodecahedron. Somewhere in Italy there is an elaborate floor on which a picture of this polyhedron appears, nicely drawn in 1420 as a mosaic (Fig. 1-21). So the small stellated dodecahedron may have been discovered by Paolo Uccello, two centuries before Kepler. Figure 3-3 shows a different view of the same polyhedron.

Kepler seems to have understood the nature of *reciprocation:* the idea that the cube and the octahedron are reciprocal, and the dodecahedron and the icosahedron are reciprocal. Figure 3-4 shows a shaded cube and white octahedron with corresponding edges crossing each other at right angles. You see that each corner of the cube emerges through a face of the octahedron and vice versa. Curiously enough, although Kepler had that idea, he did not pursue it in the matter of the star polyhedra. If you think of a Schläfli symbol, whereby the cube is called {4,3} and the reciprocal octahedron {3,4}, then it would be natural to call the small stellated dodecahedron {5/2,5} because the faces are star pentagons corresponding to the fraction 5/2, and there are five at each corner. And similarly the great stellated dodecahedron is {5/2,3}. The faces are pentagrams again and there are three of these at every corner. Turn these symbols around and you have {5,5/2} and {3,5/2}. {3,5/2}, the great iscosahedron, is shown in Fig. 3-5. These were not discovered until two hundred years later, by the Frenchman Louis Poinsot, but when they finally were there, it became clear they were reciprocal to Kepler's {5/2,5} and {5/2,3}.

I have a little more to say about Kepler, because of his interest in mystical connections between the Platonic solids and astronomy. Of course, he was very interested in astronomy. He had a curious idea about the orbits of the planets which is very nicely described in a book by Arthur Koestler,[3] the man who wrote *Darkness at Noon*. Koestler said: "Into the orbit or sphere of Saturn he inscribed a cube." Now let's say exactly what that means. You think of the orbit of Saturn as the equator of a big sphere and similarly the orbit of Jupiter as the equator of a smaller sphere inside. Kepler was a sufficiently good astronomer to know that the orbits were not really circles, but more like ellipses, so that there is a minimum distance and a maximum distance from the sun in each case. What he did was to imagine a sphere in space that was made as a shell, not a mathematical sphere but a solid shell with an outer radius corresponding to the maximum distance of the planet and an inner radius to the minimum distance. So it is a hollow sphere. And he would take the minimum distance of Saturn and divide it by the maximum distance of Jupiter to get the ratio of the circumradius and inradius of a cube. For the six planets known to Kepler, Table 3-1 gives the distances from the sun in millions of miles for comparison with the circumradius ${}_0R$ and inradius ${}_2R$ of each Platonic solid. For instance, in the case of Saturn and Jupiter the astronomical result is 1.69 as an approximation to the

Table 3-1. Distances of the planets from the sun in millions of miles, together with ratios needed for examining Kepler's theory of the solar system.

Planet	Maximal distance	Minimal distance	Ratio	Polyhedron	${}_0R/{}_2R$
Saturn	935	837			
			$\frac{837}{507} = 1.69$	Cube	$\sqrt{3} = 1.73$
Jupiter	507	459			
			$\frac{459}{155} = 2.98$	Tetrahedron	3
Mars	155	128			
			$\frac{128}{94\frac{1}{2}} = 1.35$	Dodecahedron	$\sqrt{(15 - 6\sqrt{5})} = 1.26$
Earth	$94\frac{1}{2}$	$91\frac{1}{2}$			
			$\frac{91\frac{1}{2}}{68} = 1.35$	Icosahedron	$\sqrt{(15 - 6\sqrt{5})} = 1.26$
Venus	68	67			
			$\frac{67}{43} = 1.54$	Octahedron	$\sqrt{3} = 1.73$
Mercury	43	28			

square root of 3. So let me go on quoting from Koestler:

Into the orbit, or sphere, of Saturn he inscribed a cube, and into the cube another sphere which was that of Jupiter. Inscribed in that was the tetrahedron, and inscribed in that the sphere of Mars. Between the spheres of Mars and Earth came the dodecahedron, between the Earth and Venus, the icosahedron, between Venus and Mercury, the octahedron. Eureka!. . . This is the ultimate fascination of Kepler, both as an individual and as a case history. For Kepler's misguided belief in the five perfect bodies was not a passing fancy, but remained with him in a modified version to the end of his life, showing all the symptoms of a paranoid delusion; and yet it functioned as the *vigor matrix*, the spur, of his immortal achievements.

I think it is rather nice to see how Koestler acknowledged that, although all this is nonsense, if Kepler hadn't had these curious fantasies he might never have gone on to do all the great things that he did.

Symmetry

One of the most remarkable things about the regular and Archimedean solids is their symmetrical nature. In his little book *Symmetry*, Hermann Weyl[4] mentions that the old problem of enumerating the five kinds of Platonic solids was superseded in the 1870s by the problem of enumerating the five kinds of rotation groups. I would like just to run through the details because this is a beautiful piece of pure mathematics that Felix Klein did a hundred years ago, in his book *Lectures on the Icosahedron*.[5]

Klein considered rotations of the sphere into itself, the way an eyeball rotates in its socket, but continued through a whole turn. There are rotations of various periods. A half turn is a rotation of period 2; a quarter turn is a rotation of period 4, and so on. Suppose that you have a rotation of period p, greater than or equal to 2. The axis of rotation penetrates the sphere at two opposite *poles* of that same period p. Working on a sphere, you may have poles of various periods, and a given group of rotations will transform various poles into one another; you get a certain class of equivalent poles. If the total order of the group (the total number of rotations altogether, including the identity) is N, then in each class of equivalent poles, if they are p-gonal poles, there will be N/p poles. That is because if you take a point close to a pole and just move it by that rotation, you get a little p-gon around that point. It is rotated by a rotation of period p. And so all the rotations in the group, when applied to this point near one of the poles, will give you a lot of p-gons all over the place. As there are N of these points altogether, there are N/p equivalent poles in each class.

The next thing to observe is (thinking of all the rotations in the group), for each axis of period p there are $p - 1$ rotations not including the identity. Turn through one pth of a turn and then two $p$ths and so on. There are $p - 1$ rotations for each axis. And as there are two poles at opposite ends of every axis, there are $(p - 1)/2$ for each pole. Now as there are N/p poles in each class of equivalent poles, the number of rotations for each class is

$$\frac{N}{p}\frac{(p-1)}{2} = \frac{N}{2}\left(1 - \frac{1}{p}\right).$$

Not counting the identity, the whole rotation group consists of $N - 1$ rotations, so you simply put them together. Summing over the various classes of equivalent poles, you have

$$\frac{N}{2}\sum\left(1 - \frac{1}{p}\right) = N - 1,$$

where N is the total number of rotations in the group. Just twist that over by a little more algebra and you get

$$2 - \frac{2}{N} = \sum\left(1 - \frac{1}{p}\right).$$

This is summed over the classes of equivalent poles.

How many will there be? Well, if $N = 1$ that is a trivial case, of course, where there is no pole. So from now on we can suppose the number of rotations in the whole group is greater than or equal to 2. And if that is so, then you get this inequality:

$$1 \leq (2 - 2/N) < 2.$$

Now we have this sum,

$$\sum (1 - 1/p).$$

Could there be only one term? No, there couldn't be only one, because $(1 - 1/p) < 1$, while the summation has to be greater than or equal to 1, and also less than 2. Could there be as many as four terms in this summation? No, because four terms, each 1/2 or more, would add up to two or more. So we know that this sum has two or three terms.

Take two terms and you have

$$2 - 2/N = (1 - 1/p) + (1 - 1/q).$$

Cancel the $2 = 1 + 1$, multiply all through by N and you get this curious little equation:

$$N/p + N/q = 2.$$

But as we saw right at the beginning, N/p is the number of poles in a class of equivalent poles. So N/p is a whole number. Similarly N/q is a whole number. And those two whole numbers add up to 2. So there is no conclusion except that they are both 1: $N = p = q$. And that means that if there are only two terms in this sum, you have a group of order p with only one axis but two poles, one at each end of that axis. That is C_p, a cyclic group, the group that you get by having a p-gonal rotation and all its powers, and that is all.

Now suppose there are three terms in this summation, say

$$(1 - 1/p) + (1 - 1/q) + (1 - 1/r) = 2 - 2/N,$$

so that

$$1/p + 1/q + 1/r = 1 + 2/N.$$

Since $1/3 + 1/3 + 1/3 = 1$, the three periods, p, q, r cannot all be 3 or more. So at least one of them is 2, say $r = 2$, and you have

$$1/p + 1/q = 1/2 + 2/N.$$

Multiplying through by $2pq$, you get

$$2q + 2p = pq + 4pq/N,$$

whence

$$(p - 2)(q - 2) = 4 - 4pq/N.$$

So $(p - 2)$ and $(q - 2)$ are two nonnegative integers whose product, being less than 4, can only be 0 or 1 or 2 or 3. Assuming, for convenience, that $p \geq q$, you may have $(q - 2) = 0$, but otherwise the "two nonnegative integers" can only be 1 and 1, or 2 and 1, or 3 and 1. It follows that, apart from the cyclic group C_p, the only finite rotation groups (p,q,r) are the

- Dihedral group $(p,2,2)$ with $N = 2p$
- Tetrahedral group $(3,3,2)$ with $N = 12$
- Octahedral group $(4,3,2)$ with $N = 24$
- Icosahedral group $(5,3,2)$ with $N = 60$

If you think of figures possessing this kind of symmetry, you soon see that each $(p,q,2)$ is

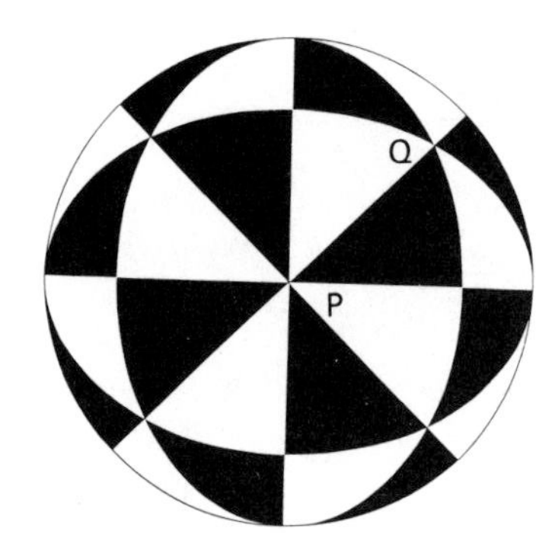

Fig. 3-6. The nine circles of symmetry of the cube.

the group of rotatory symmetry operations of the polyhedron $\{p,q\}$ and of its reciprocal $\{q,p\}$. The inequality

$$(p - 2)(q - 2) < 4$$

gives you at the same time a proof that there are only five Platonic solids: {3,3}, {4,3}, {3,4}, {5,3}, and {3,5}.

The planes of symmetry of such a solid $\{p,q\}$ decompose its circumsphere into a pattern of $2N$ spherical triangles (with angles π/p, π/q, $\pi/2$) which may be vividly distinguished by blackening N of them. The rotation group permutes the N triangles of either color. For instance, (4,3,2) yields Fig. 3-6, which is really simpler than it looks at first sight. Simply draw a circle to indicate the sphere, put in one ellipse, then another; then draw their major axes and two diagonal lines bisecting the angles between them. And now you see that the sphere has been divided into little triangles with angles $\pi/4$, $\pi/3$, $\pi/2$: 24 white and 24 black, half of them visible and the rest hidden behind. The angle at P is 45° because it belongs to four triangles of each color. Six such points (including the antipodes of P) are the vertices of the octahedron {3,4}. The angle at Q, which belongs to three triangles of each color, is 60°. Eight such points are the vertices of the reciprocal cube {4,3}. The twelve points where right angles occur are the vertices of the *cuboctahedron,* an Archimedean solid whose faces consist of six squares and eight triangles.

The *Archimedean solids* are polyhedra which have regular faces of two or three kinds, while all the vertices are transformed into one another by one of the rotation groups described above.* Thus the cycle of faces round a vertex is the same for all the vertices of each solid, and the numbers of sides of the polygons in this cycle provide a concise symbol. For instance, the symbol for the cuboctahedron is $3 \cdot 4 \cdot 3 \cdot 4$ or $(3 \cdot 4)^2$, because each vertex belongs to two triangles and two squares, arranged alternately. The p-gonal prism is $4^2 \cdot p$ and the p-gonal antiprism is $3^3 \cdot p$. The books of Archimedes on this subject were lost. So it was left to Kepler to give names for them. The name "cuboctahedron" is rather natural: it is a combination of the words cube and octahedron. Similarly, he called the common part of the dodecahedron and the icosahedron an *icosidodecahedron*. It has twenty triangles and twelve pentagons. Four others are shown in Figs. 3-7–3-10. The whole list of thirteen is as follows:

- The truncated tetrahedron $3 \cdot 6^2$
- The truncated cube $3 \cdot 8^2$
- The truncated octahedron $4 \cdot 6^2$
- The truncated dodecahedron $3 \cdot 10^2$
- The truncated icosahedron $5 \cdot 6^2$
- The cuboctahedron $(3 \cdot 4)^2$
- The icosidodecahedron $(3 \cdot 5)^2$
- The rhombicuboctahedron $3 \cdot 4^3$
- The rhombicosidodecahedron $3 \cdot 4 \cdot 5 \cdot 4$
- The truncated cuboctahedron $4 \cdot 6 \cdot 8$
- The truncated icosidodecahedron $4 \cdot 6 \cdot 10$
- The snub cube (*cubus simus*) $3^4 \cdot 4$
- The snub dodecahedron (*dodecahedron simum*) $3^4 \cdot 5$

Before Klein there was another German, A. F. Möbius, who observed that the use of Fig. 3-6 (as above) to construct the vertices of the cube, octahedron, and cuboctahedron can be extended to yield all the Platonic and Archimedean solids. One of the black triangles in Fig. 3-6 is marked PQR in Fig. 3-11. Its angles have been bisected so as to yield points S (equidistant from the great circles RP, PQ), T (equidistant from PQ, QR), U (equidistant from QR, RP), and V (equidistant from all three sides of the triangle PQR, so that V is the center of the inscribed small circle). When the planes of the great circles RP and PQ are taken to be mirrors, the images of S in this two-

* Prisms and antiprisms satisfy this definition, but are not usually considered Archimedean.

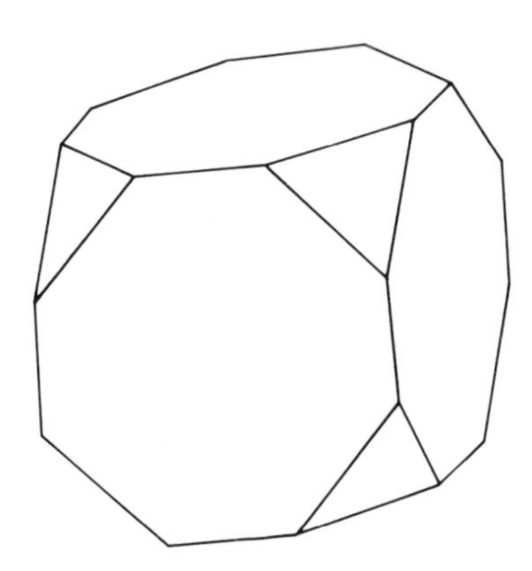

Fig. 3-7. The truncated cube $3 \cdot 8^2$.

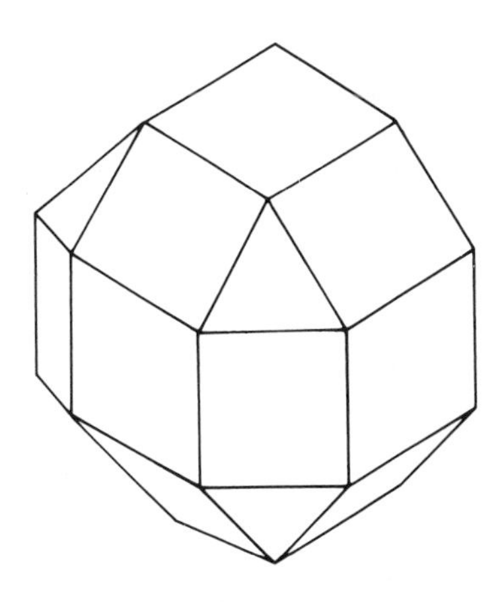

Fig. 3-8. The rhombicuboctahedron $3 \cdot 4^3$.

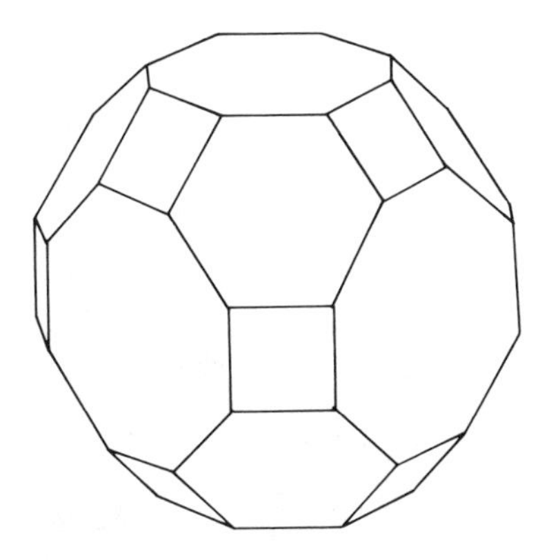

Fig. 3-9. The truncated cuboctahedron $4 \cdot 6 \cdot 8$.

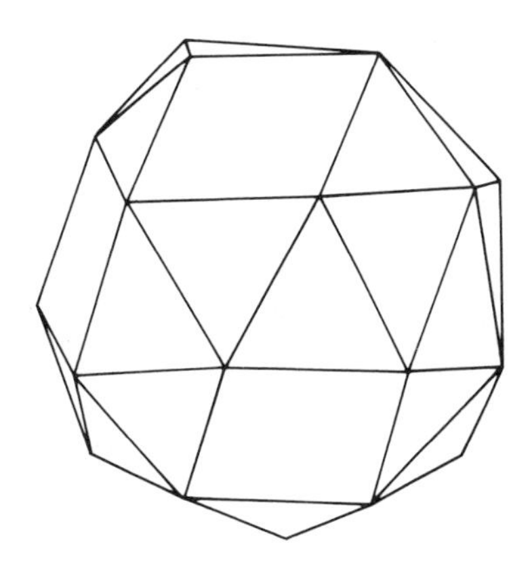

Fig. 3-10. The snub cube $3^4 \cdot 4$.

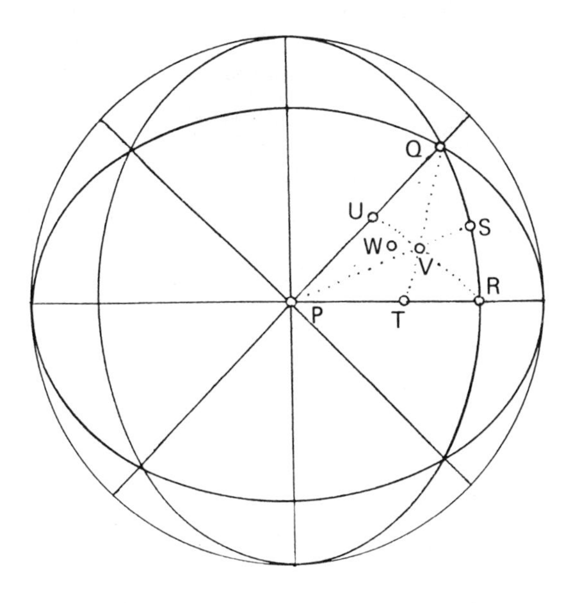

Fig. 3-11. Typical vertices of seven polyhedra.

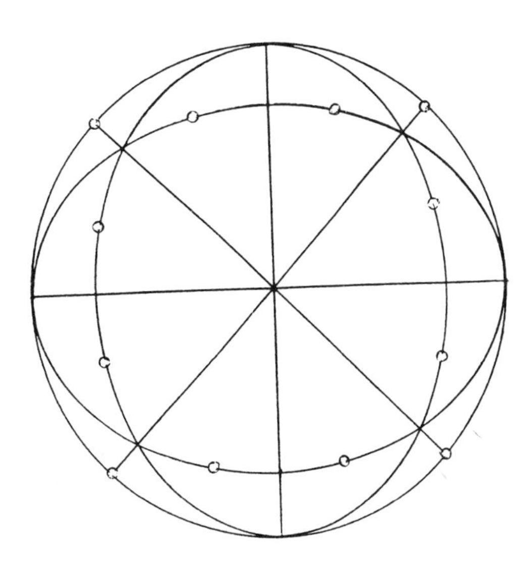

Fig. 3-12. Vertices S of the truncated cube.

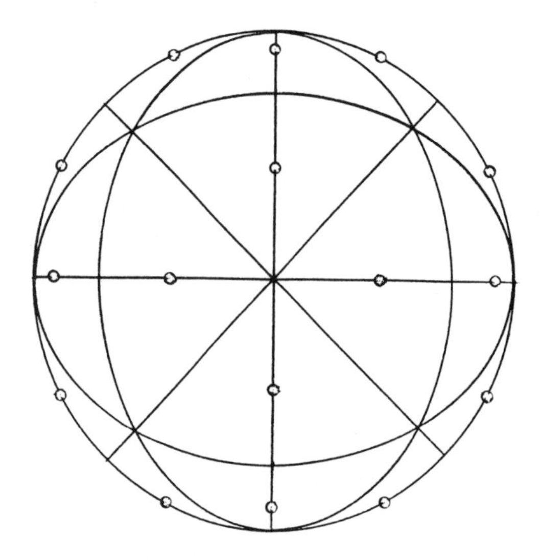

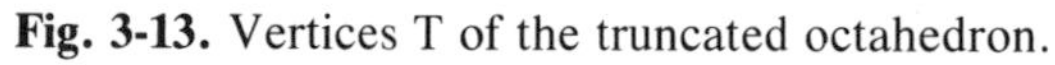

Fig. 3-13. Vertices T of the truncated octahedron.

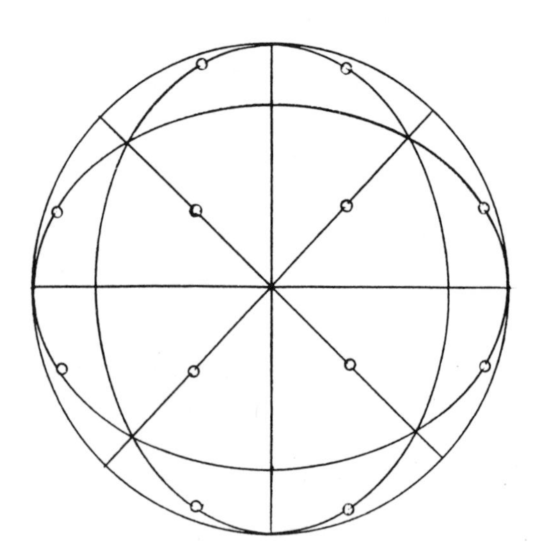

Fig. 3-14. Vertices U of the rhombicuboctahedron.

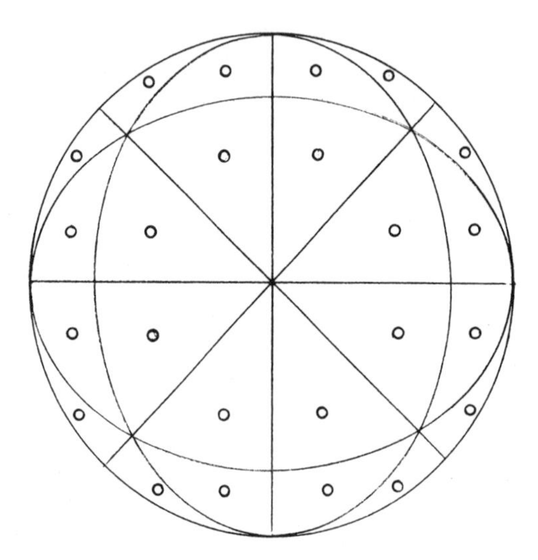

Fig. 3-15. Vertices V of the truncated cuboctahedron.

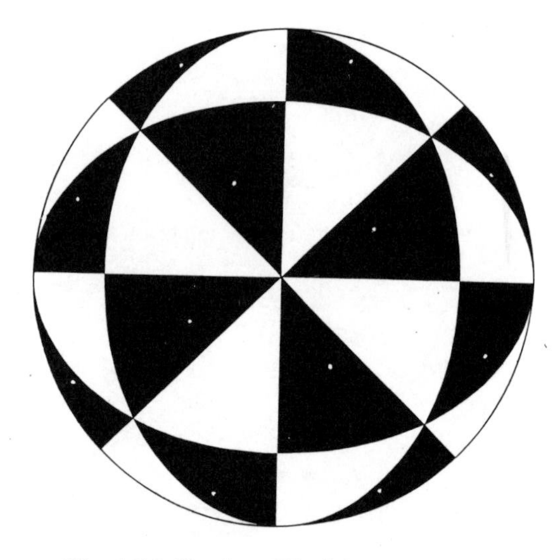

Fig. 3-16. Vertices W of the snub cube.

mirror kaleidoscope are the vertices of a regular octagon. Introducing a third mirror QR, we get Möbius's three-mirror kaleidoscope in which the images of S are the vertices of the truncated cube, as in Fig. 3-12. Other Archimedean solids can be derived similarly from the points T, U, V, as in Figs. 3-13–3-15. Figure 3-16 shows points derived from a point W, different from V, by applying the octahedral rotation group (4,3,2) of order 24, which is a subgroup of index 2 in the kaleidoscopic group of order 48. In other words, the 24 vertices of the snub cube $3^4 \cdot 4$ are points situated like W in all the *black* triangles.

Figure 3-17 shows six of the nine great circles in Fig. 3-6 or 3-18; they yield the tetrahe-

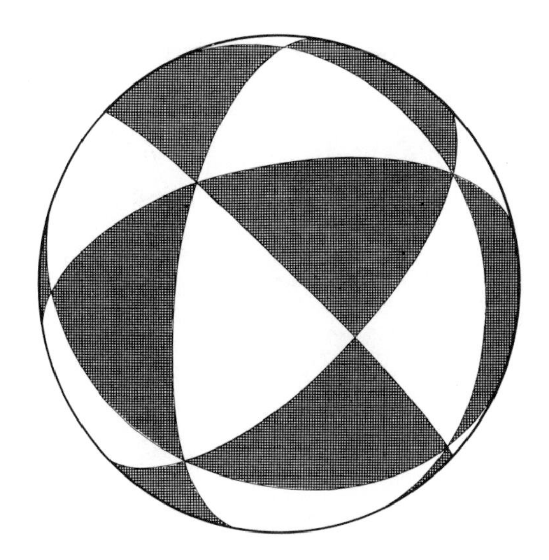

Fig. 3-17. Six of the great circles of Fig. 3-6, related to the group (3,3,2). Drawn by Patrick DuVal for his book *Homographies, Quaternions and Rotations* (London: Oxford University Press, (1964)).

Fig. 3-18. Great circles related to the group (4,3,2). Drawn by Patrick DuVal for his book *Homographies, Quaternions and Rotations* (London: Oxford University Press, (1964)).

Fig. 3-19. Great circles related to the group (5,3,2). Drawn by Patrick DuVal for his book *Homographies, Quaternions and Rotations* (London: Oxford University Press, (1964)).

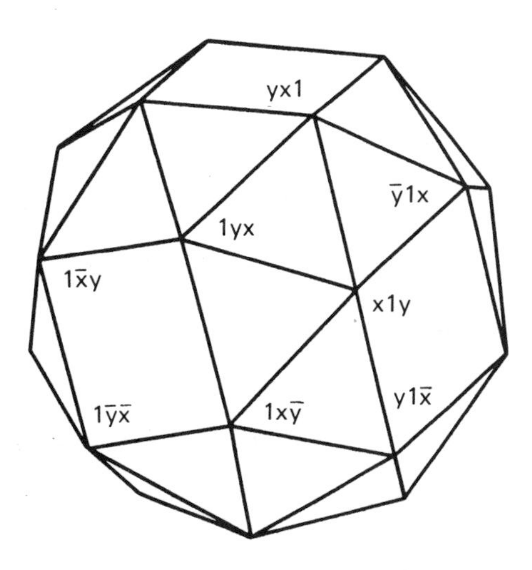

Fig. 3-20. The snub cube $3^4 \cdot 4$.

dral rotation group (3,3,2) and the truncated tetrahedron. Figure 3-19 shows the analogous set of great circles related to the icosahedral group (5,3,2). This "icosahedral kaleidoscope" yields the complicated polyhedra; for instance, points suitably placed in all the black triangles now yield the snub dodecahedron.

It is interesting to see how one could work out coordinates for the vertices of a snub cube. Let us begin with a large cube in its natural position for Cartesian coordinates, so that the vertices of the cube, are, shall we say, (1,1,1), and the same with various changes of sign: one vertex is (1,1,1) and another one is (−1,1,1) and so on. In a concise notation, the eight vertices are $(\pm 1, \pm 1, \pm 1)$. A smaller

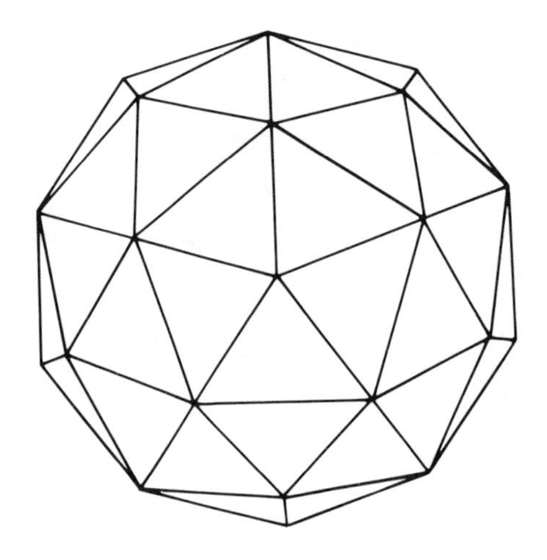

Fig. 3-21. Pentakisdodecahedron, $\{3,5+\}_{1,1}$.

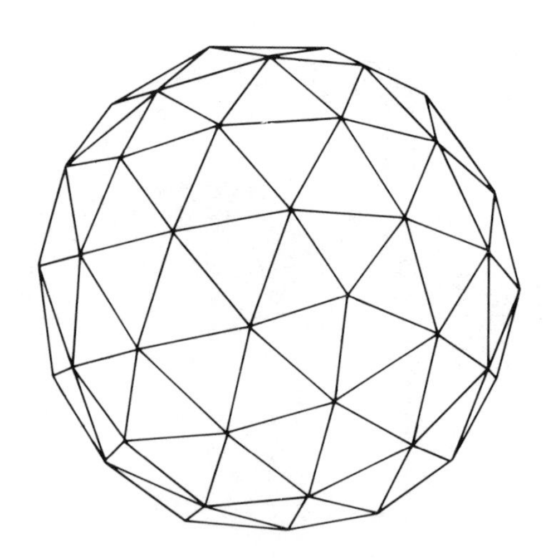

Fig. 3-22. $\{3,5+\}_{2,1}$

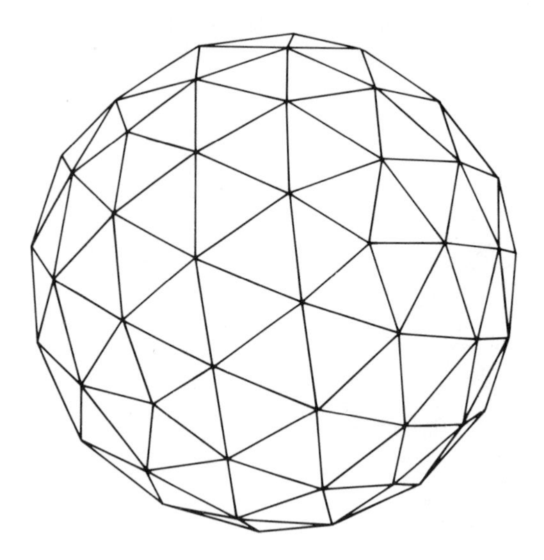

Fig. 3-23. $\{3,5+\}_{3,0}$

square inside the face $(1,\pm1,\pm1)$ of the cube may be supposed to have vertices $(1,y,x)$, $(1,x,-y)$, $(1,-y,-x)$, $(1,-x,y)$, where $x > y$, this being in the face where the first coordinate is 1. (In Fig. 3-20, $(1,-x,y)$ appears as $1\bar{x}y$.) These four points are the vertices of a square that is inside one square face of the cube, but twisted around through a certain angle. And you ask that these points and others analogously situated in other faces of the cube should be at the same distance from their neighbors, so that the distance from one point to another is the same in those various pairs. One pair gives you $xy + y + x = 1$, whence

$$x = (1 - y)/(1 + y).$$

Another yields

$$x = y(1 + y)/(1 - y).$$

Multiplying these together, you get

$$x^2 = y.$$

Substituting x^2 for y in $xy + y + x = 1$, you are left with the nice cubic equation

$$x^3 + x^2 + x = 1.$$

You can work this out in various ways and find x to be about 0.543689. Then you have to take the point with coordinates $(1,x^2,x)$, apply all the cyclic permutations, put in an even number of minus signs, and put them in a different order with an odd number of minus

Fig. 3-24. H. S. M. Coxeter. Photograph by Stan Sherer.

signs; that gives you all the 24 vertices of the snub cube (Fig. 3-20).

The late R. Buckminster Fuller, a great engineer and architect, was very interested in the structures that he called *geodesic domes* and they were often made by modifications of the icosahedron and dodecahedron. In Fig. 3-21 you see one of those, where you simply take a dodecahedron and put a small pentagonal pyramid onto each of its faces, so that you have altogether $5 \times 12 = 60$ triangles. Although they are not equilateral, they are all congruent and so you get an attempt toward finding a sixth regular solid. Of course it is not regular, because there are five triangles around some points, and six triangles at other points. But at least this is the sphere covered with a large number of triangles that are all nearly equilateral and nearly alike. And Fuller went on doing this in more and more elaborate ways. Figure 3-22 is a slightly different one of the same sort, with nearly equilateral triangles, nearly the same number around each corner. If you look closely you see that at some corners there are six triangles coming together and at others there are five. And so you can classify these polyhedra by seeing how you can go from one place where there are five triangles to another place where there are also five triangles. To go from one to the other is a sort of modified knight's move: two steps this way and then one step that way, and you get

Fig. 3-25. A truncated octahedron, in the Campus School Exhibit. Photograph by Stan Sherer.

Fig. 3-26. Professor Coxeter showing Figure 3-19 during his lecture. Photograph by Stan Sherer.

Fig. 3-27. Professor Coxeter talking with members of the audience after his lecture. Photograph by Stan Sherer.

another pentagonal point. Everywhere in between there are six triangles at a vertex. It is a rather nice consequence of the theory of polyhedra that just as the icosahedron has twelve vertices, each one of which belongs to five triangles, even after you have put in all these extra triangles it is still true: there are just twelve points on the sphere where the number of triangles is only five instead of six. And that essentially, is a consequence of Euler's formula. The number of vertices minus the number of edges plus the number of faces of a polyhedron is two. If you fiddle with that formula a little you can see it always must be true: that if you have a polyhedron whose faces consist entirely of triangles, six coming together at some vertices and five at others, then the number of vertices where there are five triangles coming together is exactly twelve. You might ask the question, how many of the other kind will there be? That was answered by a very able geometer, Branko Grünbaum, who showed that if a convex polyhedron has only triangles for faces, five or six round each vertex, then the number of vertices where you've got six triangles coming together may be any number except one.[6] It can be two or anything greater. Such a polyhedron is a remarkable generalization of the Platonic solids.

Notes

[1] B. L. van der Waerden, *Science Awakening*, rev. ed. (Leiden: Noordhoff International Publishing; New York: Oxford University Press, 1974).

[2] P. J. Federico, *Descartes on Polyhedra* (New York: Springer-Verlag, 1982).

[3] A. Koestler, *The Watershed* (Garden City, N.Y.: Anchor Books, 1960), from Koestler's larger book, *The Sleepwalkers*.

[4] H. Weyl, *Symmetry* (Princeton, N.J.: Princeton University Press, 1952).

[5] F. Klein, *Lectures on the Icosahedron*, English trans., 2nd ed. (New York: Dover Publications, 1956.)

[6] B. Grünbaum and T. S. Motzkin, "The Number of Hexagons and the Simplicity of Geodesics on Certain Polyhedra," *Canadian Journal of Mathematics 15*, (1963):744–51.

4

Milestones in the History of Polyhedra

JOSEPH MALKEVITCH

Considering the fact that polyhedra have been studied for so long, it is rather surprising that there has been no exhaustive study of their history. But we are very lucky that the four modern classics on the theory of polyhedra have had authors (Brückner, Coxeter, Fejes-Tóth and Grünbaum) who were interested in historical information and provided detailed historical notes in their books. I propose to present an outline of the milestones in the history of the subject, putting together the thread of what happened as the theory developed. I will pay special attention to regularity concepts.

I would like you to imagine what might happen if, a thousand years from now, someone at an archaeological site should find a Rubik's cube. What would the archaeologists, historians, and mathematicians of that future time deduce about our knowledge of geometry from this one object? We can get some idea from an Ed Fisher cartoon from The New Yorker (Fig. 4-1). It shows a museum statue labeled "man"; it has three legs, two heads, and misplaced feet and hands. There are two "Martian" creatures looking at it; one is saying: "It certainly is amazing what our scientists can reconstruct from just a few bones and fragments." I think that is the state of our knowledge in studying polyhedra in ancient times.

Let me begin with what perhaps are the most famous of polyhedral objects, the pyramids of Egypt (Fig. 1-24). They are awesome,

"It certainly is amazing what our scientists can reconstruct from just a few bones and fragments."

Fig. 4-1. From *The New Yorker,* January 6, 1968, p. 29. Drawing by Ed Fisher; © 1968 The New Yorker Magazine, Inc.

and the engineering accomplishment involved in having made them raises the question as to exactly what kinds of information the geometers of Egypt had about polyhedra. Our knowledge of Egyptian geometry comes down to us from two sources, from two papyri. One is called the Rhind mathematical papyrus and the other one the Moscow mathematical papyrus. There are some problems on the Rhind papyrus showing computations about the relation between what we would call today the slant height of the pyramid and the height. The Moscow papyrus has a calculation that illustrates the kind of detective work one has to do when one looks at these papyri. The authors did not spell out: "We will now do this calculation." Instead there are some symbols and sometimes an accompanying diagram, and the modern historian has to try to determine what it was that they were trying to do. In this particular instance, it seems that they may have been calculating the volume of the truncated pyramid. This is the *first milestone* in the history of polyhedra.

Volume of Truncated Pyramid

1. (c. 1890 B.C.) Problem 14 of the Moscow Papyrus suggests Egyptians may have known how to compute the volume of a truncated square pyramid, using:

$$V = \frac{h}{3}(a^2 + ab + b^2)$$

(where a and b are sides of the square bases and h is the height of the pyramid).

The calculation seems to follow the modern formula for the volume of the frustrum of a pyramid. There is quite a bit of scholarly debate as to whether this formula was actually known to the Egyptians, but I will not get involved in that particular thicket. Part of the reason for the controversy is that we have no real basis for trying to determine how the Egyptian geometers might have arrived at that result. Furthermore, the papyri do not contain any problem which calculates the volume of an untruncated pyramid! They did not seem to be in possession of anything like calculus. How else might they have found the formula? We know from modern work that you cannot find the volume of a tetrahedron by the method of cutting it up into a finite number of pieces and reassembling them into something whose volume can be computed easily. (This follows from Max Dehn's solution to Hilbert's problem on equidecomposability.) Various proposals have been made for what the Egyptians did do; all are very speculative.

Let me briefly describe what actually seems to be the history of the development of the theory of the volume of a pyramid, because it is certainly not an intuitive result. It was quite an accomplishment for ancient peoples; we can consider it to be the *second milestone*. Archimedes referred to the fact that Democritus, who flourished at the end of the fifth century B.C., knew that the volume of the pyramid was one-third the area of the base times the height, and that the proof was devised by Eudoxus. Eudoxus's method is known as the method of "exhaustion" and his approach is the one Euclid follows in the *Elements*.

Volume of a Pyramid

1. Democritus (fl. end 5th century B.C.) discovers that volume of a pyramid is equal to $\frac{1}{3}$(area of base)(height).
2. Eudoxus (c. 409–c. 356 B.C.) proves above result using method of "exhaustion."

(Achimedes confirms Eudoxus' role).

Let us turn now to the origin of regularity concepts. More specifically, we might ask the extent to which ancient peoples had knowledge of regular solids and theories of regular solids. Before attempting to answer this question, let me clarify what I mean by "regular." As the theory of polyhedra has developed, the meaning of the word "regular" has broadened. There are two major uses of the word. One approach to regularity is "local": it requires congruence of faces and/or vertex-figures (that is, the pattern of faces at a vertex) and/or edge-figures. The other approach considers the polyhedron as a whole: a polyhedron is regular if its symmetry operations act transitively on its vertices, edges, or faces. For example, vertex transitivity would mean that any vertex of the polyhedron could be moved to any other vertex by a symmetry operation of the polyhedron. In either approach we are primarily interested in the case where the faces are regular polygons. The second approach is very modern (nineteenth century) and almost certainly was not the approach used by the Egyptians or Greeks. Henceforth, when the word *regular* is used, it will refer to the first approach to regularity. As near as I am able to determine, there was no knowledge of individual regular solids, and no theory of them in ancient Egypt; I will have some more comments about that later.

As we know from Professor Chieh's essay on crystallography (Chapter 5), many very beautiful polyhedral forms occur in nature as crystals. Fluorite crystals often grow as octahedra and there are pyrite (combinatorial) dodecahedra. Examples of such crystals must have been known in ancient times. The oldest man-made dodecahedral object is generally attributed to pre-Pythagorean times. It is a dodecahedral shape with incised markings on it, discovered in an excavation on Mt. Loffa in Italy near Padua. Nobody knows what the symbols actually mean. There is some linguistic evidence that the dodecahedron was not known by the name *dodecahedron* in early days, but instead was referred to as "the sphere of 12 pentagons." That may have had something to do with how the theory actually developed.

There is a tradition, which goes back into Greek history, that assigns knowledge of the five "Platonic solids" to the Pythagoreans. Eudemus of Rhodes referred to Pythagoras himself as having discovered the five regular polyhedra. But modern scholars seem to discount this. In a scholarly thesis in 1917, Eva Sachs gave some cogent arguments suggesting that, in fact, the Pythagoreans did not know all of the five Platonic solids. Many modern scholars argue that the proper history is based on a scholium or comment in an extended edition of Euclid's *Elements* which reads as follows: "In this book, the 13th [the thirteenth book of Euclid], are constructed five figures called Platonic, which do not however belong to Plato. Three of these figures, the cube, pyramid and dodecahedron, belong to the Pythagoreans, while the octahedron and icosahedron belong to Theatetus." But the history is complicated by an understanding of what "belong to" means. The Greeks were very interested, as we know, in ruler-and-compass constructions. The question arises as to whether this quotation means that Theatetus had found ruler and compass constructions for these polyhedra. Perhaps all five solids were known, as objects, earlier.

Professor Waterhouse gives[1] some very interesting linguistic and other arguments in favor of a later date than the Pythagoreans for the origins of actually thinking of the regular solids as a *family* and singling them out for study. Theatetus (415–369 B.C.) seems to have looked at this collection of solids not merely as isolated objects; he considered the question of discussing them as part of a theory (*Milestone 3*). On the other hand, as we have seen, these polyhedra are often referred to as the Platonic solids. Plato, who was a friend of Theatetus, built them into his cosmology in the dialogue *Timaeus* (*Milestone 4*). This is important for the history of polyhedra because one of the threads that kept polyhedra alive during the Renaissance was the renewed interest in classical studies. (See, for example, Raphael's painting *The School of Athens,* in which he showed the geometers at work.) The association of Plato with these solids probably helped keep this knowledge alive for a long period of time.

Theatetus (c. 415–369 B.C.)

1. Develops a general theory of regular solids, specifically adding the octahedron and icosahedron to solids known earlier.

Plato (427–347 B.C.)

1. In the dialogue *Timaeus* Plato incorporates his knowledge of the five regular polyhedra into his philosophical system.

(His popularity and influence result in their becoming known as "Platonic solids.")

By the time we get to Euclid, we already have a fairly full-blown theory of solid geometry. In Book XI of *Elements,* Euclid gave a full treatment of metric properties of polyhedra, in Book XII he discussed the volume of prisms and pyramids including Eudoxus' proof, and in Book XIII he showed how to construct the regular convex polyhedra and "proved" that there are only five of them (*Milestone* 5). I use quotation marks because Euclid never told us what a polyhedron was. This raises the question of what people at various times have had in mind when they used the word "polyhedron." It is my contention that throughout the history of the development of regularity concepts, the notions of polygon and polyhedron have diversified. This diversification has been the driving force behind creating a lot of the theory that we know today.

Euclid (fl. 323–285 B.C.)

1. Book XI of the *Elements* treats metric properties of polyhedra.
2. Book XII of the *Elements* discusses the volume of prisms and pyramids.
3. Book XIII of the *Elements* treats the five regular polyhedra, concluding with a "proof" that there are exactly five.

The sixth milestone is the description by Archimedes, in a manuscript that now is lost, of what we today call the semiregular or Archimedean solids (*Milestone* 6). By some miracle or another, Pappus (whose works unfortunately don't appear to exist in English) gave an account (*Milestone 7*) of the lost book of Archimedes which dealt with the semiregular solids. It is significant that he explicitly mentioned that there are thirteen of them, and described them in terms of how many polygons each has with various numbers of sides at a vertex. We will see that this turns out to be rather significant at a later time.

Archimedes (c. 287–212 B.C.)

1. Describes 13 "semi-regular" solids in a now lost treatise.

Pappus (fl. 4th century)

1. (320 A.D.) Pappus' *Collection* (Book V) gives an account of the 13 "semiregular" solids discovered by Archimedes.

In the period between the time that Archimedes lived and through the time of Pappus, polyhedral objects of various kinds and types were made. Fortunately some of these objects can still be seen today. I had the big thrill of examining some of them personally at the Metropolitan Museum of Art in New York. There I saw regular icosahedral objects incised with the first 20 letters of the Greek alphabet (Figs. 4-2 and 4-3). Four similar icosahedra used to be on display in the Egyptian rooms of the British Museum in London. The exact origins and provenance of these objects is not known. It is, however, fairly certain that they are not indigenous to Egypt, that they are probably not even of Greek origin. They may date from the Roman period. Claims in the literature about Egyptian knowledge of regular solids appear to be false extrapolation from the assumption that these icosahedra were made in Egypt.

There are also other objects, from perhaps a somewhat later period, that have been cited by scholars. Most of these were found in England, France, and Italy, and have dodecahedral shapes. Figure 4-4 shows a bronze dodecahedron dug up in 1768 in Carmarthen, typical of about fifty which have been found in the northwestern provinces of the Roman empire. It has been suggested that they were used as surveying instruments.[2] There is considerable controversy about their origins and uses; the best guess is that they were candle holders.[3] I have never actually seen an original of one of these, but from photographs some of them appear to be quite handsome objects. A large collection of such polyhedral objects has been described,[4] including a rhombic triacontahedron, opening to question the claims that nobody had seen such a thing until Kepler's time. The best estimate for when these objects were made is about 500 A.D.

There does not seem to have been any systematic account of polyhedra from the time of

Fig. 4-2. Steatite icosahedron with Greek letters incised on the faces. The Metropolitan Museum of Art, New York. 27.122.5, Fletcher Fund, 1927.

Fig. 4-3. Faïence icosahedron with Greek letters incised on the faces. The Metropolitan Museum of Art, New York. 37.11.3, Museum purchase, 1937.

Fig. 4-4. Bronze dodecahedron found in Carmarthen 1768, and now in the possession of the Society of Antiquaries of London. Overall height $4\frac{1}{8}$ inches. Photograph reproduced by courtesy of the Society of Antiquaries of London.

Pappus until the Renaissance. As we know, a lot of mathematical traditions died during that period. As I mentioned earlier, what ultimately resuscitated these ideas was the renewed interest in Plato. So I will jump to the Renaissance period. I would like to single out the work of Albrecht Dürer for *Milestone 8*. This is the concept of studying polyhedra by drawing what are today called nets. By folding a planar piece of cardboard along prescribed lines and joining the edges of the figure, the net becomes a polyhedron. Dürer made nets for the dodecahedron and for other regular and semiregular solids.

Albrecht Dürer (1471–1528)

1. Invents the concept of the "net" of a polyhedron.

(A net of the cube.)

The situation during the Renaissance is extremely complicated. A great many scholars, artists, and artisans discovered and rediscovered various Platonic and Archimedean solids (*Milestone 9*). Some of them drew what appear to be star-shaped solids; others drew compounds; others drew convex polyhedral solids. In discussing this period it is very difficult to reach any firm conclusions about who discovered what and when. For example there is a picture of a solid in the famous Jamnitzer pictures (published in 1568) that some say was discovered by Kepler about 1619. Professor Coxeter alluded to the book of Luca Pacioli's work on the regular and semiregular polyhedra which was illustrated by Leonardo da Vinci. There are indications that Pacioli actually made polyhedron models of glass; there is a painting which shows Pacioli with a picture of a glass model of the rhombicuboctahedron. The important thing to realize in discussing this period is that these ideas were very much in the air. The real issue is not who discovered a particular star polytope first. I am not sure that that approach is particularly profitable. I think it is more interesting to try to understand why it was that all of this activity was going on at this particular time. Why was there so much interest in polyhedra, and why did the subject flourish during that period? The answer seems related to the emergent study of perspective and the renewed interest in the Greek classics.

Renaissance artists, architects, artisans and scholars such as:

Paolo Uccello
Wentzel Jamnitzer
Lorenz Stoer
Daniel Barbaro
Piero della Francesca
Luca Pacioli
Leonard da Vinci
Albrecht Dürer
Simon Stevin
François de Foix
R. Bombelli

"Discover" and "rediscover" various Platonic and Archimedean solids, star-polyhedra, compounds, and other polyhedral objects.

The *tenth milestone* in our history is the work of Johannes Kepler, which Professor Coxeter discusses in Chapter 3. Kepler drew tilings of the plane in which he used nonconvex and nonsimple polygons. He also gave a very elaborate, case-by-case, proof that there are thirteen semiregular polyhedra. It is interesting that Pappus didn't describe the prisms and the antiprisms which, as we know, are two infinite families which are also semiregular polyhedra. But Kepler explicitly both drew and discussed them. It is also interesting that in modern times we refer to thirteen Archimedean solids because we adopt the regularity definition based on symmetry, although almost certainly this was not the definition used by Archimedes and Pappus nor by Kepler. With the congruent vertex figure definition, Archimedes and Pappus missed one: the pseudo-rhombicuboctahedron. Kepler's detailed work on the semiregular polyhedra appears in Book II of *Harmonices Mundi*. Here he finds only thirteen semiregular solids (plus the prisms and antiprisms). However, in an offhand remark in the "Six-Cornered Snowflake" he refers to fourteen semiregular solids! No supporting detail is given, however.

Johannes Kepler (1571–1630)

1. Studies tilings of the plane using convex, non-convex, non-simple polygons.
2. Proves there are 13 "semiregular" polyhedra (plus two infinite families, prisms, and anti-prisms).
3. Constructs:
   (A) Stella octangula
   (B) Two (regular) stellated polyhedra
   (C) Rhombic dodecahedron
   (D) Rhombic triacontahedron.

Kepler also constructed the rhombic dodecahedron and the rhombic triacontahedron and, as Professor Coxeter mentioned, two "new" "regular" (star) polyhedra. At least, today we call them regular; it is hard to tell whether Kepler really thought of them as being regular. It is my impression that perhaps he did not. I also have some questions as to whether he really understood fully reciprocation and duality notions, as has been claimed. Since he was so careful, one might guess that if he understood these matters, he would have also constructed the two remaining regular star polyhedra and the duals of the Archimedean polyhedra. But that is my purely subjective view.

Unfortunately, there appears to be a pattern for many of the great discoveries about polyhedra, even for geometry in general: all too often people do very great things and then the work "goes to sleep" for long periods. Euclid's work went to sleep, the work of Archimedes went to sleep; Kepler's work, as modern in spirit as it seems to be, in fact, wasn't looked at seriously until relatively recently. Although people constantly refer to Kepler's work, it does not really seem to have affected the history of polyhedra in any direct way.

Another person who made major contributions to the theory of polyhedra was René Descartes. Professor Coxeter talked about the famous story of the lost manuscript (Chapter 3). In that manuscript (*Milestone 11*) there appears a tantalizing theorem about polyhedra: the sum of the defects of the vertices is 4π, where the defect of a vertex is defined to be 180° minus the sum of the face angles at the vertex. (Of course no one knows what kind of polyhedra Descartes was really taking about, but presumably he had convex three-dimensional polyhedra in mind.)

René Descartes (1596–1650)

1. (1619–1620) (Manuscript lost, but copy by Leibniz published in 1820).

Main Theorem:
If P is a (convex) 3-dimensional polyhedron, then the sum of the defects at the vertices is 360°. (*Defect* of a vertex is 180° minus the sum of the face angles at the vertex.)

It turns out that one can very quickly get from that theorem of Descartes to the very famous polyhedral formula of Euler, $V - E + F = 2$, and vice versa. The fact that Descartes' theorem is logically equivalent to Euler's formula has created the widespread impression that Descartes actually knew the formula, although scholars over and over again have said that this is not so. You can find in the papers of Lebesgue from the 1920s that he does not believe Descartes knew Euler's formula. You can find the same statement in Pólya,[5] and in a very nice paper by Peter Hilton and Jean Pedersen.[6] More recently Federico has written a whole book[7] on the contribution of Descartes to the theory of polyhedra, including an English translation of the manuscript and an extensive summary of this debate. In this connection it is helpful to remember remarks of Jacques Hadamard. He wrote, concerning some of his own work, that "two theorems, important to the subject, were such obvious and immediate consequences of the ideas contained therein that, years later, other authors imputed them to me, and I am obliged to confess that, as evident as they were, I had not perceived them."[8]

Evidently there is a strong tendency on the part of many people who know of Descartes' theorem to assume that if Descartes had only gone one tiny step further, he would have discovered this or that. But if you look carefully at the work of Descartes, it is very clear that he did not think of polyhedra as *combinatorial* objects. It was not a tiny step that was needed, but a big one. That great leap forward was made in part by Euler.

I might mention that Descartes is another example of a person whose work went to sleep. (His work was literally lost.)

Aside from the material that fuels the Euler–Descartes controversy, Descartes talked about the semiregular polyhedra. He didn't enumerate all thirteen of them, only the eleven that can be obtained from the Platonic solids by truncation. Nor did any of those artists and artisans, who were obviously bright and talented people, discover all thirteen of the semiregular solids. On the other hand, Kepler explicitly referred to the fact that these objects were Archimedean solids. So we know that he had seen Pappus's work describing what Archimedes had done. Kepler had a very big assist in knowing that thirteen solids existed. One wonders what would have happened during this period if people had had wide access to Pappus's work. Of course, one can only speculate.

I was originally going to refer to landmarks rather than milestones, and I think that *Milestone 12* really deserves the title *landmark*: Euler's letter to Christian Goldbach in 1750 in which he referred to his discovery of the fact that the number of vertices of a polyhedron minus the number of edges plus the number of faces equals two. Like Euclid and Descartes, however, he did not say what kinds of polyhedral objects he had in mind, and that omission has created a long list of further controversies about the history of this subject. Suffice it to say that Euler, although he found his formula, was not successful in proving it.

Leonard Euler (1707–1783)

1. Euler discovers that polyhedra obey:

   Vertices + Faces − Edges = 2

Let me briefly point to the long list of very distinguished and very interesting work that was done on Euler's polyhedral formula (*Milestone 13*). The first proof was provided by Legendre. (There are claims that Meyer Hirsch gave a correct proof prior to Legendre. I believe that this is an error that came about due to a misreading of something in Max Brückner's book). Many people contributed to the theory of the formula by figuring out what happens for different types of polyhedra and providing different proofs. What is important about this development is, first, that it provided the roots of modern topology, and it was clearly the interest in Euler's polyhedral formula, and the more general idea of the Euler characteristic, that brought about important developments. Second, much of the impetus for studying higher dimensional polyhedra grew out of the work on the Euler polyhedral formula.

Development of the Theory of Euler's Polyhedral Formula by:

1. Adrian-Marie Legendre (1752–1833)
   First proof
2. Augustin-Louis Cauchy (1789–1857)
3. J. D. Gergonne (1771–1859)
4. S. Lhuilier (1750–1840)
5. J. Steiner (1796–1863)
6. Von Staudt (1798–1867)
7. Many others!

The next milestone, *Milestone 14*, is Poinsot's 1810 discovery of the four regular stellated polyhedra. It is clear that Poinsot understood that there was a sense in which these were regular polyhedra. Poinsot's work on the star polyhedra grew out of his work on star polygons. He seems neither to have looked at Kepler's original work nor to have been aware of Kepler's discovery of two star polyhedra. (There have been allegations, however, that Poinsot plagiarized Kepler.) Other contributors to the theory of star polyhedra were A. Cauchy (1811), J. Bertrand (1858), and A. Calyey (1859).

Louis Poinsot (1777–1859)

1. In his 1810 *Mémoire*, Poinsot discovers four "regular" stellated polyhedra, using both star-shaped vertices (i.e. $\{5,\frac{5}{2}\}$ and $\{3,\frac{5}{2}\}$) and star-shaped faces (i.e. $\{\frac{5}{2},5\}$ and $\{\frac{5}{2},3\}$—already known to Kepler).

Milestone 15 followed about a year later, when Cauchy made major contributions to the theory of polyhedra. He gave what is the most common proof of Euler's formula using graph-theoretic ideas, and he also proved his famous result about simplicial 3-polytopes being rigid. He also gives a "proof" that there are no regular star polyhedra other than those found by Kepler and Poinsot.

Augustin-Louis Cauchy (1789–1857)

1. Cauchy "proves" that simplical 3-polytopes are rigid.
2. Gives a graph-theoretic approach to proving Euler's formula.
3. Shows there are 9 "regular" polyhedra.

The first systematic account of duality of polyhedra that I have found is in the work of Catalan in 1865 (*Milestone 16*). In a long article he described, very explicitly, the duals of the semiregular polyhedra. It is curious that researchers never cited this paper. In other words, this was a paper that also went to sleep. One finds Catalan's work mentioned only in historical footnotes of books written in the twentieth century. Nobody earlier seems to have paid any attention to it.

Eugene Charles Catalan (1814–1894)

1. Catalan gives a systematic account of the duals of the Archimedean solids.

In the middle to late 1800s there was a tremendous flourishing of geometric activity. Max Brückner published a book (*Milestone 17*) in which he summarized all of what was known at the time and also gave some extensive historical notes on the subject. It has very beautiful pictures of uniform polyhedra, which served as an inspiration to people later. (The uniform polyhedra are those—not necessarily convex—that have regular polygons as faces and symmetries that are transitive on the vertices.)

Max Brückner (1860–1934)

1. Brückner publishes an extensive summary of the known results on polygons and polyhedra, with historical notes.

Let me breeze through more recent work. I have indicated that perhaps Kepler had known the pseudo rhombi-cuboctahedron. This is certainly possible. This polyhedron is often referred to as Miller's solid; sometimes the Russian mathematician Ashkanazy is given credit for discovering it. But George Martin was kind enough to call my attention to a paper in 1905 of D. M. Y. Sommerville (you may know his work on n-dimensional space) in which there are Schlegel diagrams for both the rhombi-cuboctahedron and the pseudo one (*Milestone 18*). The significance of this milestone is that many proofs in this area of geometry require a delicate interplay of both theory and case-by-case analysis. Here is one of many situations where earlier work was not fully correct.

D. M. Y. Sommerville (1879–1934)

1. Sommerville describes the pseudo rhombi-cuboctahedron.

By far the most important early twentieth-century contributor to the theory of polyhedra was Ernst Steinitz. Steinitz, about 1916, developed a combinatorial characterization of convex three-dimensional polyhedra (*Milestone 19*). This work appeared in an encyclopedia of mathematics that was published in German and in French translation. Steinitz also wrote a book on polyhedra, which was almost finished at the time of his death; it was completed by Rademacher and published in 1934. As is typical of our subject, although Steinitz's main result is extremely important, there were almost no references to it before 1963. Why does it happen so often that important work in geometry attracts so little attention?

Ernst Steinitz (1871–1928)

1. Characterizes 3-polytopes combinatorially.
2. Rademacher's completion of Stenitz's almost completed book on polyhedra is published.

Next we come to Professor Coxeter's very important work on regularity (*Milestone 20*). He and various others did work on the stellated icosahedra, and he developed some very important regularity concepts which allowed skew, nonplanar, and infinite polygons. His famous work on uniform polyhedra has been referred to several times. He also combined algebraic with geometrical techniques in polyhedral group theory, which led to many important results in several branches of mathematics in addition to polyhedra theory. Much of this work is summarized in his *Regular Polytopes*.

H. S. M. Coxeter (1907–)

1. Coxeter (et al.) develops "regularity" concepts for polyhedra allowing skew and infinite polygons as faces.
2. Coxeter (et al.) conjectures that there are 75 "uniform polyhedra" (+ classical examples).
3. Coxeter pioneers work in "polyhedral" group theory.
4. *Regular Polytopes* summarizes all known work, explores new material (emphasizes higher dimensions).

A somewhat overlooked contribution to the theory of polyhedra is George Dantzig's discovery of the simplex method in 1947 (*Milestone 21*). Dantzig's work resulted in a big explosion of attempts to study the path structure on polyhedra which was very important to the development of the combinatorial theory.

George Dantzig (1914–)

1. Develops the "simplex method" for solving linear programming problem, which stimulates interest in path problems for polyhedra.

Then there is a surprisingly neglected subject. What convex polyhedra exist all of whose faces are regular convex polygons? (There is the classical work that we have discussed here, and the rediscovery by Freudenthal and van der Waerden that there are eight convex polyhedra with equilateral triangles for faces.) Norman Johnson, about 1960, conjectured that there are 92 such solids, in addition to the Platonic and Archimedean ones, including the prisms and antiprisms. In a series of papers Johnson, Grünbaum and Zalgaller proved the conjecture. The final version in English of that proof appeared in 1969! This work, taken together, is *Milestone 22*.

Regular Faced Polyhedra

1. Classical work.
2. H. Freudenthal and B. L. van der Waerden show there are eight 3-polytopes with equilateral triangles for faces.
3. N. W. Johnson conjectures there are 92 regular faced 3-polytopes (+ prism + antiprisms + platonic and Archimedean solids).
4. Johnson, Grünbaum, J. A. Zalgaller et al. prove Johnson's conjecture.

Finally, let me mention some of Grünbaum's major contributions to our subject (*Milestone 23*). The extremely important work of Steinitz was resurrected by Grünbaum in about 1962 when he realized that he could rephrase Steinitz's work in graph-theory terminology, making it possible to do all the combinatorial theory of three-dimensional polytopes in the plane. This means that those of you who can't see things in 3-space and are interested only in the combinatorial theory can study anything you want in the plane. Grünbaum summarized what he knew on the subject of convex polytopes in 1967 in his beautiful book. Then, building on work of Professor Coxeter, he published an article in 1977 in which he described a very general notion of regular polyhedra which allowed very general kinds of "regular" polygons as faces, not necessarily ones that can be spanned by membranes of any sort. Just within the last two years Andreas Dress proved that, aside from a small omission, the list of regular polyhedra that Grünbaum gave is complete.

Branko Grünbaum (1929–)

1. Restates Steinitz's characterization of 3-polytopes: a graph G is 3-polytopal if and only if *G* is planar and 3-connected.
2. Publishes *Convex Polytopes,* an exhaustive account of the combinatorial theory of polytopes.
3. Publishes a very general framework to study "regular" polyhedra, building on ideas of Coxeter.

I think we can see by the success of this conference that the subject will not go to sleep again in the way it all too often has in the past.

Notes

Acknowledgments: I wish to thank D. M. Bailey of the British Museum and Dr. Maxwell Anderson of the Metropolitan Museum of Art for their cooperation in obtaining access to the information about early man-made polyhedra. J. Wills (Siegen) provided me with a copy of Lindemann's paper and Robert Machalow (York College Library) helped obtain copies of many articles, obscure and otherwise.

[1] William C. Waterhouse, "The Discovery of the Regular Solids," *Archive for History of Exact Sciences 9* (1972):212–21.

[2] Friedrich Kurzweil, "Das Pentagondodekaeder des Museum Carnuntinun und seine Zweckbestimmung," *Carnuntum Jahrbuch 1956* (1957):23–29.

[3] F. H. Thompson, "Dodecahedrons Again," *The Antiquaries Journal 50,* pt. I (1970):93–96.

[4] F. Lindemann, "Zur Geschichte der Polyeder und der Zahlzeichen," *Sitzungsberichte der Mathematisch-Physikalische Klasse der Koeniglich Akademie der Wissenschaften [Munich] 26* (1890):625–783 and 9 plates.

[5] George Pólya, *Mathematical Discovery* (New York: John Wiley and Sons, 1981), combined ed., vol. 2, p. 154.

[6] P. Hilton and J. Pedersen, "Descartes, Euler, Poincaré, Pólya and Polyhedra," *L'Enseignment Mathématique 27* (1981):327–43.

[7] P. J. Federico, *Descartes on Polyhedra* (New York: Springer-Verlag, 1982).

[8] Jacques Hadamard, *An Essay on the Psychology of Invention in the Mathematical Field* (Princeton, N.J.: Princeton University Press, 1945), 51.

Bibliography

Artman, B. "Hippasos und das Dodehaeder." *Mitteilungen, Mathematisches Seminar, Universität Giessen 163* (1984):103–21.

Berman, M. "Regular-faced Convex Polyhedra," *Journal of the Franklin Institute 291* (1971):329–52.

Boltianskii, V. *Hilbert's Third Problem.* New York: John Wiley and Sons, 1975.

Brückner, M. *Vielecke und Vielfläche.* Leipzig: Teubner, 1900.

Catalan, M. E. "Mémoire sur la théorie des polyèdres." *Journal de l'École Imp. Polytechnique 41* (1865):1–71.

Cauchy, A. L. "Recherches sur les polyèdres." *Journal de l'École Polytechnique 9* (1813):68–86; *Œuvres Complètes,* ser. 2, vol. 1, pp. 1–25, Paris: 1905).

Cauchy, A. L. "Sur les polygones et les polyèdres." *J. de l'École Polytechnique 9* (1813):87–98.

Cayley, A. "On Poinsot's Four New Regular Solids." *The London, Edinburgh, and Dublin Philosophical Magazine and Journal of Science,* ser. 4, *17* (1859):123–28.

Chamier. "Nouvelles Hypotheses sur les dodécaèdres Gallo-Romains." *Revue Archéologique de l'Est et du Centre-Est 16* (1965):143–59.

Claggett, Marshall. *Archimedes in the Middle Ages*. Madison: University of Wisconsin Press, 1964.

Claggett, Marshall. *Greek Science in Antiquity*. New York: Abel Schuman, 1959; 3rd ed., New York: Collier, 1969.

Conze. *Westdeutsche Zeitschrift für Geschichte und Kunst 11* (1892):204–10.

Coxeter, H. S. M. *Regular Polytopes*. 3rd ed. New York: Dover Publications, 1973.

Coxeter, H. S. M. "Kepler and Mathematics," ch. 11.3 in Arthur Beer and Peter Beer, eds., *Kepler: Four Hundred Years, Vistas in Astronomy,* vol. 18 (1974), pp. 661–70.

Coxeter, H. S. M. *Regular Complex Polytopes*. London: Cambridge University Press, 1974.

Coxeter, H. S. M., P. du Val, H. T. Flather, and J. F. Petrie. *The Fifty-nine Icosahedra*. New York: Springer-Verlag, 1982 (reprint of 1938 edition).

Coxeter, H. S. M., M. S. Longuet-Higgins, and J. C. P. Miller. *Uniform Polyhedra. Philosophical Transactions of the Royal Society* [London], sec. A, *246* (1953/56):401–50.

Descarques, P. *Perspective*. New York: Van Nostrand Reinhold, 1982.

Dijksterhuis, E. J. *Archimedes*. Copenhagen: Munksgaard, 1956.

Dress, Andreas W. M. "A Combinatorial Theory of Grünbaum's New Regular Polyhedra. Part I. Grünbaum's New Regular Polyhedra and Their Automorphism Groups." *Aequationes Mathematicae 23* (1981):252–65.

Dress, Andreas W. M. "A Combinatorial Theory of Grünbaum's New Regular Polyhedra. Part II. Complete Enumeration." (to appear).

Dürer, Albrecht. *Unterweysung der Messung mit dem Zyrkel und Rychtscheyd*. Nürnberg: 1525.

Euler, Leonhard. "Elementa Doctrinae Solidorum." *Novi Commentarii Academiae Scientiarum Petropolitanae 4* (1752–53):109–40 (*Opera Mathematica,* vol. 26, pp. 71–93).

Euler, Leonhard. "Demonstratio Nonnullarum Insignium Proprietatus Quibus Solida Hedris Planis Inclusa Sunt Praedita." *Novi Commentarii Academiae Scientiarum Petropolitanae 4* (1752–53):140–60 (*Opera Mathematica,* vol. 26, pp. 94–108).

Federico, P. J. *Descartes on Polyhedra*. New York: Springer-Verlag, 1982.

Fejes Tóth, L. *Regular Figures*. New York: Pergamon Press, 1964.

Field, J. V. "Kepler's Star Polyhedra." *Vistas in Astronomy 23* (1979):109–41.

Gillings, R. J. "The Volume of a Truncated Pyramid in Ancient Egyptian Papyri." *The Mathematics Teacher 57* (1964):552–55.

Gillings, R. J. *Mathematics in the Time of the Pharaohs*. Cambridge, Mass.: MIT Press, 1972.

Grünbaum, Branko. *Convex Polytopes*. New York: John Wiley and Sons, 1967.

Grünbaum, Branko, and N. W. Johnson. "The Faces of a Regular-faced Polyhedron." *Journal of the London Mathematical Society 40* (1965):577–86.

Grünbaum, Branko. "Regular Polyhedra—Old and New." *Aequationes Mathematicae 16* (1977):1–20.

Günther, S. *Vermischte Untersuchungen zur Geschichte der Mathematischen Wissenschaften*. Leipzig: Teubner, 1876.

Haussner, Robert, ed. *Abhandlungen über die regelmässigen Sternkörper, Ostwald's Klassiker der Exacten Wissenschaften,* no. 151. Leipzig: Engelmann, 1906.

Heath, T. L. *The Thirteen Books of Euclid's Elements*. Cambridge, England: Cambridge University Press, 1925; New York: Dover Publications, 1956.

Hilton, Peter, and Jean Pedersen. "Descartes, Euler, Poincaré, Pólya and Polyhedra." *L'Enseignement Mathématique 27* (1981):327–43.

Hirsch, M. *Sammlung Geometrischer Aufgaben*. part 2, Berlin, 1807.

Johnson, N. W. "Convex Polyhedra with Regular Faces." *Canadian Journal of Mathematics 18* (1966):169–200.

Jordan, C. "Recherches sur les polyèdres." *Journal für die Reine und Angewandte Mathematik 66* (1966):22–85.

Kepler, Johannes. *Harmonices Mundi*. Linz, 1619; *Opera Omnia,* vol. 5, pp. 75–334, Frankfort: Heyden und Zimmer, 1864; M. Caspar, ed., *Johannes Kepler Gesammelte Werke,* vol. 6, Munich: Beck, 1938.

Lakatos, I. *Proofs and Refutations*. New York: Cambridge University Press, 1976.

Lebesgue, H. "Remarques sur les deux premières demonstrations du théorème d'Euler, rélatif aux polyèdres." *Bulletin de la Société Mathématique de France 52* (1924):315–36.

Legendre, A. *Élements de géometrie*. 1st ed. Paris: Firmin Didot, 1794.

Lindemann, F. "Zur Geschichte der Polyeder und der Zahlzeichen, *Sitzungsberichte der Mathematisch-Physikalischen Klasse der Koeniglich. Bayerischen Akademie der Wissenschaften* [Munich] *26,* pp. 625–783 and 9 plates (1890).

Lyusternik, L. A. *Convex Figures and Polyhedra*. Boston: D. C. Heath, 1966.

Malkevitch, Joseph. "The First Proof of Euler's Theorem." *Mitteilungen, Mathematisches Seminar, Universität Giessen 165* (1986):77–82.

Pacioli, L. *Divina Proportione*. Milan: 1509; Milan: Fontes Ambrosioni, 1956.

Pappus. *La Collection Mathématique*. P. Van Eecke, trans. Paris: De Brouwer/Blanchard, 1933.

Poinsot, L. "Mémoire sur les polygones et les polyèdres." *Journal de l'École Polytechnique 10* (1810):16–48.

Poinsot, L. "Note sur la théorie des polyèdres." *Comptes Rendus 46* (1858):65–79.

Pottage, J. *Geometrical Investigations Illustrating the Art of Discovery in the Mathematical Field*. Reading, Mass.: Addison-Wesley, Advanced Book Program, 1983.

Roman, Tiberiu. *Reguläre und Halbreguläre Polyeder*. Berlin: Deutscher Verlag, 1968.

Sachs, E. *Die Fünf Platonischen Körper*. Berlin, 1917.

Skilling, J. "The Complete Set of Uniform Polyhedra." *Philosophical Transactions of the Royal Society* [London], ser. A, *278,* (1975):111–35.

Sommerville, D. M. Y. "Semi-regular Networks of the Plane in Absolute Geometry." *Transactions of the Royal Society* [Edinburgh] *41* (1905):725–47 and plates I–XII.

Steinitz, Ernst. "Polyeder und Raumeinteilungen." In W. F. Meyerand and H. Mohrmann, eds. *Encyklopädie der mathematischen Wissenschaften,* vol. 3. Leipzig: B. G. Teubner, 1914–31.

Steinitz, Ernst, and H. Rademacher. *Vorlesungen über die Theorie des Polyeder*. Berlin: Springer-Verlag, 1934.

Thevenot, E. "La mystique des nombres chez les Gallo-Romains, dodécaèdres boulets et taureauxtricomes." *Revue Archéolgique de l'Est et du Centre-Est 6* (1955):291–95.

Thompson, F. H. "Dodecahedrons Again." *The Antiquaries Journal 50,* pt. I (1970):93–96.

Valet, M. "Des polyèdres semi-réguliers, dits solides d'Archimède." *Mem. de la Soc. des Sci. phys. et math. de Bordeaux 5* (1867):319–69.

van der Waerden, B. L. *Science Awakening*. New York: Oxford University Press, 1961.

Waterhouse, William C. "The Discovery of the Regular Solids." *Archive for History of Exact Sciences 9* (1972):212–21.

Wiener, C. *Über Vielecke und Vielfläche*. Leipzig, 1864.

Zalgaller, V. A. *Convex Polyhedra with Regular Faces*. New York: Consultants Bureau, 1969.

(Unsigned). "Roman Dodecahedron from Wales." *The Antiquaries Journal 4* (1924):273–74.

5

Polyhedra and Crystal Structures

CHUNG CHIEH

I have long been interested in searching for interesting relationships between polyhedra and crystal structures, especially with the application of polyhedra as units for crystal structures. Crystallography uses geometry as a foundation. As a crystal scientist, I am interested in understanding how and why certain crystal structures are built the way they are, particularly from a geometric veiwpoint. I am also constantly searching for relationships among the various crystal structures.

Shapes, colors, and geometry have historically been subjects of great interest to philosophers, mathematicians, and scientists. We are here to understand, to construct, to design, to create, to appreciate, and to love the shapes and forms of various kind. Among various shapes, perhaps polyhedra are especially interesting to us all, and we may also develop an affection for the aesthetic shapes of some very nice crystals such as those shown in Figs. 5-1–5-4. My interest is in the arrangements of atoms, molecules, and ions in those crystals, and how the arrangements are related to geometry and polyhedra.

The crystalline state is the most common form of all matter at sufficiently low temperatures. In modern terms, crystals consist of at-

Fig. 5-1. A wulfenite crystal: an orange flattened octahedron.

Fig. 5-2. Crystals of vanadinite: red hexagonal prisms.

Fig. 5-3. Hematite-stained quartzoids: polyhedra with threefold symmetry.

Fig. 5-4. Chrome alum: perfect octahedron, weighing 867 grams.

oms, molecules, or ions arranged in a periodic, repeated manner. For a crystal visible to the naked eye, there may be more than a million repeating units in each of the three directions. Those periodic arrangements may be described by symmetry operations such as one-, two-, three-, four-, and sixfold rotations, mirror and glide planes, screw axes and centers of inversion.[1] Their structures are fascinating from both the architectural and geometric viewpoints. However we look at them, the crystal structures are beautiful, three-dimensional repetitive patterns.

Figures 5-1–5-4 show some representative crystals. Their shapes are certainly related to some familiar polyhedra. The almost-perfect octahedral alum crystal was grown by a high school student, and it weights 867 grams. After the publication of a picture of this crystal by the *Chem 13 News* (a monthly publication of the University of Waterloo Chemistry Department), the editor received pictures of even larger alum crystals from which coffee tables, seats, and other interesting things had been made. (These crystals stand up to normal use when their surfaces are protected by varnish.)

The interesting external shapes of crystals must certainly be related to their internal structures. What are the basic units from which these wonderful and geometrically interesting crystals are built? Perhaps inspired by the beauty of crystals, the philosopher Plato (427–347 B.C.) associated the regular polyhedra with the *primal substances* from which everything is derived. Aristotle agreed that earth, air, fire, and water are the primal substances; however, he disagreed with Plato's associating these substances with four of the regular solids. During the seventeenth century there was lively discussion about the basic units of crystals. Johannes Kepler (1571–1630), Erasmus Bartholin (1625–98), René Descartes (1596–1650), and Robert Hooke (1635–1703) suggested that spheres are the ultimate particles. Today, packings of spheres are used as *models* for the discussion of crystals made up of atoms, but no one knows the real shape of atoms. Certainly the electronic configuration of each constituent atom has the symmetry of the electrostatic environment of that atom. In a crystal that environment is never spherical.

Fig. 5-5. Transformation of a packing of circles in a plane to a packing of triangles. The equivalent three-dimensional transformation is between spheres and polyhedra.

Figure 5-5 shows the packing of circular disks in a two-dimensional space, and the gradual transformation from packing of circles to that of triangles. In three-dimensional space, the transformation can be made from spheres to polyhedra. In Fig. 5-6, we see the packing of spheres as a model of the crystal structure of common table salt. This is also a model of the structures of many binary compounds. Yet, we still like to think of the unit as a little cube. We tend to find interesting the relationship between the packing of spheres and the packing of polyhedra. Are crystal structures really packings of spheres, or are they packings of polyhedra? The choice for crystal science is very much a matter of convenience and a matter of aesthetics. The partition of space into shapes, even within a crystal, is a subject of interdisciplinary interest, involving art, mathematics, and science.

Some highly symmetrical polyhedra such as the Archimedean truncated octahedra have been used by crystallographers and mineralogists to represent complicated crystal structures. Zeolites are natural architectures sometimes employed as chemical ion exchangers and molecular sieves. These natural three-di-

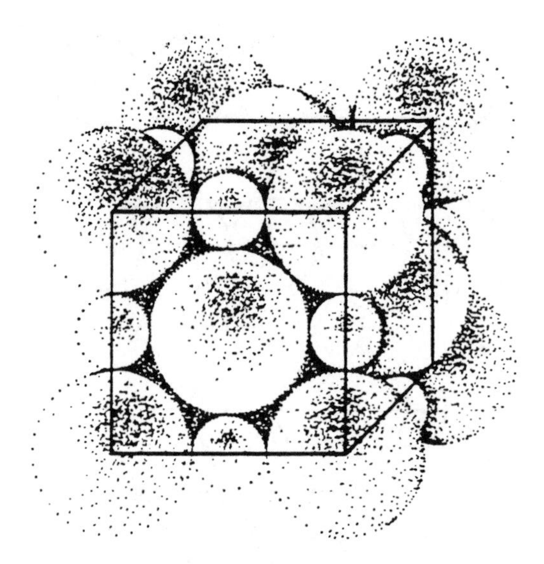

Fig. 5-6. The packing of spheres as a model of crystalline sodium chloride, common table salt. The small spheres represent sodium cations, and the large spheres represent chloride anions. The unit cell, a cube, is sketched with straight lines.

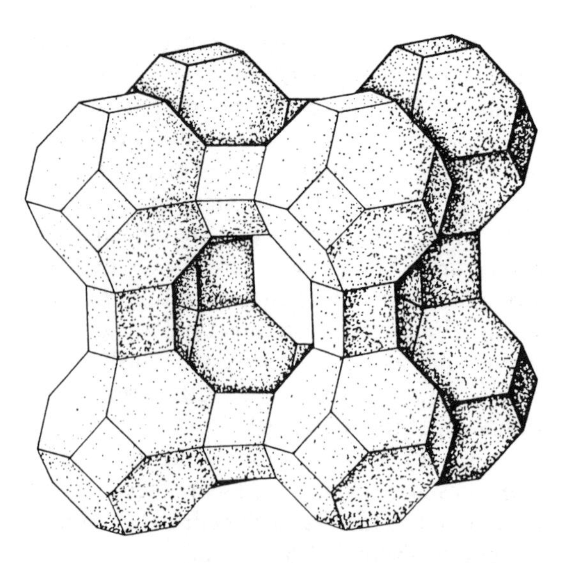

Fig. 5-7. A portion of the open-framework packing of Archimedean truncated octahedra, a model that represents the structures of many zeolites.

mensional structures are beautiful in their own right. They are silicates with some silicon atoms replaced by aluminum atoms with a general formula $(Al,Si)_nO_{2n}$. We can easily identify a cagelike unit formed by connecting $(SiO_{4/2})$ groups of atoms. The silicon atoms in the crystal structure are located at the vertices of Archimedean truncated octahedra. These large cages are interconnected in many ways; Fig. 5-7 shows one of the open packings or connections of them. The square faces are separated at distances equal to the length of the edges of these polyhedra, thus making the square faces the faces of cubes. Of course, the structure may also be considered to be an open packing of cubes. Figure 5-7 shows only part of the framework, and there are millions of these truncated octahedra in any direction in a real crystal. The possible ways that these polyhedra may interconnect is an interesting topological problem.

Mathematicians have contributed greatly to the study of crystallography, and their methods are used for the description of crystal structures.[2] Because of the periodic nature of a crystal, description of the large structure is simplified by considering the crystal to be built up by repetition, in all directions, of the structure enclosed within a parallelepiped (the *unit cell*). Although there are standard conventions for selecting the unit cell, the choice is not unique. As a crystallographer, I am interested in finding out how a particular structure is formed, what are the basic units (not necessarily the unit cells) that build a specific structure, and why a structure type is of common occurrence. I would like to find a general geometric scheme by which crystal structures are formed. Because crystal structures are three-dimensional patterns, they are too complicated to illustrate my approach to the problem of finding the basic units. Thus I shall use some two-dimensional artworks to demonstrate my search. Let us start by looking at one of Escher's drawings, *Fish and Birds,* reproduced as Fig. 5-8.

There are many ways to choose a basic unit that can be used to build these beautiful patterns of fish and birds. Let me show how crystallographers would choose unit cells from a pattern like this. There are various choices as indicated in Fig. 5-9. Choices 5-9a and 5-9b are arbitrary, and 5-9c is somewhat obscure, yet each is legitimate because each is a parallelogram. None of these choices is unique. But there is a unique way of defining a different

Fig. 5-8. *Fish and Birds* by M. C. Escher. © M. C. Escher Heirs c/o Cordon Art–Baarn–Holland.

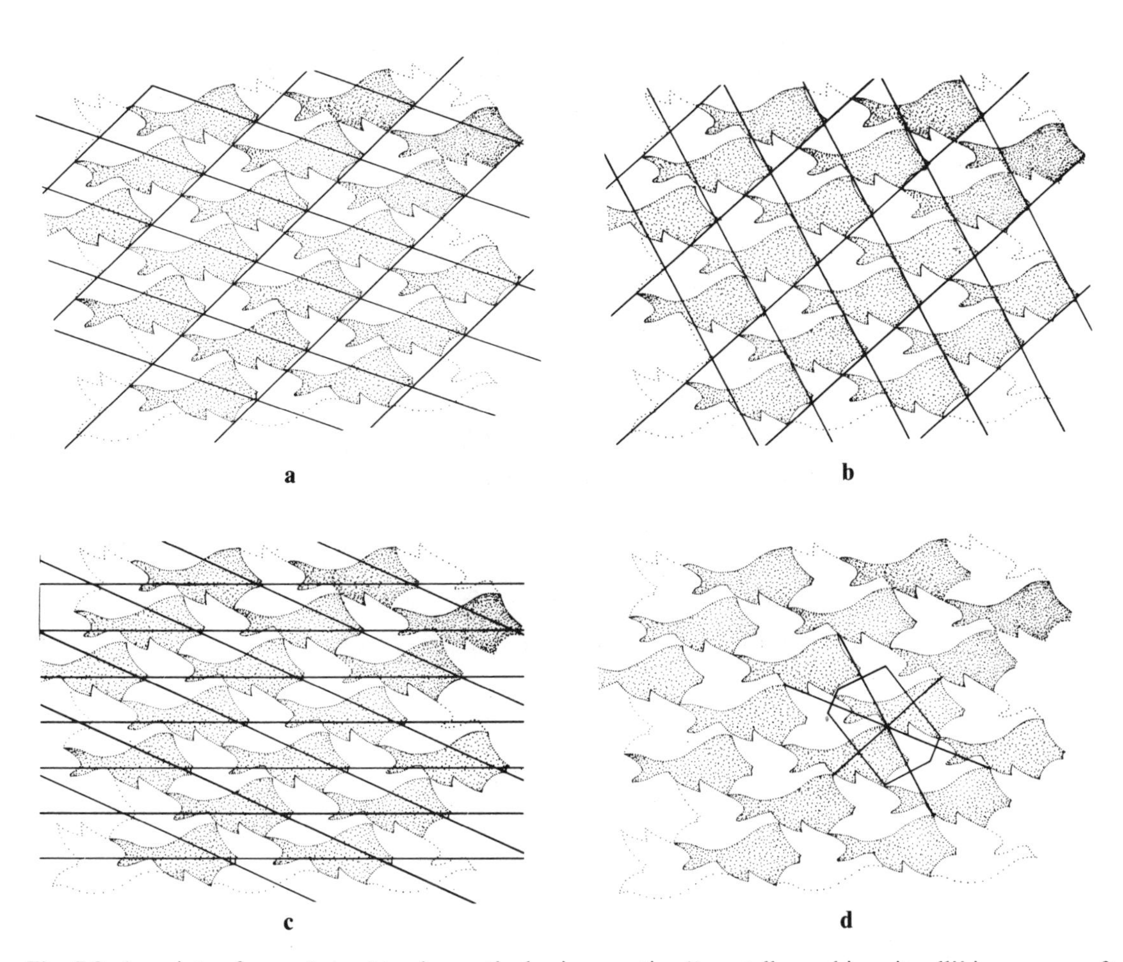

Fig. 5-9. A variety of ways (a to c) to choose the basic repeating "crystallographic unit cell" in an array of fishes and birds. The Dirichlet domain (d) of that array.

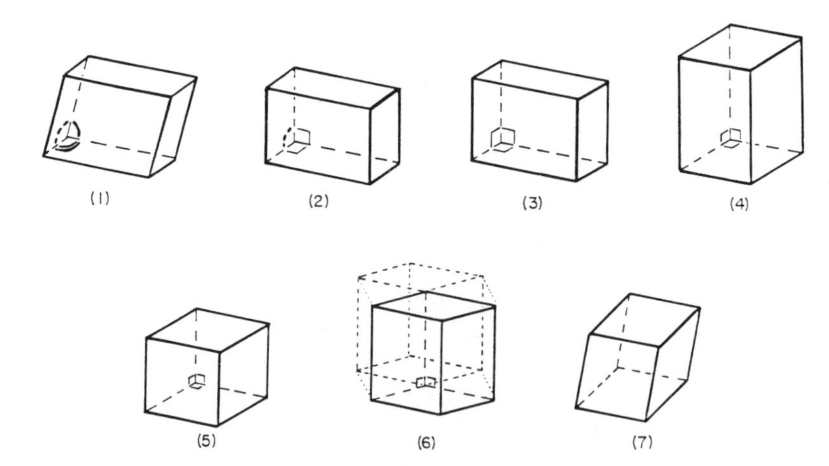

Fig. 5-10. The seven polyhedra used as crystallographic unit cells for the seven crystal systems. Unique parameters: (1) Triclinic—a, b, c, α, β, γ; (2) Monoclinic—a, b, c, β or γ depending on choice; (3) Orthorhombic—a, b, c; (4) Tetragonal—a, c; (5) Cubic—a; (6) Hexagonal—a, c (γ = 120°); (7) Rhombohedral—a, α.

Fig. 5-11. *Black and White Knights* by M. C. Escher. © M. C. Escher Heirs c/o Cordon Art–Baarn–Holland.

kind of basic unit. It was first suggested[3] by the German mathematician G. Lejeune Dirichlet (1805–59). In his method, a particular point in the pattern is chosen, for example the eye of a bird, and then it is connected to all other similar points (eyes of birds). We then draw bisection lines of these vectors, and the smallest area enclosed by the bisection lines is a unique unit called the Dirichlet domain. In two-dimensional space these domains are polygons (Fig. 5-9d), whereas in three-dimensional space they are polyhedra. Note that the Dirichlet domain for a two-dimensional pattern usually is not a parallelogram, and that of

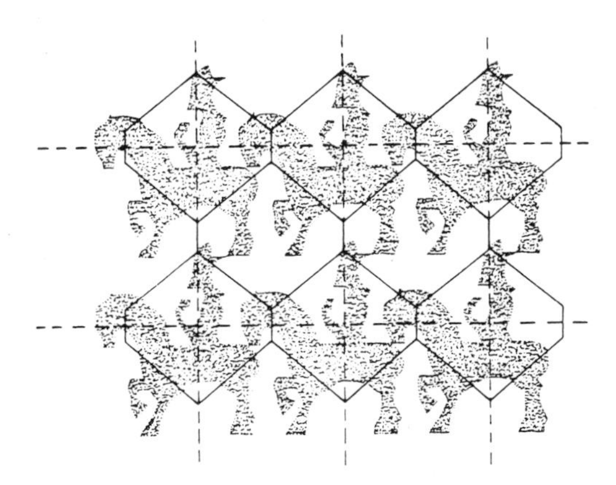

Fig. 5-12. The geometric plan for Escher's *Black and White Knights* as depicted by Dirichlet domains.

a three-dimensional crystal structure is usually not a parallelopiped.

The eyes of the birds (or any other set of translationally equivalent points) constitute a two-dimensional *lattice*. Similarly, a three-dimensional lattice is a set of points generated by three noncoplanar vectors. Classified by symmetry, there are seven types of coordinate systems. They are often depicted by unit cells (see Fig. 5-10). In the nineteenth century, Bravais studied the symmetries of poiyhedra and of lattice points, and he came to the remarkable conclusion that there are only fourteen symmetry types of point lattices. Nowadays, we take the fourteen lattices for granted, and in many books, the fourteen Bravais lattices and their relation to the unit cells are displayed together. (There are fourteen Bravais lattices and only seven symmetry types of unit cells; seven of the unit cells contain more than one lattice point.) Therefore, we ask the question: What are the units if we divide the crystal structures according to the fourteen lattices, instead of by unit cell shape? The Dirichlet method gives unique shapes, but the difficulty is that there are more than fourteen different polyhedra because different axis ratios of the same lattice type give rise to Dirichlet domains with various shapes. I shall return to this problem later.

Let us return to two dimensions. Applying the Dirichlet technique, I shall illustrate the geometric plan for Escher's *Black and White Knights,* shown in Fig. 5-11. If we choose a point along a certain line (equivalent to a glide line) on this artwork, we can see that the drawing is made up of two types of units, a black knight and a white knight. There is just one catch, which was pointed out to me by a little boy looking over my shoulder when I placed the Dirichlet domains over the *Black and White Knights*: if the center of the polygon is an arbitrarily chosen point, then the patterns within the polygon (arrangement of atoms in the polyhedron in case of crystal structures) may no longer be related (by crystallograhic and color symmetry).

Let us return to the very beautiful Archimedean truncated octahedron, one of my favorite polyhedra, as the unit for the cubic crystal system. This is a system in which the unit cell is a cube. There are three Bravais lattices in this system. In one case, the "primitive" lattice, the cell does not contain any lattice point in the cube, but the vertices are marked by the lattice points. In the "body-centered" lattice there is a lattice point at each cube center as well as at the vertices. In the "face-centered" cubic lattice, there is a lattice point in the center of each cube face, but none in the center of the cube.

For the body-centered cubic lattice, the Dirichlet domain is the Archimedean truncated octahedron. When we used it to represent the structures of zeolites, we did not emphasize the fact that they pack together to fill the entire space, leaving no gaps. But they do. Figure 5-12 illustrates the packing of these semiregular polyhedra. Suppose we make transparent polyhedra of the same size and

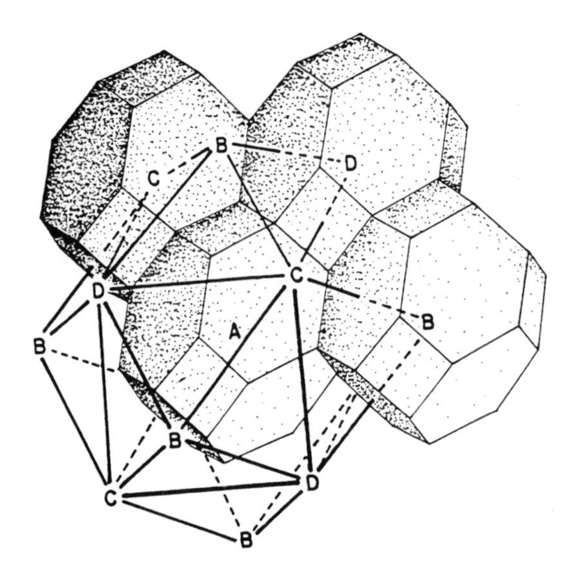

Fig. 5-13. Close packing of four "different" types (different, perhaps, because of color) of Archimedean truncated octahedra in the formation of a face-centered lattice.

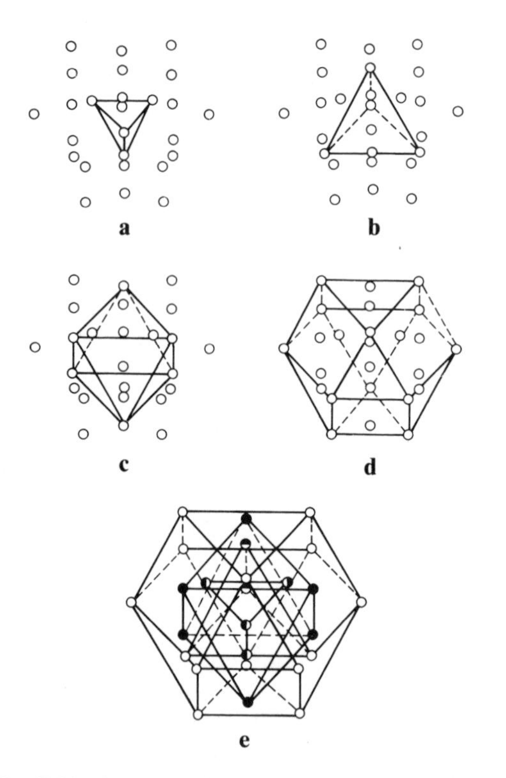

Fig. 5-14. A geometric unit of the γ-brass crystal structures, consisting of inner and outer tetrahedra (a and b), an octahedron (c), and a cuboctahedron (d). The composite of all these is shown in (e).

shape, and put a structural feature in each. Then we may use these polyhedra to build three-dimensional structures, as Escher put fish and birds together to make patterns. For the discussion of crystal structures, the polyhedra are only conceptual units; they represent clusters of packed atoms, ions, or molecules. A packing of one type of Archimedean truncated octahedron gives rise to a cubic body-centered lattice, a packing of two types according to a specific order gives a primitive lattice, and a packing of four types as shown in Fig. 5-13 gives a face-centered cubic lattice. If we put some configurations of certain symmetries into these transparent polyhedra, we can build structures having various kinds of symmetry compatible with the symmetry of the cubic lattice.

By assuming that we have one, two, or four types (due to the enclosed configuration of atoms) of units all having the shape of Archimedean truncated octahedra, we are able to classify all cubic space groups, and eventually all cubic crystal structures. The details of the classification have been published.[4]

We shall now see some examples of atomic arrangements of the cubic crystals. Let us look at the atomic arrangement within one of these polyhedra, within the structure of one geometric unit. This unit comes from the crystal structure of a γ brass, an alloy.[5] Starting from the center of the unit, there are four atoms arranged tetrahedrally as outlined in Fig. 5-14a. Given the symmetry, or point group, there should be a limited number of ways to build a geometric unit from the center, keeping in mind that atoms in most metallic crystals maintain definite equilibrium distances among each other. A beautiful way to add atoms to the unit is to place them at each face of the existing small tetrahedron, resulting in a larger one. This is followed by arranging six atoms in an octahedron outside the two tetrahedra, and finally arranging twelve atoms in a cuboctahedron to complete the unit. The model (using spheres as atoms) in Fig. 5-15 shows how these units are fitted together. In one, the two units are slightly separated for clarity. It should be pointed out that these units stack in a three-dimensional fashion, rather than linearly.

Fig. 5-15. Packing of the geometric units in γ brasses, isolated (left) and close-packed (right).

All cubic γ brasses belong to one of only three space groups, and they have one, two, and four types of units respectively. The atomic arrangements in these units are very similar from a purely geometric viewpoint; the differences arise because of the elements which occupy the vertices of the tetrahedra, octahedra, or cuboctahedra.

Metallic crystals are not the only structures which can be described or represented by the idea that the units actually have the shape of an Archimedean truncated octahedron. Organic molecules such as hexamethylenetetramine, $C_6H_{12}N_4$ (see Fig. 5-16), are natural units. These units pack in a cubic space group, and the Dirichlet domain for the molecule as a whole is an Archimedean truncated octahedron. This does not mean that the shape of this molecule is that of an Archimedean truncated octahedron; molecules do have bumps and craters on the surface. The birds and fish, or black and white knights in Escher's drawings, are not polygons either, but they do fit together forming a two-dimensional "crystal" in the way polygons do.

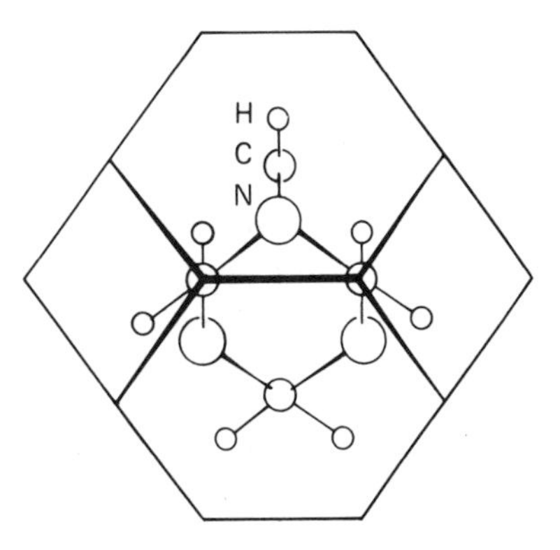

Fig. 5-16. The shape of an organic molecule, hexamethylenetetramine. The volume occupied by this molecule in its crystal approximates that of an Archimedean truncated octahedron.

Let us turn to other crystal systems, and at this point look at some Dirichlet domains of tetragonal, rhombohedral, and hexagonal lattices.[6] Polyhedra for Dirichlet domains of tetragonal lattices depend on the shape of the unit cell. A tetragonal lattice may be described by the lengths of two vectors a and c. Three of the four possible shapes of the Dirchlet domains are shown in Fig. 5-17a ($c > \sqrt{2}a$), 5-17b ($c = \sqrt{2}a$) and 5-17c ($c < \sqrt{2}a$). Actually, Fig. 5-17b shows a cubic face-centered lattice,

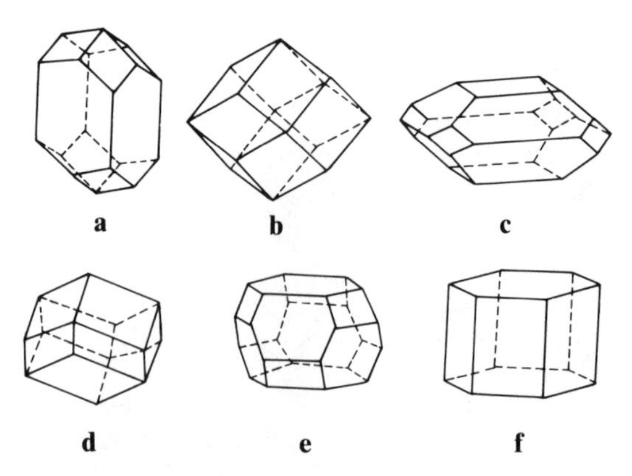

Fig. 5-17. Dirichlet domains of tetragonal (a–c), rhombohedral (d and e), and hexagonal (f) lattices.

a special case of both rhombohedral and tetragonal lattices. The fourth shape is the Archimedean truncated octahedron described earlier, where the tetragonal system may be metrically the same as that of a body-centered cubic lattice. Depending on the rhombohedral angle, there are two types of polyhedra. The Dirichlet domain for a rhombohedral lattice in which the angle is less than 60° is shown in Fig. 5-17d, whereas Fig. 5-17e is the Dirichlet domain for one whose rhombohedral angle is greater than 60°. Of course, the special cases when the angle is 60° or 90° are included in the cubic system. The Dirichlet domain for a hexagonal lattice is the hexagonal prism in Fig. 5-17f, but the conventional unit cells are of course parallelopiped.

Something becomes apparent when the Dirichlet domains of various types of lattice are depicted. A three-dimensional point lattice is a set of points generated by translations defined by three noncoplanar vectors. It has six parameters, three each of angles and magnitudes of the vectors. The variations of these parameters generate an infinite number of lattices, but they fall into the 14 Bravais symmetry types. Thus, each of these six parameters may vary independently, but the six-dimensional space may be divided into 14 regions, within each of which the symmetry of the Dirichlet domains is the same. However, there is still a variation of the shapes of the polyhedra representing the Dirichlet domains in each region.

The partition of a crystal structure into units according to Dirichlet domains is an interesting strategy for their study. However, there are many ways to choose the points from which the Dirichlet domains are derived. For example, the lattice points (which are not unique) could be chosen as the centers of atoms, or the gravitational centers of molecules. A reasonable method should keep all the units for a structure the same shape and size; the packing patterns of these units should apply to perhaps many structures for easy memory. Furthermore, the arrangement of atoms or molecules in this unit takes advantage of the symmetry properties, and all units for a structure have the same symmetry. Keeping these criteria in mind, we may derive the Dirichlet domains from points of the highest symmetry in a structure and call them *geometric units*. I have been concerned about the ways these geometric units pack, and the possibility of classifying crystal structures using geometric units and their packing patterns. This concern led me to study the space groups. I tried to classify them according to the packing of geometric units, since they represent hypothetical crystal structures. I termed the study of this scheme "geometric properties or geometric plans of space groups." Space groups theoretically classify all crystal structures according to their symmetry, and the study of space groups for their geometric plans is therefore a study of crystal structures for the same.

Using Dirichlet polyhedra derived from sites of highest symmetry of the tetragonal crystal system, we may proceed in a similar way to those of the cubic space groups to work out the geometric plan of the tetragonal space groups. The polyhedra used as units may vary in shape due to the axial ratio of the tetragonal system. The arrangements of units are shown using a plane perpendicular to the $x + y$, and x directions respectively (represented by (110) and (100)) as given in Fig. 5-18. The details of classification of these space groups have been published in the form of a table.[7] I note that there are nine types of arrangement for 68 space groups, and the nine packing patterns are also given in Fig. 5-18.

As an illustration, I will choose a series of organometallic compounds,[8] tetraphenyl derivatives of the group IV elements C, Si, Ge, and Pb. The molecules of these compounds belong to the same point group, and they oc-

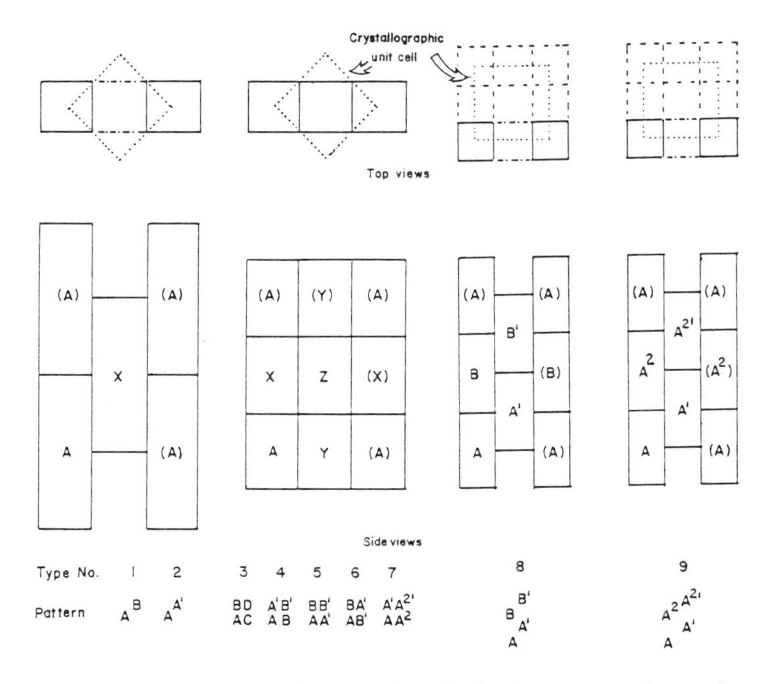

Fig. 5-18. Nine packing types of geometric units in the tetragonal crystal system.

cupy sites of the same symmetry in the space group. The molecules are natural geometric units. The packing of these molecules in solid state is shown in Fig. 5-20.

After working on the analysis of spatial arrangements in the tetragonal system, we have been able to apply the knowledge gained for the solution of an interesting tetragonal crystal structure, that of anhydrous zinc bromide. The geometric units can be seen as cubes or large tetrahedra, $Zn_4Br_6Br_{4/2}$ (see Fig. 5-19). The real structure belongs to a space group that is too complicated for theoretical calculation in the analysis of its infrared spectrum. At this point, I realized that these units may be packed together in more than one way, as exemplified by illustration in Fig. 5-19, which shows two types of stacking related to two very common crystal structures, ice and diamond. In ice, the tetrahedron is made up of an oxygen atom in the center, and four shared hydrogen atoms at the vertices $OH_{4/2}$, whereas in diamond, each tetrahedron represents a carbon atom, which is connected to four others in a tetrahedral fashion. The ice structure belongs to a hexagonal lattice, but the diamond structure belongs to a cubic one. Let us return to the $ZnBr_2$ structure. By a simple change in the orientation of these large tetrahedral units, $Zn_4Br_6Br_{4/2}$, we manage to reduce the complexity of the calculation by using another space group, to which the structure approximates.

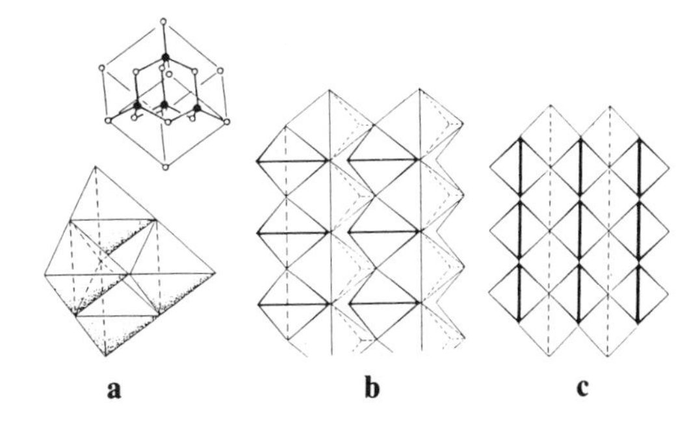

Fig. 5-19. The crystal structure of $ZnBr_2$. The geometric unit of this compound consists of a large unit with a formula $Zn_4Br_6Br_{4/2}$ that can be viewed as a large tetrahedron made up of four small ones (a). There are many ways to connect tetrahedra, and crystal structures may be represented by these connections: (b) ice, (snow), (c) diamond.

Further variations of packing of tetrahedra are indicated in Fig. 5-21. The interconnected tetrahedra at three of their vertices form a layer 5-21a; the fourth vertex and every three-connected point are for interlayer connections. If another layer of the same type is used

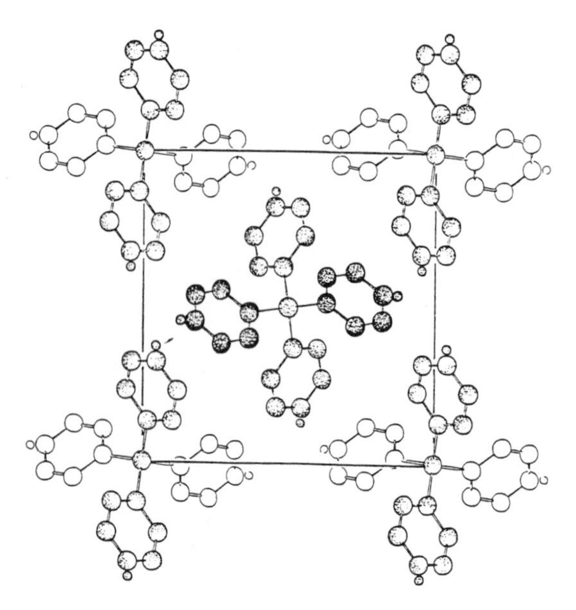

Fig. 5-20. Molecular packing of tetraphenyl derivatives of the group IV elements (C, Si, Ge, and Pb). The molecule at the center of the diagram is moved up by a half period in the direction perpendicular to the paper, and because of the orientation difference of this molecule with respect to one at the origin, the structure belongs to a primitive lattice.

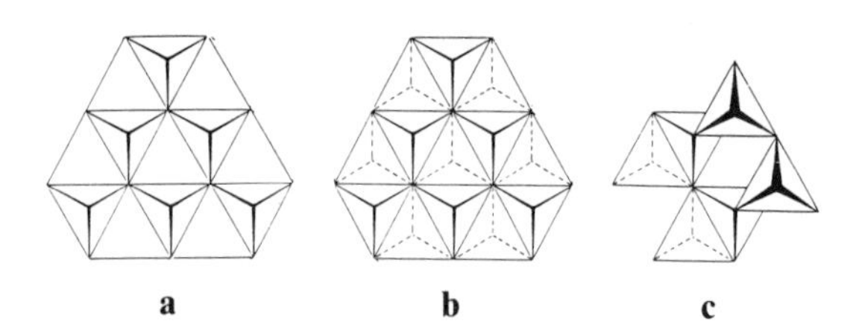

Fig. 5-21. Layer packing of interconnected tetrahedra: (a) a layer of interconnected tetrahedra, (b) a double layer, and (c) a triple layer.

but turned upside-down, then we have a double layer 5-21b. The ice structure is simply a back-to-back stacking of these double layers on top of each other. However, a third layer may be added to the double layer in a fashion shown in Fig. 5-21c. By varying the positions of interlayer connections, we can stack a sequence of any length. This aspect of geometry is richly demonstrated in the natural crystal structures.

The foregoing discussion about crystal structures indicates that I agree with Plato's argument that *all matter is the result of combinations and permutations of a few basic (polyhedral) units*. Nowadays, we know a lot more about crystal structures than Plato did. My association of these structures with polyhedra is partly for the ease of recognition and partly for providing a minimum inventory to get the maximum diversity in application. The intricate shapes of crystals stimulate us to study geometry, but geometry is the most important tool for the understanding and systematic classification of crystal structures.

My heartbeat increases whenever I see art by M. C. Escher. I have tried to find out how some of his exciting works were created. Escher has fantasized geometry and symmetry into visually stimulating forms. It can also be said that crystal structures are the artwork of God or nature. I am as curious about the formation of crystal structures, and about the geometric design of those structures, as I am about Escher's art.

In conclusion, I am excited to see so many people enthusiastically making contributions in terms of models, in terms of educational materials, and in terms of teaching me how to understand and appreciate geometric units. Your effort has made it a little easier for me to understand the crystal structures or natural three-dimensional patterns in terms of their geometric plan, something that I wanted to comprehend.

Notes

Acknowledgment: Thanks to Nancy McLean for redoing some of the drawings.

[1] Geometric considerations show that crystals cannot have fivefold rotational symmetry, but the discovery of diffraction patterns showing fivefold symmetry by D. Shechtman, I. Gratias, and J. W. Cahn ["Metallic Phase with Long-range Orientational Order and No Translational Symmetry," *Physical Review Letters 53* (1984):1951–53] indicates its existence in the solid state of pseudocrystals or quasicrystals.

[2] See, for example, A. L. Loeb, "A Systematic Survey of Cubic Crystal Structures," *Journal of Solid State Chemistry 1* (1970):237–67.

[3] G. Lejeune Dirichlet, "Über die Reduction der positiven quadratischen Formen mit drei unbestimmten ganzen Zahlen," *Journal für die reine und angewandte Mathematik 40* (1850):209–27.

[4] Chung Chieh, "The Archimedean Truncated Octahedron, and Packing of Geometric Units in Cubic Crystal Structures," *Acta Crystallographica,* sec. A, *35* (1979):946–52; "The Archimedean Truncated Octahedron II. Crystal Structures with Geometric Units of Symmetry $\bar{4}3m$," *Acta Crystallographica,* sec. A, *36* (1980):819–26; "The Archimedean Truncated Octahedron III. Crystal Structures with Geometric Units of Symmetry m3m," *Acta Crystallographica,* sec. A, *38* (1982):346–49. Chung Chieh, Hans Burzlaff, and Helmuth Zimmerman, "Comments on the Relationship between 'The Archimedean Truncated Octahedron, and Packing of Geometric Units in Cubic Crystal Structures' and 'On the Choice of Origins in the Description of Space Groups' " *Acta Crystallographica,* sec. A, *38* (1982):746–47.

[5] M. H. Booth, J. K. Brandon, R. Y. Brizard, C. Chieh, and W. B. Pearson, "γ Brasses with F Cells," *Acta Crystallographica,* sec. B, *33* (1977):30–36 and references therein.

[6] Chung Chieh, "Geometric Units in Hexagonal and Rhombohedral Space Groups," *Acta Crystallographica,* sec. A, *40* (1984):567–71.

[7] Chung Chieh, "Geometric Units in Tetragonal Crystal Structures," *Acta Crystallographica,* sec. A, *39* (1983):415–21.

[8] Chung Chieh, "Crystal Chemistry of Tetraphenyl Derivatives of Group IVB Elements," *Journal of the Chemical Society* [London] *Dalton Transactions* (1972):1207–8.

6

Polyhedra: Surfaces or Solids?

Arthur L. Loeb

It is a great pleasure to note how the network of polyhedrists is growing. Many of us started in isolation and others wondered what it was all about, this matter of polyhedra. Most people still wonder, but at least we are not so singular anymore; we seem to be becoming connected.

To begin with, what is a polyhedron? Since I am especially interested in the relationship between concepts and images, I decided to approach the subject from that point of view and to try to relate mathematical concepts and images. A polyhedron is an image of many, many different concepts, some of them inconsistent with each other.

In Fig. 6-1 you see our friend Escher, whose spirit seems to hover over the Shaping Space Conference. He is contemplating the question of the apparent solid on the left and, on the right, the surface. When we talked about this print, he often said that it is very curious, it is really the reflectivity of a surface that matters. Even when material is transparent, some of the light is bounced off; some of it is transmitted, but it is modified when it is transmitted. So we really cannot tell unequivocally what goes on inside. The one on the right is totally impenetrable. Everything bounces off the white surface. We cannot tell anything about the inside. Figures 6-2 and 6-3 show that Escher, who as we know was very much concerned with plane tessellations, was very much aware of the difference between tessellating a plane and tessellating a sphere; we shall return to this later.

Escher was enormously skilled as a graphic artist and his fame rose considerably in the time of conceptual art. He was truly a concep-

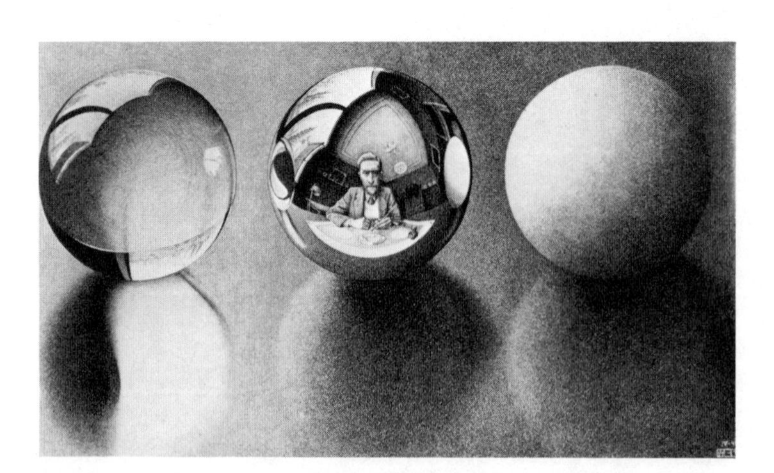

Fig. 6-1. M. C. Escher contemplating the apparent solid on the left and the surface on the right. *Three Spheres II*. Lithograph. M. C. Escher, April 1946.

Fig. 6-2. *Sphere with Angels and Devils*. Stained maple. M. C. Escher, 1941. © M. C. Escher heirs c/o Cordon Art – Baarn – Holland.

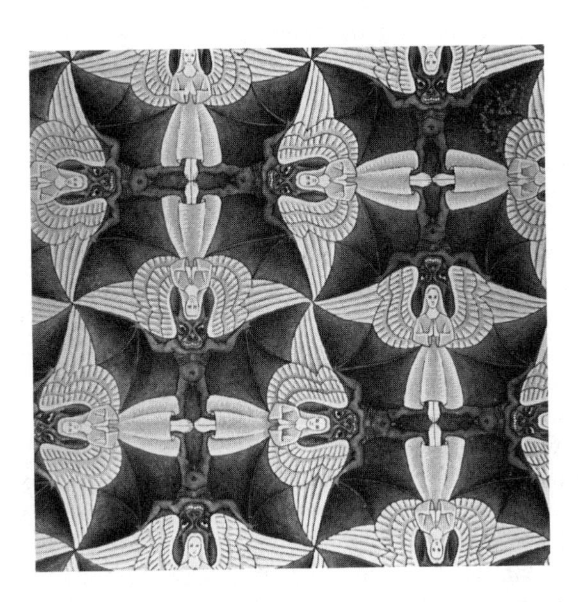

Fig. 6-3. *Angels and Devils*. Pencil, India ink, crayon, and guache. M. C. Escher, 1941. © M. C. Escher Heirs c/o Cordon Art – Baarn – Holland.

tual artist, but unlike a good many conceptual artists who had essentially become minimalists and had no more physical substance to their art, Escher had the skill to express his ideas and his concepts visually. What Escher never did (and he said he could not) was to relate the visual concepts to equations. Nevertheless, we have here a projection of a concept in a language that is not mathematical in the sense of verbal formulas of sequences of symbols, but nevertheless is very important as a visual language.

The surface is most important. The two ways I am going to try to approach polyhedra have a certain duality in the very broadest sense (though not in the strict mathematical sense). On the one hand there is the point of view of a set of connected items of different dimensionalities, and on the other that of a set of very rigorously defined points. The first gives us the connectivity point of view, the second the symmetry point of view. We can inscribe vertices of zero dimension, edges of single dimension, faces of two dimensions, on a surface. The edges then do not have to be straight and the faces do not have to be flat; that of course means we leave the domain of defining incircles and outcircles. We simply talk about networks on a surface. Then it matters very much how this surface is connected, whether we have a sphere or the analog to a sphere, or whether we have a toroid such as the hat that has been designed for this conference and that was worn in Florence many centuries ago (see Chapter 1). We call a sphere singly connected because it has no hole, a doughnut doubly connected because it has a hole. If one travels inside a doughnut, and wants to travel past its hole, one must choose one or the other of two distinct kinds of paths in order to avoid the hole. Inside a sphere one can travel between two points along an infinity of different routes, and these routes may in principle differ one from the next one by an infinitesimal amount. Inside a doughnut there is also an infinite number of routes between two points, but they divide into two distinct groups, those going around the hole on one side or on the other. It is like driving from Northampton to Cambridge, Massachusetts: there are many ways, but one must go either north or south of the Quabbin Reservoir. Similarly, a pretzel, having three holes, is quadruply connected.

One of my students, Beth Saidel, observing that connectivity relationships are the same for all singly connected surfaces regardless of their exact shapes, decided to use the Ukrainian technique of Easter-egg painting to apply some tessellations which we had discussed (see Fig. 6-4).

Consider a finite number of vertices V on a surface. Connect them by a number of lines, called edges. To be exact, an edge is a curve (not necessarily a straight line) which joins two vertices, but does not contain any vertex except at either end. No edges cross each other: their crossing would imply a vertex at the intersection. The number of edges is called E. A region of the surface surrounded by a closed circuit of alternating edges and vertices, which does not contain either edges or vertices except on its boundary, is called a *face*. The number of faces on the surface is called F. For a surface of connectivity g

$$V - E + F = 2 - 2g. \tag{6.1}$$

It is amazing how much practical information can be derived from this equation. For our purposes it will be convenient to translate it into an expression relating valencies. If we call r the number of edges coming into any one vertex and n the number of edges (hence also the number of vertices) surrounding any one face, then we can add up the total number of edges in two different ways. One way is to find the number of vertices, V_r, having valency r. The total number of edges coming into a vertex having valency r equals rV_r. If we then sum over all possible values of r we would get, not the total number of edges, but twice that amount, because every edge terminates at two vertices, hence would have been counted twice. Therefore

$$2E = \sum_r rV_r, \tag{6.2}$$

and analogously

$$2E = \sum_n nF_n. \tag{6.3}$$

Fig. 6-4. Tessellations in the style of Ukrainian Easter-egg painting, by Beth Saidel. From the Teaching Collection in the Carpenter Center for the Visual Arts, Harvard University. Reproduced with permission of the Curator.

We can define weighted averages for both r and n:

$$r_{av} = \frac{\sum_r rV_r}{V}; \tag{6.4}$$

$$n_{av} = \frac{\sum_n nF_n}{F}. \tag{6.5}$$

Dividing Eq. (6.1) by $2E$ and then substituting Eqs. (6.2)–(6.5) into Eq. (6.1) produces[1]

$$\frac{1}{r_{av}} + \frac{1}{n_{av}} = \frac{1}{2} + \frac{1}{E} - \frac{2g}{E}. \tag{6.6}$$

Both Eq. (6.1) and (6.6) will prove useful. To begin with, consider the tiling of a singly connected surface ($g = 0$) with nothing but pentagons and hexagons, three tiles meeting at each vertex; in that case $r = 3$. Therefore $2E = 3V$. Moreover, $F = F_5 + F_6$. Hence Eq. (6.1) becomes

$$F_5 + F_6 = 2 + \frac{E}{3}. \tag{6.7}$$

Counting edges by summing over all hexagons and pentagons, and remembering that every edge is shared by two faces:

$$5F_5 + 6F_6 = 2E. \tag{6.8}$$

Solving Eqs. (6.7) and (6.8) by eliminating F_6, we find that we automatically eliminate E as well and obtain $F_5 = 12$. This is a startling result: it tells us, for example, that a soccer ball must have exactly twelve pentagonal (usually black) faces. We also find that there are berries having exactly twelve pentagonal faces in the company of hexagonal faces. It has been shown[2] that the number of hexagons can be any positive integer except 1. These results are noncommittal about the number of

Table 6-1. Solutions of Eq. (6.6) for regular structures.

r	n	E	Comments
2	n	n	A polygon having n sides
r	2	r	A pumpkin-like structure having r diagonal faces join at each of two points
3	3	6	Tetrahedron
3	4	12	Cube
4	3	12	Octahedron
3	5	30	Pentagonal dodecahedron
5	3	30	Icosahedron
3	6	∞	Hexagonal tiling of the plane
6	3	∞	Triangular tiling of the plane
4	4	∞	Square tiling of the plane

Source: This table and several of the line drawings in this chapter are from Arthur L. Loeb, *Space Structures: Their Harmony and Counterpoint* (Reading, Mass. Addison-Wesley, Advanced Book Program, 1976).

hexagons, but emphatically limit the number of pentagons to twelve.

Two structures are called duals to each other if to each vertex of one there corresponds a face of the other, and vice versa. The dual of the pentagon-hexagon tessellation is dealt with by interchanging V and F, and n and r. The duals therefore must have $n = 3$, hence triangular faces. If a Fuller dome were built extending all the way around a sphere instead of being anchored in the soil, it would have exactly twelve 5-valent vertices, together with a large number of hexagons that determines the size of the dome. Accordingly, the occurrence of the number 12 in connection with berries, domes, and soccer balls is not a coincidence, but is in fact a fundamental property of the space in which we live, and a constraint with which we need to be familiar if we desire to shape that space.

Structures having all vertices equivalent to each other, as well as all faces equivalent to each other, are called *regular* (see Chapter 3). For such structures r_{av} and n_{av} are integers, as of course is E. For regular structures Eq. (6.6) has the solutions given in Table 6.1. No other solutions are possible; this table exhaustively enumerates all regular structures. Neither r nor n can exceed 6 except when n, respectively r, equals 2. The cases $n = 2$ (digonal faces) are very real once we accept the possibility of curved edges. There are many objects in nature which are digonal polyhedra (for instance pumpkins, which have r digonal faces meeting at the stem and at the bottom), and there are many pods which are digonal trihedra. Five of the solutions correspond to the Platonic solids. Interesting are the three solutions having infinitely many edges. If the faces are to be finite in area, these solution can only be realized on a sphere having infinite radius. Such a sphere would be experienced as a plane, much as we experience the surface of our globe in our immediate environment as flat. Note that there is no solution having n equal to five and E infinite; the implication is that there can be no regular pentagonal tessellation of the Euclidean plane.

Note in Table 6-1 that interchange of n and r transforms a regular structure into its dual; the symmetry of Eq. (6.6) in r and n implies that, if a structure represents a solution of Eq. (6.6), then so will its dual.

For doubly connected surfaces ($g = 1$), we derive from Eq. (6.1):

$$\frac{1}{r_{av}} + \frac{1}{n_{av}} = \frac{1}{2}. \tag{6.9}$$

When Eqs. (6.6) and (6.9) are compared, it is observed that the solutions to Eq. (6.9) are just those of Eq. (6.6) with E equal to infinity. This means that the toroid (the Florentine hat), for example, may be tessellated just like the plane, which is not true of a sphere having a finite radius.

Besides the regular structures there are the semiregular structures and their duals, which have either all faces mutually equivalent or all vertices mutually equivalent, but not both. It has been shown[3] that Eq. (6.6) still yields an enumerable set of solutions when either r_{av} or n_{av} is an integer, which is characteristic of these structures. Once more we find a number of solutions corresponding to infinitely many edges, which again may be interpreted as plane tessellations. It is remarkable that so many of the structures shown in this book correspond to the solutions of the remarkably simple Eq. (6.6), and can be so listed and classified.

These structures may all be represented on a planar surface by means of a Schlegel diagram. You can think of Schlegel diagrams visually; if you hold a polyhedron so close to your face that one of its faces frames the entire polyhedron so that you see all the edges besides that frame just receding inside that frame, then you have a Schlegel diagram. Another way of looking at a Schlegel diagram is this. You can think of, say, a truncated octahedron beautifully inscribed on a spherical blackboard and then you suddenly realize that it didn't matter where you put the vertices and edges, it was just how they were connected. You could then think of this as a peculiar kind of string bag—you could slide everything over to one side of your spherical blackboard, contracting all edges into a very tiny figure. If you realize that our whole earth is really a gigantic sphere of very large radius, you could draw a gigantic truncated octahedron on the surface of the earth, and shrink it into a portion of that gigantic sphere which would just happen to be a blackboard. Then you would have a Schlegel diagram on your blackboard. Since n and r are only the valencies of the vertices and faces, it doesn't matter where these elements are located; we can say from our point of view of connectivity that a Schlegel diagram is entirely equivalent to the polyhedron itself. There is no difference because we don't care where the vertices are. And so I will show you a number of solutions of Eq. (6.6) in the form of Schlegel diagrams.

In Fig. 6-5, on the left, is the Schlegel diagram of a square antiprism: a square face has become extremely extended and frames the rest. All the other connections are there and you could almost solve these equations graphically just by taking a point, taking the proper valency, putting the lines in and then extending those lines until a face is created; interestingly enough, by following that procedure and not even thinking about what the polyhedron looks like, you can get your solutions directly as Schlegel diagrams. Now the question is: What about the dual? The polygon that frames the whole Schlegel diagram really represents a face corresponding to everything else in the plane—the entire universe in the plane. The reason? Remember we slid the polyhedron over to one side. But all the rest of the huge sphere is still a face, and if we now take a dual graphically, we have to put a vertex in each face (Fig. 6-5, right). Emanating from that vertex must be the same number of edges as surround the original face, and each edge has to cross one of its companions. Then what happens when we get to the outer polygon? We have a whole cycle of faces just inside it, each of them in dualizing becoming a vertex which then has to be connected to a vertex corresponding to that outer face. Notice that what I have done is to put arrows across the outer framing. Those arrows indicate that somewhere in the universe there is a vertex way out at the other pole of our infinitely large sphere and that is where the arrows will connect. I could have put it off center within this figure and connected everything, but it makes for a very unattractive, very ugly unsymmetrical kind of dual Schlegel diagram. I call these diagrams dual Schlegel diagrams because they are not Schlegel diagrams; they do not have everything framed by an outside face. We have in the dual Schlegel diagram a representation in which we have a vertex, which represents a real vertex, outside of that frame. But it is perfectly easy to deal with those; you often can visualize much better what the dual polyhedron looks like by imagining the polyhedron flattened out, and the faces which meet at the "backside" vertex folded out.

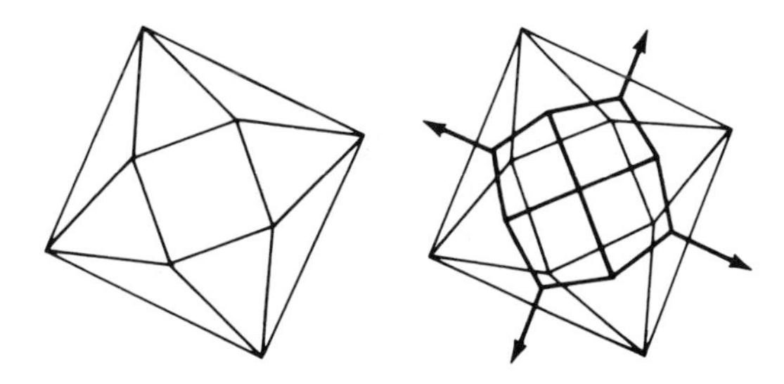

Fig. 6-5. Schlegel diagram of a square antiprism (*left*) and of its dual (*right*).

Figure 6-6 shows a Schlegel diagram of a snub figure. It is the same one that Professor Coxeter showed (Chapter 3), in which he calculated the coordinates when the square was rotated. In this case, I happened to orient it so that the outer square is the rotated one and you see the whole tier of triangles surrounding each of those squares. So this one is the snub cube, one of the solutions of that Diophantine equation. In Fig. 6-7 you see it again in the upper right, but here I put it in the company of

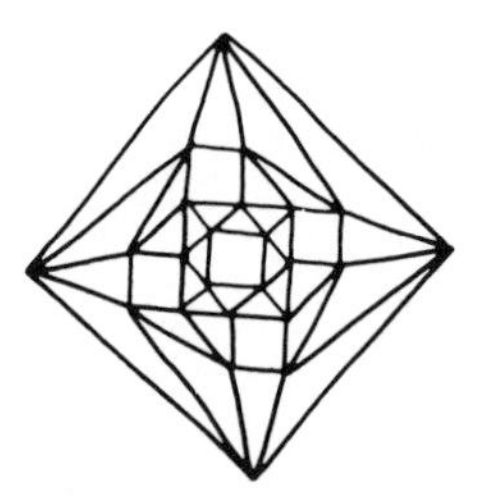

Fig. 6-6. Schlegel diagram of a snub cube.

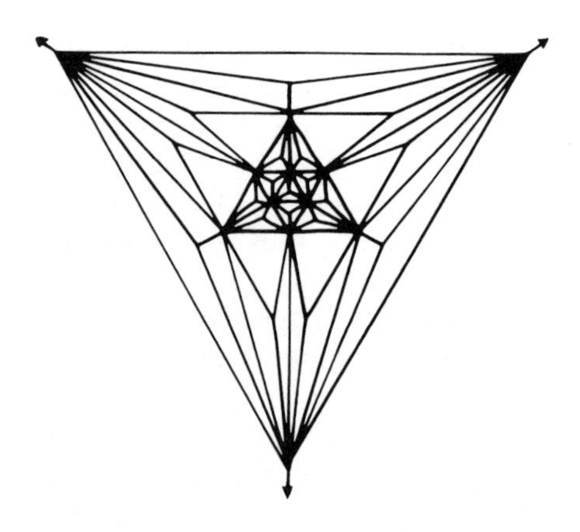

Fig. 6-9. Schlegel diagram of a stellated icosahedron.

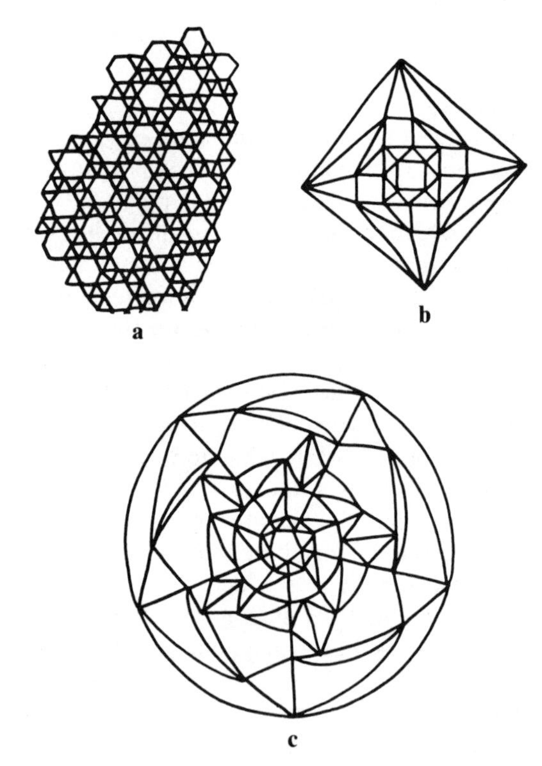

Fig. 6-7. (a) Snub tessellation of triangles and hexagons, having 5-valent vertices. (b) Snub cube. (c) Snub dodecahedron.

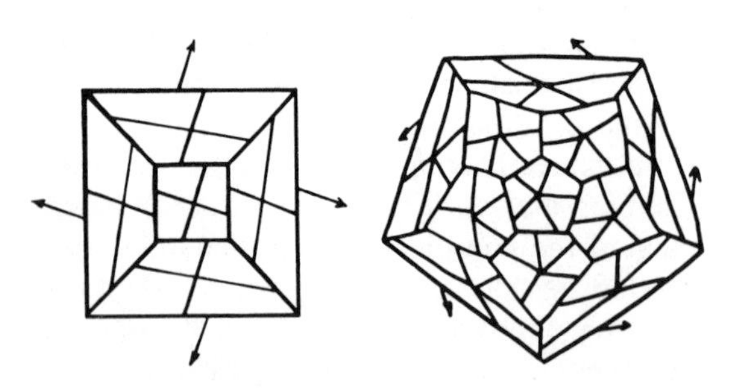

Fig. 6-8. Schlegel diagrams of (*left*) a pentagonal icositetrahedron (dual of the snub cube) and (*right*) a pentagonal hexacontahedron (dual of the snub dodecahedron).

its family having increasing numbers of edges, so corresponding to the snub cube there is a snub dodecahedron with still a finite but much larger number of edges, and finally the plane tessellation. All of these figures have in common their *chirality,* a very important property. This means they exist in forms that are distinct from their mirror images. So I could have drawn either one or I could have flipped the figure and then we would have had the other form of it. In Fig. 6-8 you see that I have taken the duals of the snub cube and snub dodecahedron in the form of the dual Schlegel diagrams.

Figure 6-9 is a stellated icosahedron and again that does very well in its dual Schlegel representation; in the center you can see the structure very well, but like a polar projection it is distorted toward the outside. But if you want to build these figures, these dual Schlegel diagrams help a lot. You get your actual polyhedron from the Schlegel representation by lifting up the arrows radiating out to infinity and bringing them together.

Figure 6-10 shows a pentagonal tessellation of the sphere. This is a model made by Brett Tomlinson. Next in the family is the limiting pentagonal tessellation (Fig. 6-11) in which E equals infinity. Incidentally, as we saw previously, there can be no regular pentagonal tessellation of the plane. But here we have a semiregular pentagonal tessellation and we have valencies in this case of 6 and of 3. You can tell fairly easily that this is also a figure that has chirality. What you have to do is

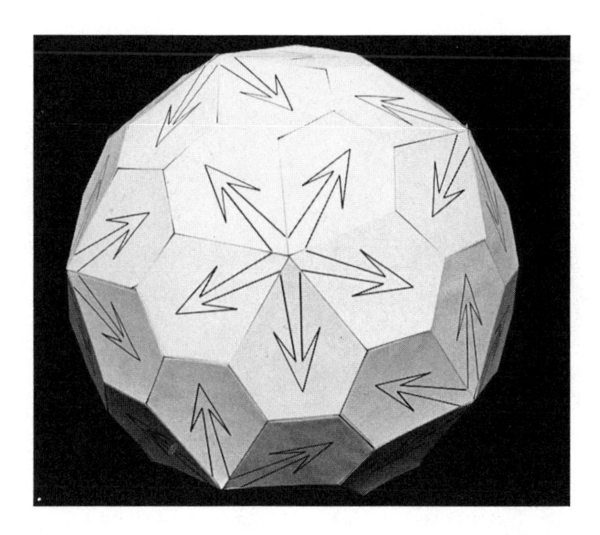

Fig. 6-10. A model of a pentagonal tessellation of a sphere, by Brett Tomlinson. From the Teaching Collection of the Carpenter Center for the Visual Arts, Harvard University. Reproduced with permission of the Curator.

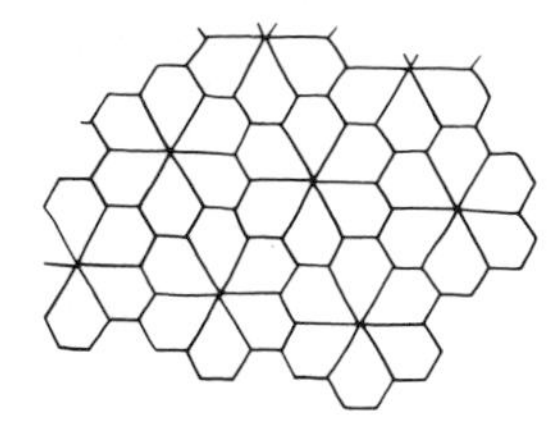

Fig. 6-11. Pentagonal tessellation of the plane with $n_1 = 4$, $n_2 = 1$, $r_1 = 3$, $r_2 = 6$.

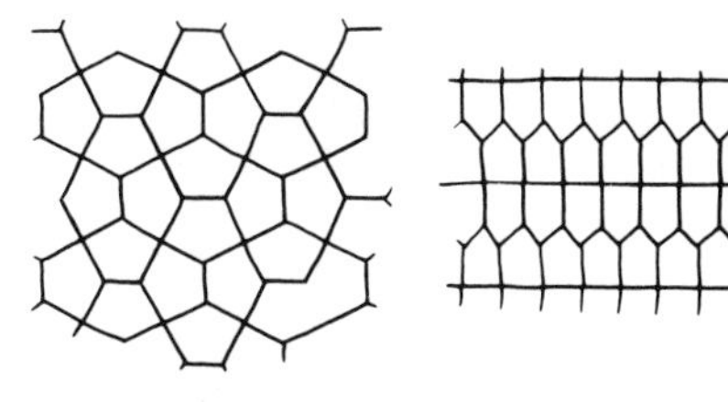

Fig. 6-12. Two pentagonal tessellations of the plane, each with $n_1 = 3$, $n_2 = 1$, $r_1 = 3$, $r_2 = 4$.

make a distinction, not only between the vertices having different valencies, but also between different 3-valent vertices. You find that some of these 3-valent vertices are connected to a 6-valent as well as to two 3-valent vertices, whereas others are connected to 3-valent vertices only. Those vertices are definitely distinct. Their contexts are different even though their valencies are the same. And so as we go around the pentagon, you will notice that we have five vertices and we have to make a choice, whether we are going to put one of each type of 3-valent vertices on the right-hand side or on the left-hand side. That means that we are forced to create a tessellation that has chirality, because we have these different types to distribute. That choice has to be made and depending on how we make it, we get this tessellation or its mirror image. Figure 6-12 shows two of the pentagonal tessellations about which Doris Schattschneider is an expert. You see again that we are dealing with a family of E increasing toward infinity. Polyhedra and tessellations are very closely related.

Now I am going to take the symmetry point of view. That is a totally different concept which again leads to the images of a polyhedron. We have already heard a great deal about symmetry. I want to talk a bit about how we can tell symmetry to a computer; this is becoming very important. We have to face the fact that computers do not easily visualize. And we should not try to teach the computers to visualize in the way that we visualize. We should develop languages; actually this is how I started to get into this field 25 years ago, when I tried to develop a language for the description of crystal structures. We probably should modify our language somewhat so that communication with the computer becomes more convenient.

Fig. 6-13. A collection of wooden polyhedra, each with the symmetry of a cube. Each polyhedron has been oriented with a threefold axis vertical.

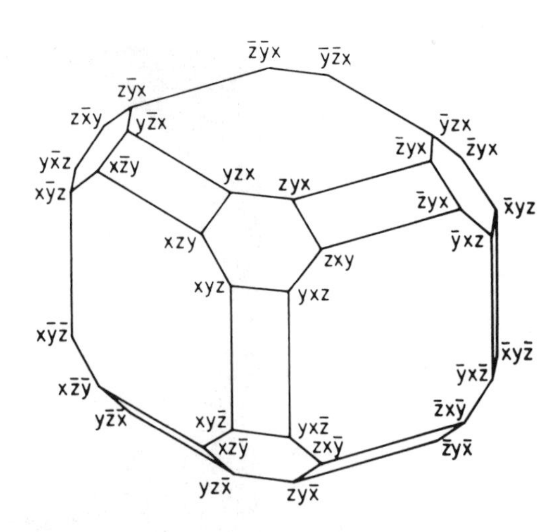

Fig. 6-14. The greater rhombicuboctahedron.

Let us first think about the board in Fig. 6-13. You see on this board a number of polyhedra, all of which have exactly the same symmetry: they all have the symmetry of a cube. The cube is placed in a rather interesting orientation, having its threefold axis vertical; that is done on purpose. Each polyhedron has been put there with a threefold axis vertical. We now want to think about the question: If they all have the same symmetry, then how are we going to distinguish among them?

One way of distinguishing is by the process of truncation. When we talk about this we can think of ordinary knives slicing pieces of cheese. But it is very difficult to tell a computer how to slice cheese. It just does not know about slicing cheese and I don't think it is too useful to teach it to do this. So it would be nice if we could now get a different point of view of these truncations. I am going to look at the coordinates of these vertices. It is these configurations, made by the atoms and ions in the crystals, that give us the so called coordination polyhedra.

Suppose that we have a point whose coordinates are x, y, and z. If this point is part of a structure having threefold rotational symmetry, then there must be two other points whose coordinates are cyclic permutations of x, y, and z; and which are related to the point (xyz) by threefold rotational symmetry. The three points form the vertices of an equilateral triangle whose coordinates are, respectively: xyz, yzx, and zxy.

Cubic symmetry moreover implies mirrors diagonally through the cartesian axes, hence an additional triplet whose coordinates are zyx, xzy, and yxz. Reflection of these six points in each of the coordinate planes produces 48 points whose coordinates are those

Table 6-2. Special cases in which the symmetry elements of a cube do not produce 48 distinct vertices.

Coordinates of generating point	Special condition	Number of vertices	Polyhedron generated	Figure
xyz	None	48	Greater rhombicuboctahedron	6.14
xxy, $y > x$	$x = y$	24	Lesser rhombicuboctahedron	6.15
$xy0$	$z = 0$	24	Truncated octahedron	6.16
xxy, $y < x$	$y = z$	24	Truncated cube	6.17
$xx0$	$x = y$, $z = 0$	12	Cuboctahedron	6.18
xxx	$x = y = z$	8	Cube	6.19
$x00$	$y = z = 0$	6	Octahedron	6.20
000	$x = y = z = 0$	1	A single point	

of the above six combined with the eight possible combinations of plus and minus. These 48 points constitute the vertices of a greater rhombicuboctahedron (Fig. 6-14).

There are special circumstances under which the symmetry elements of the cube do not generate a full complement of 48 distinct vertices. This happens when x, y, or z have special values which cause these vertices to lie precisely on a symmetry element. For instance, a point lying on one of the threefold axes would have $x = y = z$, with the result that the six points whose coordinates were all the permutations of x, y, and z have fused into a single point, which, when reflected into the cartesian planes, produces merely the eight vertices of a cube. All special cases are listed in Table 6-2 together with the names of the polyhedra whose vertices are defined by the resulting special combinations of coordinates. These polyhedra are shown in Figs. 6-14–6-20. Here we call the initial point on which the cubic symmetry elements act to generate the entire point complex the "generating point."

The computer can very quickly perform all the permutations inherent in the cubic symmetry, regardless of the specific values of x, y, and z, and then determine how many distinct points are generated. Table 6-1 translates that number into the name of the appropriate polyhedron; the computer recognizes the polyhedron on the basis of the number of distinct vertices, a task at which it is much more adept than that of slicing cheese.

Table 6-2 shows three different polyhedra having 24 vertices. It is easy enough for the computer to distinguish them: if one of the coordinates is zero, then the polyhedron is a truncated octahedron, otherwise it is either a lesser rhombicuboctahedron or a truncated cube, depending on the relative magnitudes of u and w. This latter distinction appeared to me rather subtle for two apparently so different polyhedra and hence prompted me to compare the two. As a result, I came upon a previously unrecognized relationship between the two forms. This relationship is based on the circuits traced by the edges of either polyhedron on a (spherical) surface on which it may be projected. The lesser rhombicuboctahedron has triangular, rectangular, and square faces, and truncated cube triangular and octagonal

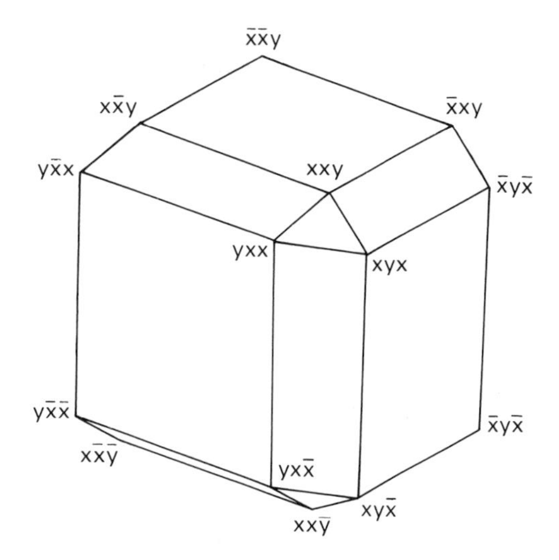

Fig. 6-15. Lesser rhombicuboctahedron.

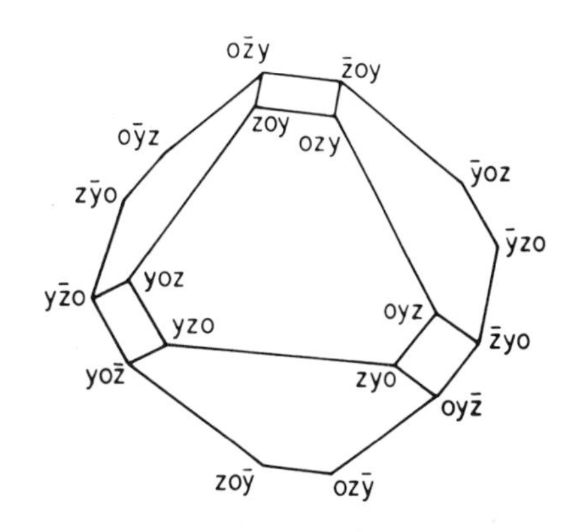

Fig. 6-16. Truncated octahedron.

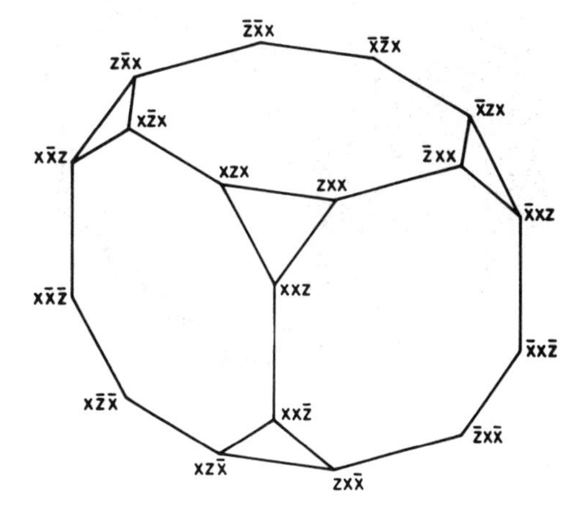

Fig. 6-17. Truncated cube.

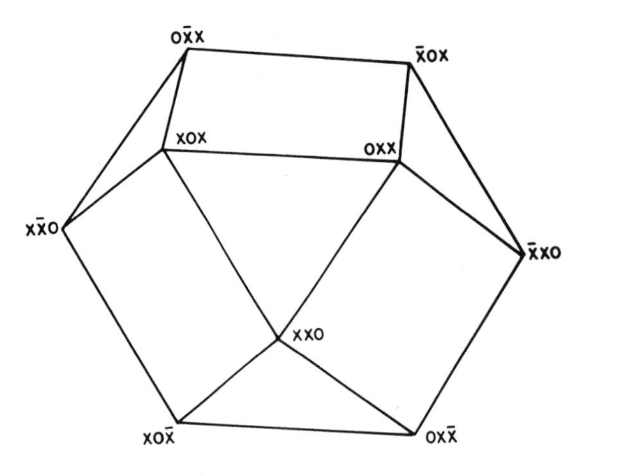

Fig. 6-18. Cuboctahedron.

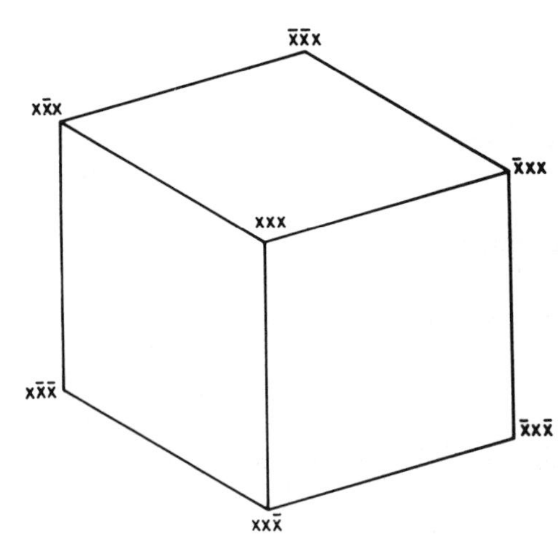

Fig. 6-19. Cube.

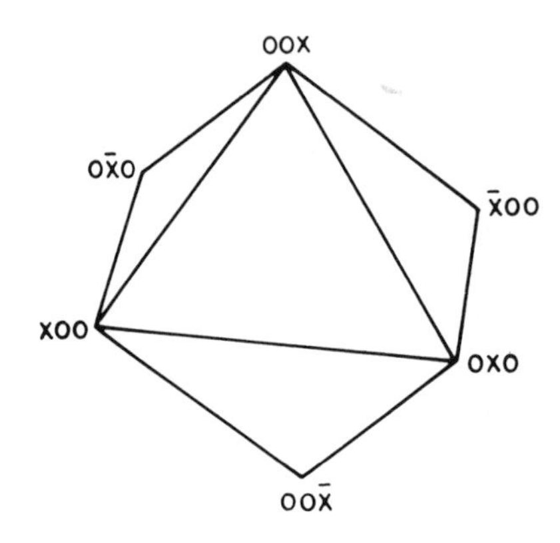

Fig. 6-20. Octahedron.

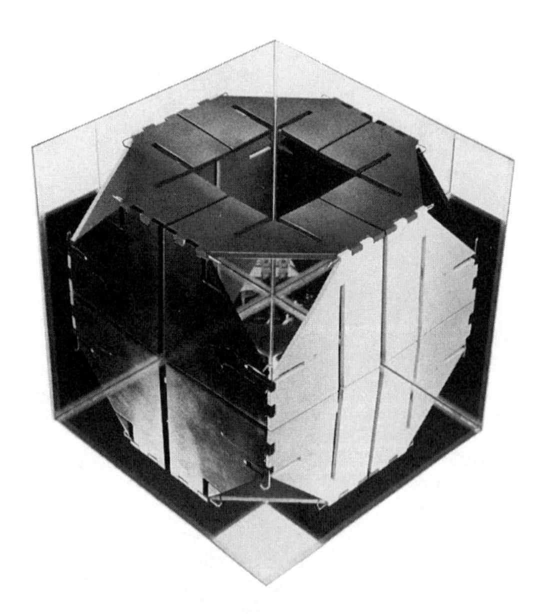

Fig. 6-21 Model of truncated cube, designed and constructed by Jonathan Lesserson. Photograph by C. Todd Stuart. From the Teaching Collection of the Carpenter Center for the Visual Arts, Harvard University. Reproduced with permission of the Curator.

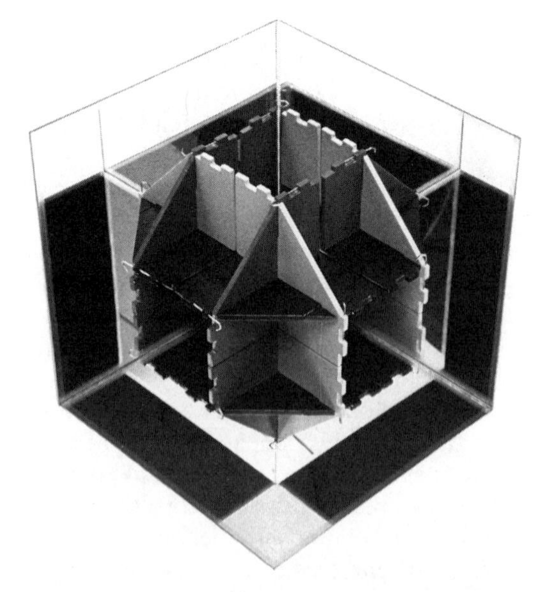

Fig. 6-22. Model of lesser rhombicuboctahedron, designed and constructed by Jonathan Lesserson. Photograph by C. Todd Stuart. From the Teaching Collection of the Carpenter Center for the Visual Arts, Harvard University. Reproduced with permission of the Curator.

faces. If, however, we look at a square face on the former, we note that its four vertices are also vertices of four triangles whose bases combine with four additional edges to form an octahedron. If we take the eight triangular faces on either form and flip them upside down, changing the relationship between u and w, then we interchange the squares and octagons, and transform one form into each other. A model of this transformation has been built, confirming the close relationship between the two forms (Figs. 6-21 and 6-22).

Notes

[1] The division of both sides of Eq. (6-1) by the summation is based on the assumption that r_{av} and n_{av} remain finite even in the limit of infinite number of vertices and edges. The detailed justification of this assumption is beyond the scope of this paper, but it is limited to surfaces in which the density of vertices and edges (averaged over a region whose area is large compared to that of a face) is reasonably uniform throughout the structure. Obviously this assumption is valid for periodically repeating structures.

[2] B. Grünbaum and T. S. Motzkin, "The Number of Hexagons and the Simplicity of Geodesics on Certain Polyhedra," *Canadian Journal of Mathematics 15* (1963):744–51.

[3] Arthur L. Loeb, *Space Structures: Their Harmony and Counterpoint* (Reading, Mass.: Addison-Wesley, Advanced Book Program, 1976).

7

Spatial Perception and Creativity

JANOS BARACS

I come from Montreal, where I belong to a group called the Structural Topology Research Group. "Structural topology" is an often criticized term, but we are stuck with it.

In their paper *Duality of Polyhedra* (this book, Chapter 13), Branko Grünbaum and Geoffrey Shephard talk about misconceptions that may arise when amateurs get mixed up in mathematics; they named the result "mathematical folklore." I like this label and I admit that I am a folklorist.

Now I will prove this point for you: I will discuss spatial perception and creativity. This term suggests some competence in the fields of psychology, philosophy, logic, and so forth, and I have none. But my experience has led to some success in understanding the creative process in morphology. We study this through a sequence of actions that are translated in geometric terms.

Let me start by defining the field with which we are dealing. When we talk about spatial perception we can talk about the physical, social, and other types of spaces. We narrow our interest to the geometrical space, in the structural and formal sense. By structural perception, I mean the combinatorial study of the topological, projective, affine, and metrical properties of configurations, of spatial models. In formal perception, we are interested in quantitative properties, such as ratios, proportions, measures, and coordinates. This part of the field is very interesting and involves working with sculptors and architects. In this chapter, however, we consider only about structural perception.

Figure 7-1 is a diagram describing the three major, distinct phases that should occur, when we start with the exposure to a spatial model and end with the perception of the spatial model. These phases are the creation of an *image,* the *imagery,* and the *imagination.* Studying the first phase, required very little. I asked an ophthalmologist to help me find out whether there are people who have some deficiencies in stereoscopic vision. He gave me a very simple tool, a little booklet with polarizing glasses. With this instrument, I was able to code everybody's stereoscopic vision. I gave the test to students, to colleagues, to everybody whom I could find. I found that an extremely small percentage of people have deficiencies in stereoscopic vision.* So the problem is not that some of us have difficulties in creating images.

The next two phases are *imagery* and *imagination.* Imagery is a phase of comprehension, of understanding space. The last phase, imagination, is the process of intervention, which is the creative process or, in our profession, the design.

For many years I have been teaching courses like descriptive geometry and structural topology, and I have also worked with architects and sculptors. Eventually—I think it is a sign of age—one starts to analyze the mental process. The diagram in Fig. 7-1 is a result of such an analysis. I divided the second phase, imagery, into three actions, and the last phase, imagination, also into three actions. I will go through a simple form in space with you and show at each step what I am proposing. We have six clearly defined geometrical actions. They are *visualization, structuration,*

* Note: R. Buckminster Fuller and the Series Editor of this book, A. L. Loeb, have been in this extremely small group having such visual deficiency.

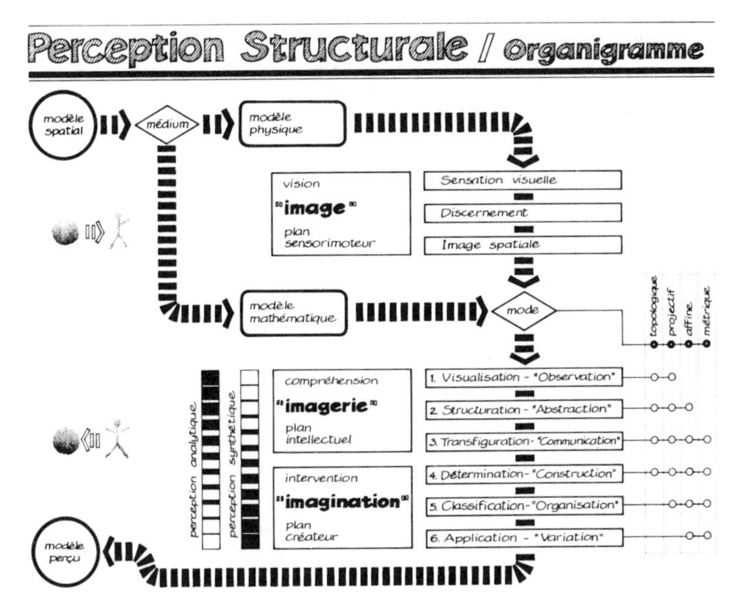

Fig. 7-1. Three phases: the creation of an image, the imagery, and the imagination.

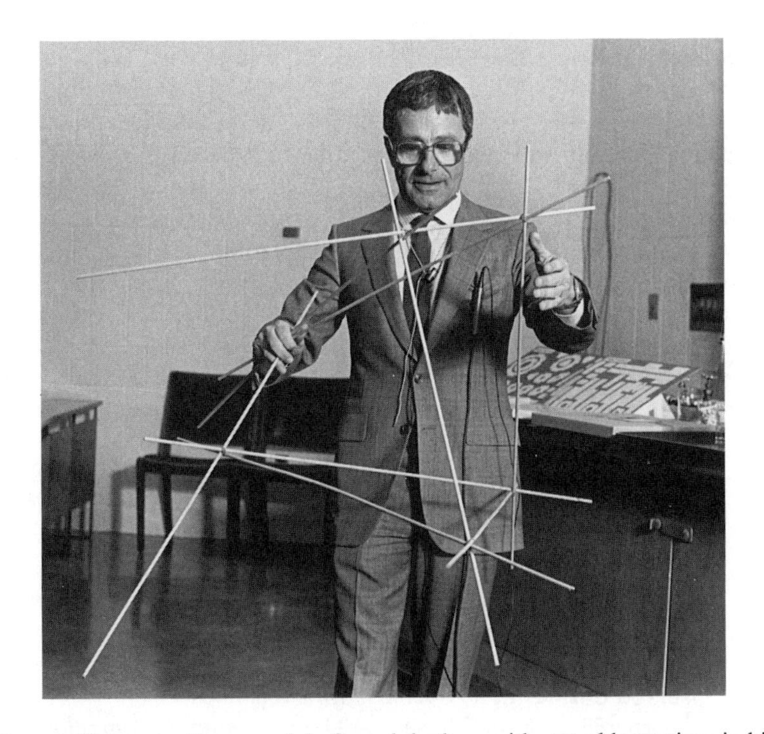

Fig. 7-2. Janos Baracs demonstrating a model of a polyhedron with movable vertices in his lecture at the Shaping Space Conference. Photograph by Stan Sherer.

transfiguration, determination, classification, and *application*. Next to the actions I have written those terms which we call skills or aptitudes. The actions to be performed are linked to these aptitudes. What we have been trying to do in the last two or three years is to devise exercises in order to introduce people, young and old, to these skills.

The model shown in Fig. 7-2 may not be as attractive as others at this conference, but it does something that other models do not do. I don't want to present the shape in a frozen,

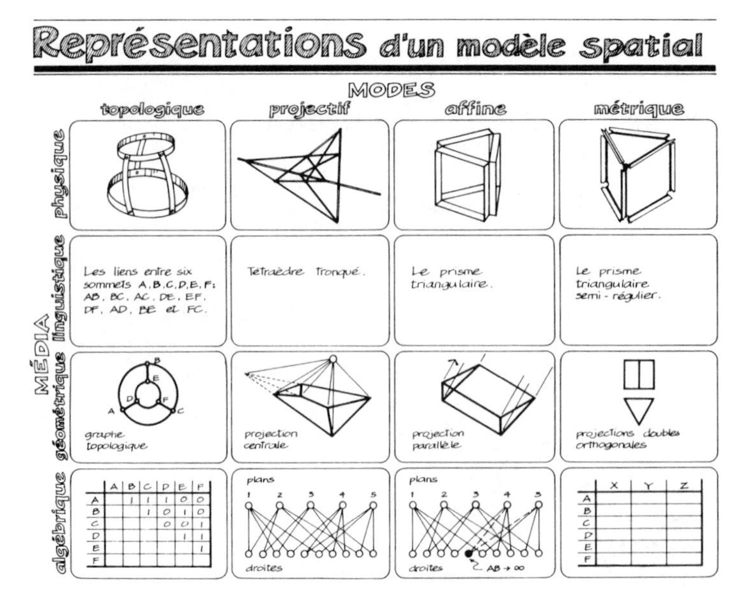

Fig. 7-3. Matrix of representations of a spatial model.

rigid form with particular metric properties; I want you to view it as I slide the vertices as a movable object which I can continuously transform. I can change lengths and angles at will, I can change symmetries, I can study many different properties.

If you can create an imagery which is movable, transformable, which you can manipulate, that is the best start for imagination and creation. This is the beginning of the voyage, an excursion in space. The model is a particular combinatorial structure composed of six vertices, nine edges, and five faces, and I will subject it to different motions.

Before describing the six actions, I should clarify the meaning of "spatial model." Figure 7.3 is a matrix of representations of a spatial model. We may use a topological model, a projective model, an affine model, or a metric model. Each model exhibits only those properties which are conserved during the proper transformations in a particular phase. These are the modes of representation. The media of representation may be physical (like the model in Fig. 7-2), linguistic (a verbal or written description), geometric (mapping or different types of projections) and finally algebraic (matrices or lattices). It is worth mentioning my surprise when I noticed in my experiments the link between linguistic abilities and the aptitude of spatial perception; students with limited verbal skills also proved to be handicapped in creating geometric imagery! We shall return to this representation matrix when we discuss action 3: transfiguration.

And now let us go on with the description of the six actions listed in Fig. 7-1.

Action 1. The first action is *visualization* (Fig. 7-4). There are two distinct steps because in architecture you are either "outside" or "inside." If you are outside you have to walk around the object to receive a complete image, while if you are inside you have to turn around, unless you have 360° vision. In both cases, you have to integrate partial images. There are various simple tests and exercises to show that this integration is not a simple process. In a cubic space the process is well exercised, but we may not be living in cubes for the rest of our lives.

The next step is to memorize images. If we cannot store a mental image of a seen object, then I think we are stuck. This can be tested easily. (People with excellent memories sometimes fail, while others who have very poor memories can be excellent here.) In the third step we are looking for composite images. For instance in Kepler's drawing, the icosahedron was shown as a pentagonal antiprism with two

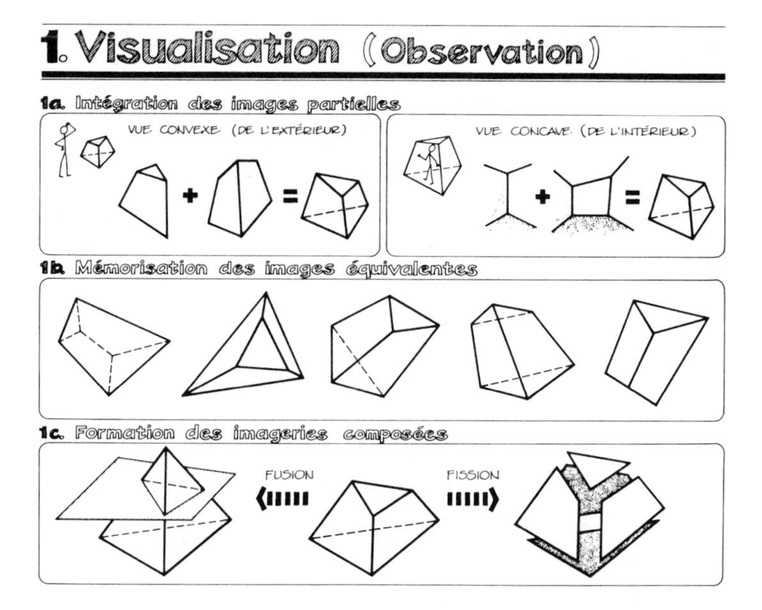

Fig. 7-4. Action 1: Visualization.

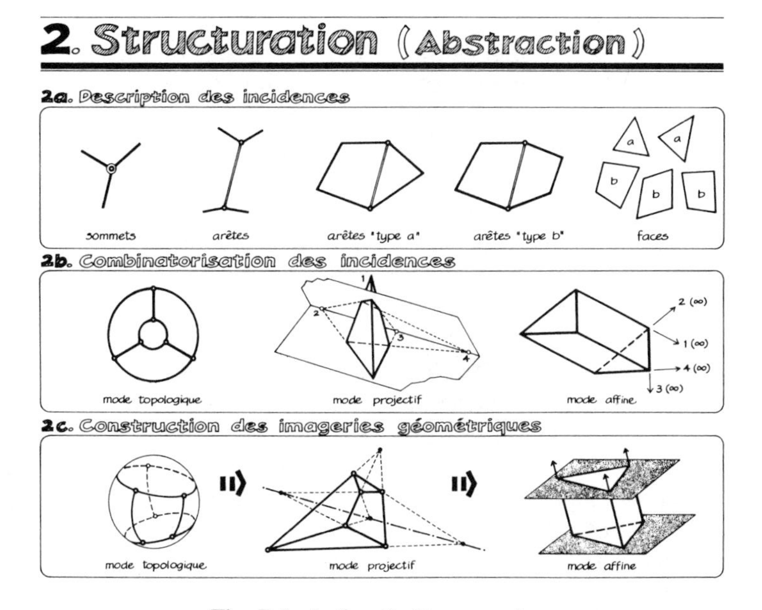

Fig. 7-5. Action 2: Structuration.

pentagonal pyramids; this is a composed image to help you memorize the structure. This completes the first action.

Action 2. The second action is *structuration* (Fig. 7-5). Here we want to study the topological, projective, and affine structures of the object. Remember that we do not measure in this phase, we do not care about angles or distances, we do not study symmetries.

As a first step, we recognize and classify incidences. In the second step, we integrate these incidences in a combinatorial structure, in the topological, projective, and affine modes. The third step is a synthesis of the two completed actions: visualization and structuration. We should now possess a geometric imagery of the spatial model in the topological, projective, and affine modes.

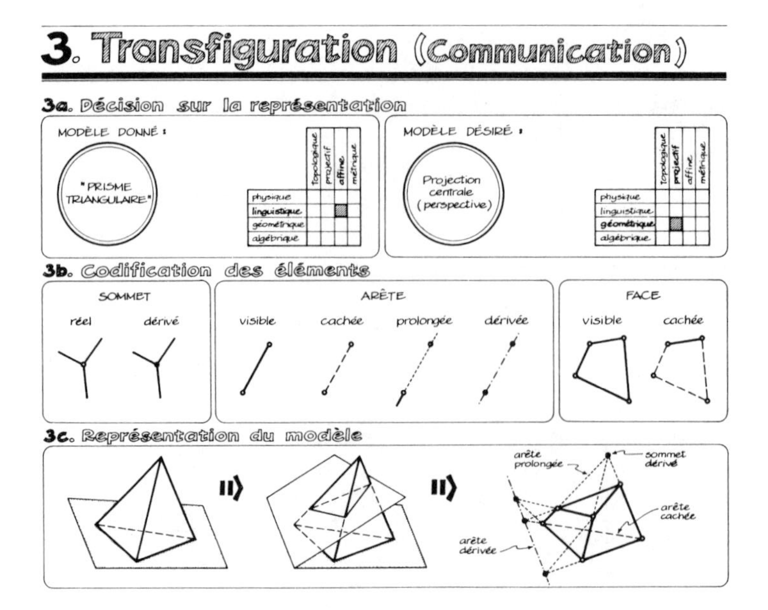

Fig. 7-6. Action 3: Transfiguration.

Action 3. Our next action is *transfiguration* (Fig. 7-6), an apprenticeship to communication. In the first step we are using the representation matrix of Fig. 7.3. A spatial model in a given mode and medium will be transferred into another (or the same) mode and medium. For instance we may ask to prepare the perspective drawing (projective mode, geometric medium) of a triangular prism (affine mode, linguistic medium). Or we may start with an incidence matrix (topological mode, algebraic medium) and decide to produce the graph of the polyhedron (topological mode, geometric medium). Essentially, we decide on a method or representation of a given model, taking into account the purpose of the representation and the type of properties we wish to exhibit.

The second step is codification (standard or created for a particular task) with a legend that allows our representation to be read by others. The third and final step is the prepared representation. This passage from our mode and medium to another mode and medium is more than mere communication; it is an essential process in order to perceive the model itself.

Action 4. Action 4 is *determination* (Fig. 7-7). The problem is rooted in combinatorial geometry and linear algebra. The example presented here is for metric determination; the same action, however, can be equally applied for projective or affine properties.

The first step is to enumerate the (metric) invariants of the spatial model. As you can see, the seven types of invariants (distances and angles) result in 148 pieces of information in the case of the truncated tetrahedron.

Now we need to know the least number of invariants that uniquely determine the polyhedron in space. If we exclude the Euclidean motions (six degrees of freedom), we need exactly nine invariants (this number, C_d, is equal to the number of edges in the case of a spherical polyhedron). The number of necessary invariants is reduced if we impose a symmetry group on the polyhedron. For instance, we have the choice of one invariant only, if we wish the truncated tetrahedron to be realized as a semiregular triangular prism.

Going back to our general position, the choice of nine invariants out of 148 elements is a terrific number (7.32×10^{13}). It is our duty in the third step to select a combination whose elements are linearly independent. This selection was done intuitively (Fig. 7.7, section 4c) in our example. Thus determination is an integral part of the perception process and also an important practical tool to design and to realize (to construct).

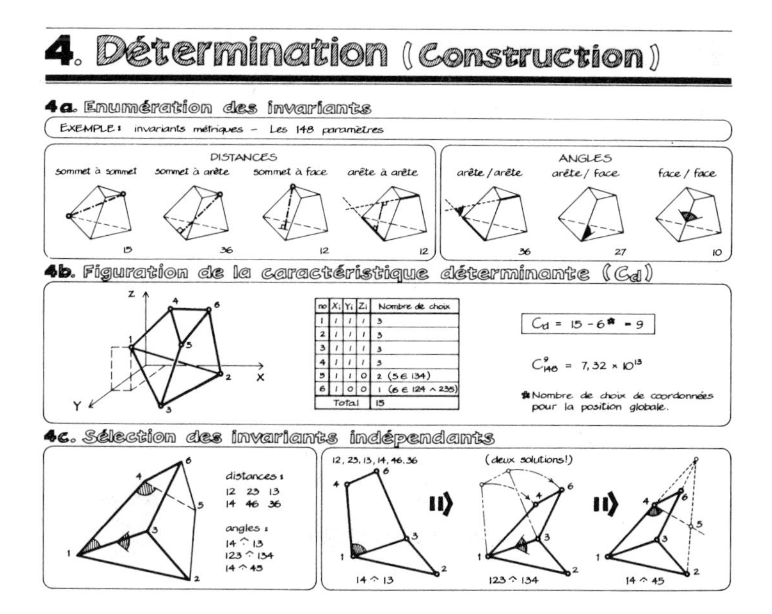

Fig. 7-7. Action 4: Determination.

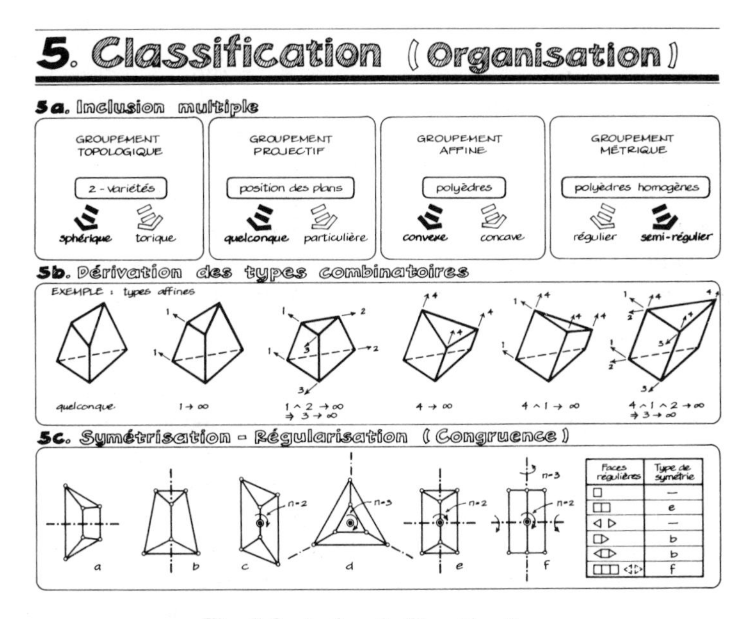

Fig. 7-8. Action 5: Classification.

Action 5. Classification is the fifth action. We now have reached a point in our actions where we are able to classify the available options of our model into multiple groupings (Fig. 7-8). In the first step we gave as examples topological, projective, affine, and metric groupings. The derivation of the combinatorial types (affine in this example) is shown in the second step, where the option of parallelism is explored.

The third step is symmetrization or regularization. Shown in the figure is the semiregular triangular prism and all its symmetry subgroups. Faces, which have to be regular polygons for each group, are also indicated.

Action 6. The last action—*application*—is the least understood and the most mysterious of all. It is the action of conception, creation, or (in my profession) design. The three steps

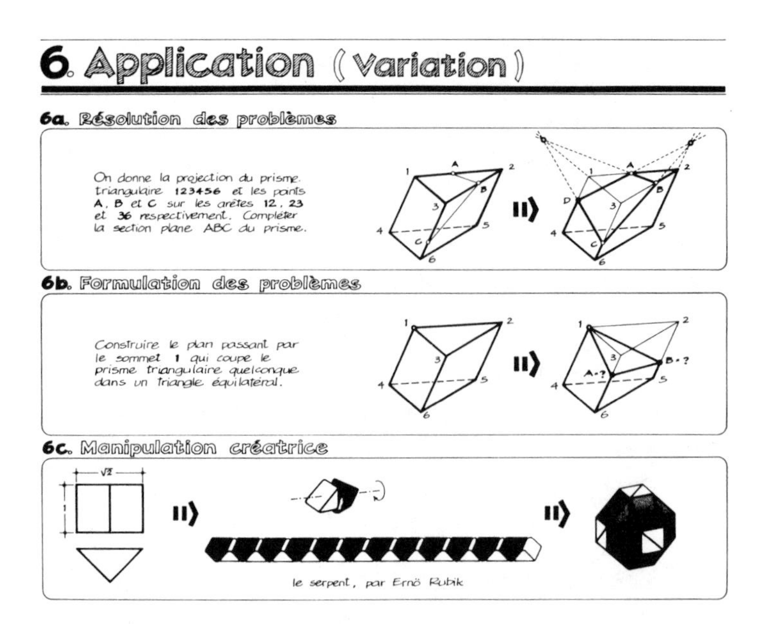

Fig. 7-9. Action 6: Application.

shown in Fig. 7-9 represent three levels of application in an increasing order of complexity.

The first level is the resolution of a given problem, which requires a certain degree of imagination. At the second level, the creative process is more advanced. Teachers will concur with me, that finding a good problem usually is harder and more rewarding than solving one.

We named the last step "creative manipulation." To demonstrate it, we use again the triangular prism, which in this example was "manipulated" into a creative toy, the serpent, by Ernö Rubik.

We propose these six actions as a logical, sequential mental process to shape space. The first three actions represent the analytic perception of a spatial model, resulting in an imagery. The last three actions, the synthetic perception of the model, provokes the imagination, ending in a creative application. During the six actions, we applied topological, projective, affine, and metric transformations in a gradual fashion to a given spatial model.

The actual process of design in my profession is somewhat different: it does not begin usually with a given spatial model. It starts with a program that describes functions, criteria and so forth. So the first step is to generate the spatial model itself, which fulfills the functions and satisfies the criteria. This model should be quite general, free of details, possessing only some essential, intrinsic geometric properties. It follows that I am proposing a topological model, to be found by enumerating the available options. We refine this model through projective, affine, and metric transformations, ending up with the desired product.

The process is illustrated in Fig. 7-10. To make this presentation brief, the example is simplified. The selected topological model is a closed curve with six labeled vertices (a, b, c, d, e, and f): no lengths, no angles, no parallelism, not even straight lines are specified in this original choice. As a result, we have a large family of figures (all the plane and spatial hexagons) by selecting a few properties, which are, however, the most intrinsic ones of this family. Let us now imagine that our model is a rubber band marked with the six vertices. While stretching this rubber band, adjacent vertices will remain adjacent and the band will stay as a closed curve. This type of transformation is called a *continuous mapping,* while the invariant properties are *adjacency* and *continuity;* we are in the realm of topology. With so few properties to scrutinize, it is surprising how rich the content of topology remains.

We now enter the second phase in shaping our hexagon. While the properties established in the topological phase are kept unchanged,

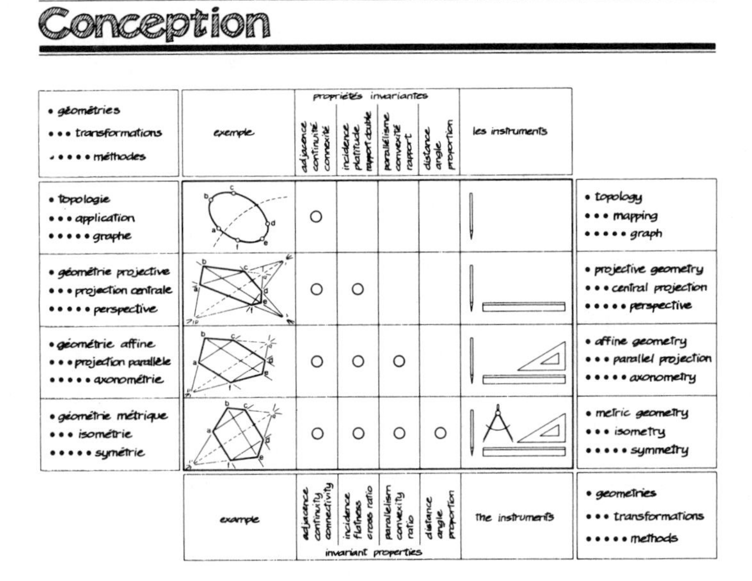

Fig. 7-10. An illustration of the actual process of design.

we decide that the curved edges shall be straight lines, all in one plane, and we choose some particular incidences among them. Let the lines af, be, and cd meet in the point s, arranged in such a way that the common point p of the lines ab and ef, the common point q of the lines bc and de, and the common point r of the lines ac and df are on one line. Many other choices and their combinations are possible, each set of choices representing a distinct hexagon with a visual impact on its shape. If we now project this figure from a point onto another plane, the projected figure will preserve all the chosen properties. We may state that in projective geometry, we study properties of a figure which are invariant under central projections. These new projective properties are incidences and flatness (straight lines and plane surfaces).

We continue to refine the shape of the hexagon by choosing new properties in the third phase, but again all previous choices—at this time topological and projective properties—are kept invariant. The option in this phase is the selection of certain lines to be parallel. Another way to phrase it is to say that certain chosen points are moved to infinity along their incident lines. For instance, we choose the points r and s to be moved to infinity. The point r is common to three lines (ac, pq, and df), which are now parallel, and similarly the lines be, af, and cd become parallel because their common point s is at infinity. If we now project this figure with parallel rays onto any other plane, the projected figure will preserve all the properties chosen so far. In affine geometry, we study properties of a figure that are invariant under parallel projection. The new affine properties are parallelism and convexity.

We have gone through three new geometries by now, but the most familiar geometric properties, such as distances and angles, have not yet been mentioned. These choices are left for the fourth and last phase of our program. Let us choose the line be to be perpendicular to the line pq. This decision will force the lines cd and af to be also perpendicular to the line pq, since these lines were parallel in the affine phase. Just as before, we do not alter choices made in a previous phase. The last choice we make now concerns distances; we want the line pq to bisect the segments be, cd, and af. An important new property has emerged in this final shape of the hexagon: if you consider the line pq as a "mirror-line," the vertices f, e, and d are the mirror-views of the vertices a, b, and c respectively or, simply, the hexagon possesses bilateral symmetry. The reflection of a figure in a line preserves distances and angles. Such an operation is called an isometric transformation. If all the distances of a figure are preserved, so are all the other geometric properties. In metric geometry, we study properties of figures, which are invariant under isometric transformations. The new metric properties are distances and angles.

We have completed a sequential approach to shape a simple plane figure. Within each of the four geometries sketched above, certain properties of a figure may be determined quite simply by counting, for instance, the number of vertices, edges, incident lines in a point, the number of lines parallel to each other, the number of equal distances and angles. These properties are the subject matter of the combinatorial theory, which has been introduced on the mathematical scene in the last fifty years.

You might say now, yes it is a nice procedure, but why not start simply at the other end and just draw a hexagon that has a mirror symmetry? This could be done for a simple example like a plane polygon. But if I had chosen a polyhedron, even one with not more than six or seven faces, you would not recognize immediately, for instance, how to manipulate the planes of the faces in order to arrange them according to certain symmetry groups; it would be an extremely difficult task. However if you go through this process, you will become aware of the available options, you will be able to control your form, you will be in charge of it. Usually, the forms are in charge of us. We have to reverse the process. We should not have to use catalogs and say, "I want this form, I want that form" as if we were shopping in a supermarket! We should be in full, complete control; we should be able to shape space, as the title of this book says.

As we progressed from the topological figure to the metric figure of the hexagon, the invariant properties have been increasing while including all the previous properties. We have been progressing from the most general toward the most specific. This concept is also supported by illustrating the necessary and sufficient drafting tools to produce the drawings in Fig. 7-10. Only a pencil is needed to draw a topological figure of the hexagon; a straight edge is required in projective geometry. To draw parallel lines in an affine figure, a new instrument is necessary, a straightedge gadget that slides on the first one. The symbol of a right triangle is used in the illustration, being the standard tool in the drafting practice (but remember that perpendicularity is irrelevant in affine geometry). Finally we have to add a compass to our repertoire of instruments to draw a metrically equivalent (isometric) figure.

The methods (practical applications in our profession) listed in Fig. 7-10—graph theory, perspective, axonometry, and symmetry operations—presently are used mostly to analyze and communicate preconceived forms. We are convinced that geometries can do much more for us; we propose to apply them in the synthetic process of conceiving forms.

I have presented the six actions to perceive a spatial model and the gradual transformations to design a product. At first glance this methodology may seem to you as one of those fashionable intellectual exercises based on personal beliefs and prejudices. That is not so! As I mentioned earlier, these methods evolved during these decades of professional practice and teaching, which gave me ample opportunity to test their didactic and practical use. I hope the following slides will convey the message: I do try to practice what I preach!

Figure 7-11 is a picture of my office. The scenery changes every year. After the yearly clean-ups, I keep some souvenirs and make room for the new projects. These models are important because I believe in the old Chinese saying: "I hear, I forget; I see, I remember; I do and I understand." But, there is a drawback. Meaningful models are painstakingly slow to build, to take apart and to transform. To solve this problem, I designed five kits, which are now in constant use at my faculty and elsewhere. They allow fast assembly, easy transformation, and high precision.

Figure 7-12 shows a kit named Poly-Form. It is a topological and projective kit; its purpose is to demonstrate through simple manipulations the links between the concepts of polyhedral graph, adjacency matrix, embeddings, projective conditions, and projective realizations. I tried Poly-Form with 8- to 12-year-old children with a result surprising to me. At first I was wary of explaining it to them; I found it hard to avoid the fancy terms. Then I was chagrinned when I realized how fast they grasped the concepts and how happily they went on to explore. College students

Fig. 7-11. ". . . I do and I understand."

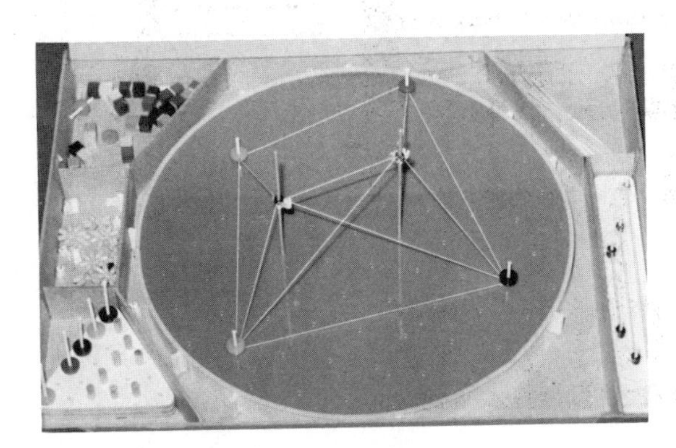

Fig. 7-12. Poly-Form: From graphs to projective polyhedra.

(ages 18 to 20) sometimes are slower than those kids and certainly are more inhibited about exploration! These and other experiments with younger children and college students led to the following proposals for teaching geometry:

1. Spatial geometry should be introduced at an early age (10 to 12 years).
2. The subject matter should be polyhedra.
3. The starting notions should be topologic and projective, to be followed later with affine and metric properties.

Figure 7-13 shows Poly-Kit No. 1. The die-cut cardboard polygons are to be attached with rubber bands to form the five regular polyhedra, the thirteen semiregular polyhedra, six of the family of prisms and antiprisms, four of the semiregular duals, the five parallelohedra, and all other polyhedra with regular faces. (The 92 convex polyhedra with regular faces enumerated by V. A. Zallgaller include the regular and semiregular polyhedra.)

Poly-Kit No 2 (Fig. 7-14) is a space-filling kit. Here the special connection between the die-cut cardboard polygons makes it possible to attach three or more faces along an edge. The circular holes in the polygons allow the user to inspect the incidence structure of the juxtaposition. Poly-Kit No. 2 allows a fast assembly of the space-fillings by the 5 parallel-

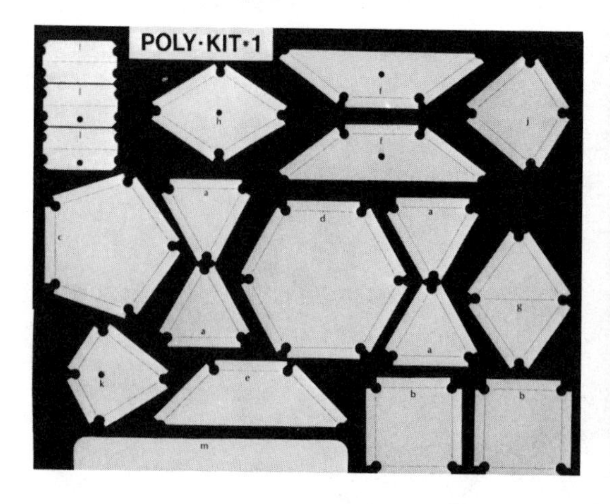

Fig. 7-13. Poly-Kit No. 1: Metric polyhedra.

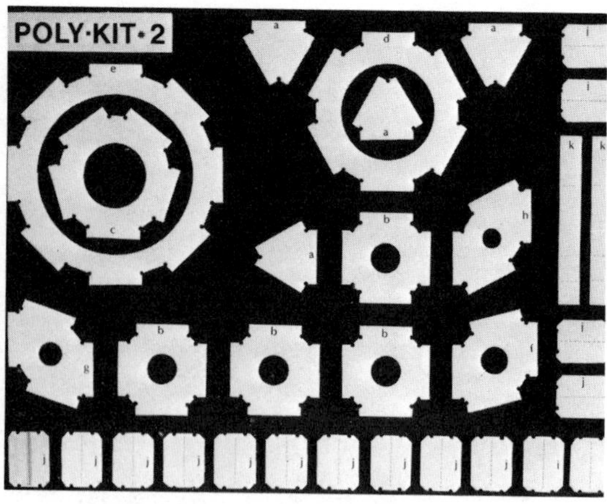

Fig. 7-14. Poly-Kit No. 2: Space-filling polyhedra.

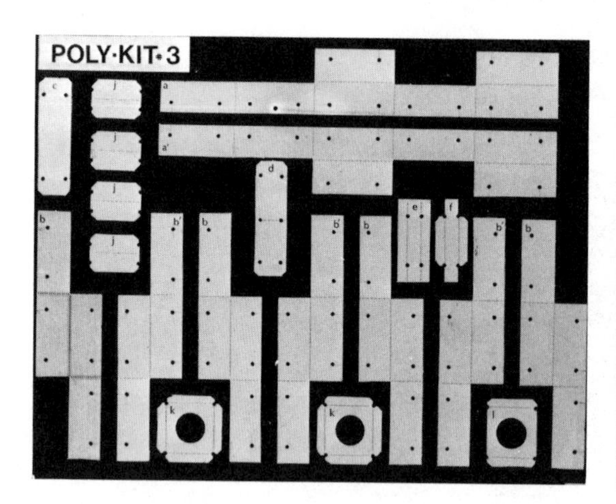

Fig. 7-15. Poly-Kit No. 3: Affine polyhedra (zonohedra).

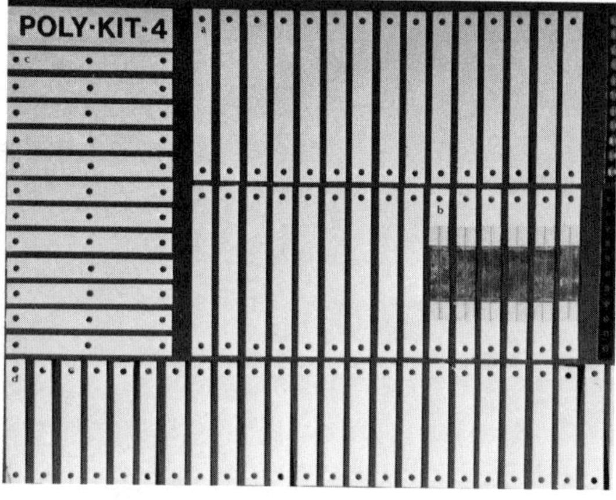

Fig. 7-16. Poly-Kit No. 4: Geometric rigidity of braced grids.

ohedra and others where the component polyhedron is composed of regular faces.

Figure 7.15 shows the die-cut cardboard bands of Poly-Kit No. 3. This kit was conceived for building a fascinating, infinite family of polyhedra called zonohedra. (A *zonohedron* is a convex polyhedron; all of its faces are parallelograms). The repetitive use of one type of band is sufficient to build any zonohedron. This is a truly affine kit, because the models that can be built allow the affine motions of the zonohedron to be demonstrated; the model retains its central symmetry during the deformations. Another type of band in the same kit may be used to build all polyhedra that can be realized with equal edge lengths (equilateral polyhedra).

Poly-Kit No. 4 (Fig. 7-16) also serves two purposes. It can be used to demonstrate the geometric rigidity of regular grids composed of bars or bars and tension members. The same kit can also be used to build polyhedra projected onto a sphere.

My students complete their apprenticeship using the different kits, then move on to build large-scale models. In this phase they are confronted with structural and technological considerations. Figures 7-17 and 7-18 show an "octet" spaceframe and its detail built with wooden bars and plastic joints, injected for

Fig. 7-17. "Octet" spaceframe (student project).

Fig. 7-18. Joint detail of "octet" spaceframe.

this particular project. In Fig. 7-19 we see a dome built with tubular aluminum rings. The geometry is based on the inscribed circles of the faces of a dualized semiregular polyhedron. The joints are all of one type and despite their articulations (they are "hinged" to each other), the dome when attached to the ground becomes rigid.

The last few figures are examples from my professional practice. I selected five projects where the geometric content is evident, and I had the good fortune to work with truly talented and adventurous architects and sculptors. My duties included proposing a geometric concept, calculating the stresses, and devising modes of fabrication and erection while respecting the ever-present budgetary constraints.

In Fig. 7-20 we see the main theme building *Man the Explorer* of Expo '67 in Montreal. It is a giant, integrated steel spaceframe, based on the juxtaposition of truncated tetrahedra and tetrahedra. Figure 7.21 is also a theme building of Expo '67 (*Man in the Community*.) Here concentric hexagonal plywood-box rings are superposed; the reduction in size results in a logarithmic silhouette. Figure 7-22 is a prefabricated concrete spaceframe with the tetrahedron-octahedron geometry, built for a shopping center in Montreal.

Fig. 7-19. Articulated ring-dome (student project).

Fig. 7-20. *Man the Explorer.* Theme building, Expo '67, Montreal. Architects: Affleck, Desbarats, Dimakopoluos, Lebensold, and Size. Structural engineers: Eskenazi, Baracs, de Stein, and associates.

Fig. 7-21. *Man in the Community.* Theme building, Expo '67, Montreal. Architects: Erickson and Massey. Structural engineer: J. J. Baracs.

The last two projects are large-scale sculptures. The first one (Fig. 7-23) is a set of juxtaposed general zonohedra built with identical hexagonal aluminum frames, articulated at their joints. The integrated lighting system adds to the visual impact for this Montreal subway station. The last figure shows a sculpture in the enclosed mall of a large public building in Quebec City (Fig. 7-24). It is interesting to note that each of these last two sculptures, despite their different appearances, is based on one of the 230 space groups: three mutually perpendicular screw motions with half-turns. In the first case, the hexagonal frames, in the second case the extruded H beams are subjected to the same symmetry operations. Even the orientations of the bolts and nuts are consistent with the motions.

The German mathematician Felix Klein gave an overall view of geometries in his famous address at the University of Erlangen in 1872. He was the first to propose transformations as criteria to distinguish geometries. This concept was applied by the Swiss psychologist Piaget. He demonstrated that growing children perceive space in a sequential fashion. They go through a topological stage until the age of 6, then progress through a projective stage (at ages 6 to 10); the perception of affine and metric properties begins around the age of 10.

The results of my own experience and research in the field of the synthetic process of creating form led me to a similar approach: in order to exploit fully the stunning richness of three-dimensional space, the initial method for conceiving form is a sequential series of combinatorial choices taken at the different levels of geometries in the order of topology, projective, affine and metric geometry. But, despite the clearly established hierarchy of the geometries, creating form can not be simplified to a linear thinking process. Certain early choices in the topological and projective level are dependent on affine or metric criteria, imposing simultaneous considerations. This linking of seemingly unconnected concepts (bisociation) is the theme of Arthur Koestler's famous book, *The Act of Creation.*

Essentially, we are trying to promote a stronger link between geometry and creative design. The only obstacles are attitudes. The widespread opinion of today's artists and architects was well described by the Soviet crystallographer Shubnikov: they have "a horror of the words law, order, symmetry, geometry; they prefer harmony, beauty, style, rhythm, unity, although the true meaning of these words differ very little from that of the former."[1] In turn mathematicians do not help either to narrow the gap, and we agree with the recent observation of the mathematicians Grünbaum and Shephard: "Current fashions in mathematics applaud abstraction for its own sake, regarding it as demeaning to work on problems related to elementary geometry . . . It seems to us to be foolish and presumptuous to believe that ours is the first generation which needs no more the inspiration that can be found in studying simple geometric objects and their mutual relations."[2]

The indifferent or hostile attitude toward geometry is probably rooted in the choice of curricula for our schools. The little geometry that is taught is limited to Euclidean geometry, giving the student an inventory of axioms, theorems, and proofs, related to the forms of static "frozen" figures. What baffles the mind is that in the last five centuries, many new, important geometries have emerged which are well-documented in our libraries, but which somehow have not penetrated into the public knowledge. The curriculum in schools should be revised to include chapters of more recent geometries based on the contrasting notions of change and conservation, motion and invariance. This approach, where the structure of forms is studied rather than the forms themselves, stimulates the imagination and appears to be more conducive to creative design. I used the term "creative design"; it may seem that the adjective "creative" is redundant, but I do not think so. Take a look at our cities, buildings, and objects: they are the results of "design," but in most cases, with little sign of creativity. Many interesting books and essays have been written on the theme of creativity, and any definition is obviously subject to debate. I like what the mathematician-philosopher D. R. Hofstadter wrote in a recent article: "Making variations on a theme is really the crux of creativity."[3] This statement confirms our geometric view on morphology: creating form is not an invention, it is a process of

Fig. 7-22. *Plaza Côtes des Neiges.* Shopping center, Montreal. Architects: Mayers and Girvan. Structural engineers: Baracs and Gunther.

Fig. 7-23. Sculpture, station Namur, Montreal. Sculptor: Pierre Granche. Structural engineer: J. J. Baracs.

Fig. 7-24. Sculpture, Palais de Justice, Quebec. Sculptor: Louis Archambault. Structural engineer: J. J. Baracs.

transformations. The same article began with a quotation of G. B. Shaw: "You see things, and you say 'why?' But I dream things that never were; and I say 'why not'?"[4] It certainly takes a poet to express so well the contrast of minds: the analytical versus the synthetic, the critic versus the designer.

Notes

[1] A. V. Shubnikov and V. A. Koptsik, *Symmetry in Science and Art* (New York: Plenum, 1974), p. 366.

[2] Branko Grünbaum and G. C. Shephard, "Tiling with Congruent Tiles," *Bulletin of the American Mathematical Society,* series 3, *3* (1980):951–73.

[3] Douglas R. Hofstadter, "Metamagical Themas: Variations on a Theme As the Essence of Imagination," *Scientific American 247,* no. 4 (October 1982): 20–29.

[4] George Bernard Shaw, *Back to Methuselath,* London: Constable (1921).

8

Why Study Polyhedra?

JEAN PEDERSEN

There are many good reasons for studying polyhedra. They are, of course, simply *interesting* in themselves. The visual perception of their symmetries delights the eye and the underlying structure stimulates the mind. We would be completely justified in studying them if these were their only attributes, though perhaps the study would not then take place in the mathematics classroom. In fact, polyhedra are incredibly rich in mathematical content, thus providing an introduction to and links among several branches of mathematics. Students who are not exposed to experiences with polyhedra, and the mathematics connected with them, have every right to feel educationally deprived.

Studying polyhedra is fun! When I talk to students about polyhedra, they don't ask, "Why do we have to study such things?" They aren't bored by exploring these wonderful ideas. When students ask, "Why do we have to do this?" they really mean, "I'm bored, teacher." In our current educational system, whenever students really enjoy mathematics and don't ask, "What's this good for?" we may have to answer to suspicious administrators and bureaucrats! Outsiders don't ask, "Why must they do this?" when mathematics students suffer by executing a dreary algorithm for days, or weeks. But when students enjoy mathematics, the principal or sometimes even a parent will say: "Your students should be doing more real mathematics. Why are you teaching them this?" The prevalence of such attitudes made me think more carefully about the totality of reasons why we study polyhedra, and to find reasons we could give to administrators and parents to justify study of such exciting mathematics.

The first reason for studying polyhedra is that they are tangible. This is particularly important to secondary students who do a lot of things that are neither mathematical nor tangible. Students who build models naturally want to know more about their creations. Just the act of constructing a model raises questions about it. This is how real mathematics begins—with questions!

Some idiosyncratic looking Platonic solids are shown in Fig. 8-1. Each was constructed by braiding together straight colored strips of paper on which appropriate fold lines had been drawn.[1] It is easy to verify[2] that

- *If* we require that every edge on the completed model is covered by at least one strip, that every point on the interior of every face must be covered by at least one thickness from the strips, and that the same total area (hence color) from each strip must show on the finished model,
- *Then* you can braid the tetrahedron from two strips, the hexahedron (cube) from three strips, the octahedron from four strips, the icosahedron from five strips, and the dodecahedron from six strips.

I do not know a general explanation for this intriguing pattern.

If we adhere to these rules, the dodecahedron we construct (shown in Fig. 8-1) is not as pleasing as we have come to expect a dodecahedron to be. The dodecahedron in Fig. 8-2 is much more attractive. It, too, is braided from six identical straight strips, but whereas the

Fig. 8-1. The Platonic solids. Clockwise, from upper left: icosahedron (five bands), dodecahedron (six bands, faces bi-colored), tetrahedron (two bands), hexahedron (three bands), octahedron (four bands).

Fig. 8-2. The golden dodecahedron, braided from six bands. The faces have pentagonal holes.

dodecahedron of Fig. 8-1 has its symmetry group reduced to that of the tetrahedron (a group with only twelve elements) because of its surface coloring, the coloring of the dodecahedron in Fig. 8-2 preserves the underlying icosahedral symmetry (its symmetry group has sixty elements). For most of us, the loss of symmetry reduces our satisfaction correspondingly.

I will eventually take these models apart to prove the assertions I have made about them—you might say it is my "proof by destruction"! One of the especially interesting things about these particular models is that their construction does not require the use of a straightedge or a compass. The pattern pieces were made simply by systematically folding straight strips of paper.

A second reason for the study of polyhedra is that they provide an interesting introduction to combinatorial geometry. An ordinary convex polyhedron would turn into the surface of a ball if its faces were made of rubber and it were inflated. It is easily observed by beginners that if V, E, and F are the number of vertices, edges, and faces, respectively, of a given convex polyhedron, then $V - E + F = 2$. It would be only a little more sophisticated for them to observe that $2E \geq 3F$ (and, likewise, that $2E \geq 3V$). Many such pretty and useful formulas can be discovered by students at the high school or beginning college level, and there are nice explanations[3] which are understandable for these students.

A third reason for studying polyhedra is that they can serve as a vivid introduction to group theory.[4]

Yet another reason for studying polyhedra is to introduce more general topological ideas.[5] We may then study manifolds whose faces are not necessarily flat or which are not necessarily homomorphic to the surface of a sphere. Figure 8-3 illustrates two such surfaces. We might also generalize this to higher dimensions.

One of the most compelling reasons for studying polyhedra is that one is led toward ideas that are deeply and intimately connected with other parts of mathematics. Peter Hilton and I have recently done some joint research of a predominantly number-theoretic nature that was originally motivated by our study of polyhedra.[6] In geometry if we speak of *generalizing* when we increase the dimensions of a given problem, then it is natural to speak of *specializing* when we reduce the dimensions of a problem. It was in this sense that Peter and I specialized the concept of constructing a regular polyhedron in space in order to look carefully at the construction of regular convex polygons[7] in the plane. We asked whether straight strips of paper could be systematically folded to construct regular convex polygons. During our investigations we used, among other things, some well-known results from number theory (including properties of the Euler φ function) and even l'Hôpital's rule. We were motivated to ask (and were able to answer) what we believe are new questions in the theory of numbers. Often it was the geometric considerations in our original problem that helped indicate the direction of proof for these purely number-theoretic results which were not directly related to our original geometric motivation. Such cross-fertilization is characteristic of creative mathematics.

Our search for a systematic algorithm for constructing all regular convex polygons was successful, and the algorithm also gave us the most efficient way to fold any regular star polygon.[8]

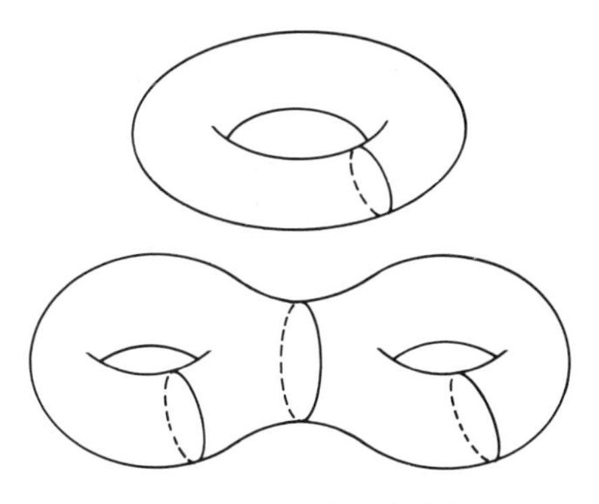

Fig. 8-3. A torus and a double torus.

I offer one more reason for studying polyhedra. The Bourbakist idea of "imposing a structure on a set" is basic to mathematics. The esthetic and intellectual appeal of a symmetric object seems to be universal, even if the further structure we impose on it is not. Sensitivity to this appeal is part of the pleasure we derive from looking at polyhedra—and this common experience motivates us to want to understand each other, even though we come from different disciplines, use different notations, and express ourselves in different terms.

Let me give some specific examples, focusing first on some group theory and combinatorics. I shall begin with a puzzle[9] made by my daughter Jennifer when she was a ninth-grade geometry student. (Construction details are given later in this chapter.) In some settings it is natural to talk about the proper rotations of objects in space. Thus, for example, let us examine this finished puzzle cube. One practical reason is that we may want to know how many ways it can be put in a box, on a shelf or into a drawer where it fits snugly. A useful definition of a *proper rotation* is any rotation of an object that leaves it occupying exactly the same space; we regard two rotations as the same if their effect on the object is the same. Let us examine all possible proper rotations for the cube.

- First we notice that there are three axes of symmetry that pass through the centers of opposite faces. Each axis is related in the same way to the rest of the cube, so we ob-

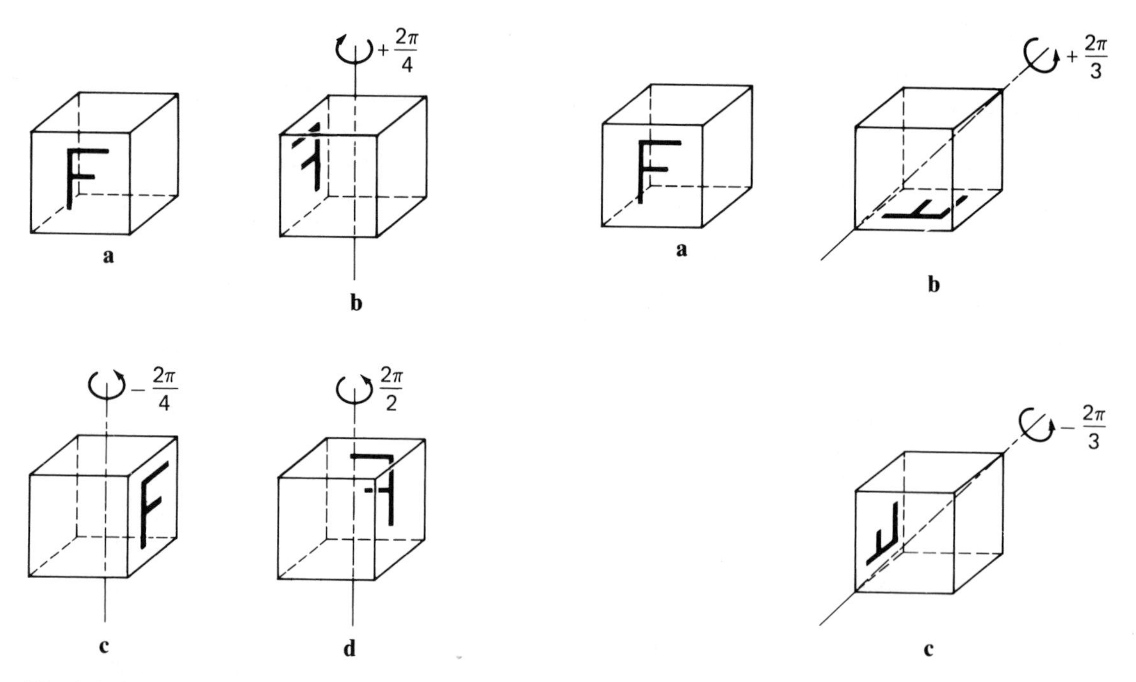

Fig. 8-4. Proper rotations of the cube about one of three axes of symmetry passing through the centers of opposite faces.

Fig. 8-5. Proper rotations of the cube about one of the four axes of symmetry passing through opposite vertices.

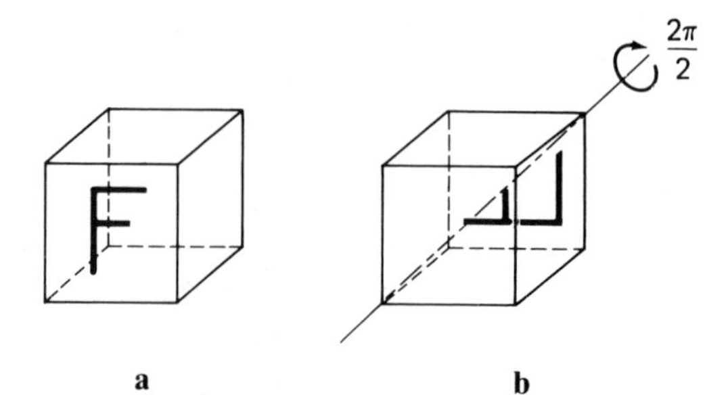

Fig. 8-6. Proper rotations of the cube about one of the six axes of symmetry passing through the centers of opposite edges.

tain a prototype by discussing the distinct proper rotations about any one of those axes. For example, we may focus on the vertical axis of symmetry as shown in Fig. 8-4. The cube can be rotated about this axis through a quarter turn ($2\pi/4$) in either direction ($\pm$) as shown in Figs. 8-4b and 8-4c so that, in Fig. 8-4b, the front face is moved to the left, while, in Fig. 8-4c, the front face is moved to the right. Thus we see that the rotations in Figs. 8-4b and 8-4c do produce distinct results. Taking into account all three axes of this type, we see that they produce a total of six distinct rotations. Furthermore, we can also rotate the cube about this same axis through a half-turn ($2\pi/2$) in either of two directions ($\pm$), but both of these rotations would produce exactly the same final result, as illustrated in Fig. 8-4d. So in this case, we regard these rotations as the same and list only one proper rotation, which we denote $2\pi/2$. This type of rotation thus accounts for a total of three distinct rotations.

- Next observe that there are four axes of symmetry passing through opposite vertices. Since each axis is related in exactly the same way to the rest of the cube, we only need discuss the proper rotations about one of these axes (see Fig. 8-5). With the cube in hand we readily verify that a proper rotation of this type is obtained by rotating through a one-third turn ($2\pi/3$) in either of two directions ($\pm$). Each direction produces a distinct position of the cube (as seen in Figs. 8-5b and 8-5c). There are four such axes, and thus eight distinct rotations of this type.

- There are also six axes of symmetry passing through opposite edges. Again, each such axis is related to the cube in exactly the same way, so we look at just one of them (see Fig. 8-6) and observe that the only proper rotation of the cube about this axis is through a half-turn ($2\pi/2$). Since positive and negative half-turns yield the same result and since there are six pairs of opposite edges, we conclude that there are six distinct rotations of this type.
- Finally, we have the easiest proper rotation of all: just let the cube sit as shown in parts (a) of Figs. 8-4–8-6. This is the lazy rotation, usually called the *identity*. Table 8-1 summarizes the results.

Table 8-1. Proper rotations of the cube.

Symmetry operation		Number of operations of this type
Rotations		
Axis passes through		
opposite faces	$\pm 2\pi/4$	6
opposite faces	$2\pi/2$	3
opposite vertices	$\pm 2\pi/3$	8
opposite edges	$2\pi/2$	6
Identity		1
	Total	24

Table 8-2. Rotations of four objects.

Number of rotations of the given type	Description of the type of rotation
6	(_ _ _ _)
3	(_ _)(_ _)
8	(_ _ _)(_)
6	(_ _)(_)(_)
1	(_)(_)(_)(_) = Identity
24	Total

There are thus 24 distinct symmetry operations for the cube. Is the fact that $24 = 4!$ an accident? Or is there something significant about 4!? It is the number of permutations of four objects—and the group of such permutations is the well-known symmetry group on four symbols, denoted S_4. Perhaps we should list the different types of permutations of four objects (which we call A, B, C, D) and see if there is any connection.

- First there are the four-cycle permutations, which we may indicate by writing (A_ _ _). Then there would be three choices for the second position, followed by two choices for the third position, and the last position is then determined. So we get six 4-cycle permutations.
- We next look at the collection of permutations of the form (_ _)(_ _)—that is, the two 2-cycle permutations. Again it is no restriction to write such a permutation as (A_)(_ _). There are three choices for the other element in the first cycle. Once that choice is made there are no other choices. Thus we see that there are precisely three permutations of the form (_ _)(_ _).
- To count the number of permutations of the form (_ _ _)(_) we may think of first selecting the element that stays fixed. This can happen in four ways. Having made this choice any one of the three remaining elements can occupy the first place in the three-cycle. The last two elements can then be arranged in exactly two ways. Thus the total number of (_ _ _)(_) permutations is seen to be eight.
- It remains to count the number of permutations of the form (_ _)(_)(_). This we do by selecting the two that are permuted. The number of ways we may do this is six.
- There is just one identity, so that we have accounted for the expected 24 permutations. The results of our count are presented in Table 8-2.

Comparison of data in Tables 8-1 and 8-2 reveals a striking resemblance between the lists of numbers. This cannot be an accident. Then what are the four things that are being permuted when we rotate the cube? *Hint:* Look at Fig. 8-7.

It is a straightforward exercise to verify that the four strips forming the "diagonal" cube in

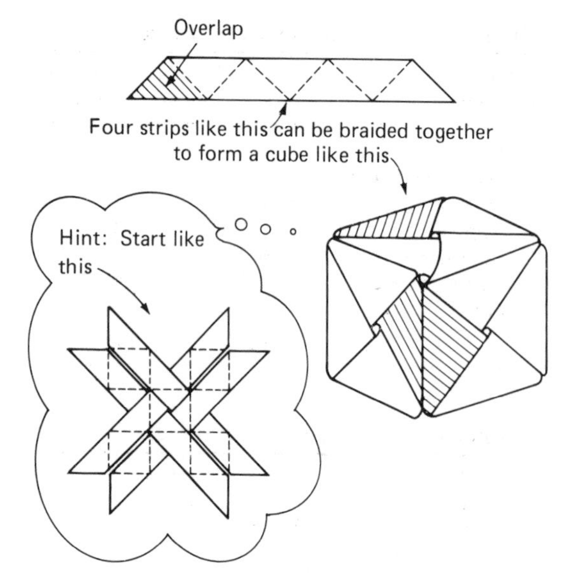

Fig. 8-7. Four strips braided together to form a diagonal cube.

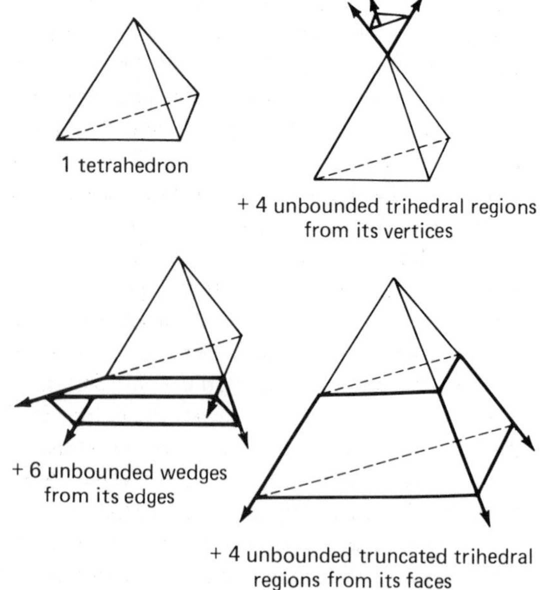

Fig. 8-8. Partition of space by extension of the face planes of a tetrahedron. There is one bounded region (the tetrahedron itself) and 14 unbounded regions.

Fig. 8-7 constitute a set of four objects that are permuted, producing exactly the types of rotation shown at the right of Table 8-2, when the cube is rotated according to the specifications in the left two columns of Table 8-1. Other sets of four objects, related to the cube, permute in this way. One particularly nice feature of the diagonal cube is that the four strips are easy to "see" (in contrast, for example, to the four *interior diagonals* of the cube, which are also permuted by any rotation).

Let us look at just one other aspect of this example. "Open" one face of the cube constructed in Jennifer's puzzle so that one of the two inscribed tetrahedra can be seen sitting inside it (with its vertices coinciding with four of the eight vertices of the cube). If we now take all of the proper rotations of the cube that leave the inscribed tetrahedron *occupying the same space* we obtain all twelve of the possible proper rotations of the tetrahedron. This is a vivid, concrete demonstration of the fact that the proper rotation group for the tetrahedron (known as the tetrahedral group) is a subgroup of the proper rotation group of the cube (known as the octahedral group). Of course the group in question is S_4, and the subgroup is A_4, of index 2 in S_4, also called the alternating group on four symbols.

Let us now turn our attention to some combinatorial aspects of the diagonal cube. Notice that the centers of the faces are surrounded by four different colors and the vertices are surrounded by three different colors. Some natural questions to ask are:

- In how many distinct ways can four colors be arranged in a cyclic order?
  Answer: Six, and each of those ways occurs on exactly one of the six faces of this cube!
- In how many distinct ways can three of four distinct colors be arranged in a cyclic order?
  Answer: Eight, and each of those ways occurs at exactly one of the eight vertices of this cube!
- Suppose you gave each student in a large class (in practice three students will generally suffice) four strips of paper (each strip being a different color) and ask them to weave the strips together to form a diagonal cube. How many different kinds of cubes could you get? Cube X is considered differ-

ent from cube Y if there does not exist a proper rotation that will make the coloring of all the faces on cube X coincide with the coloring of all faces on cube Y.
Answer: There will be just two, and they are mirror images of each other. This can be proved using the Pólya enumeration theorem, but you will be convinced that it is true if you conduct the experiment. You will find that the two cubes are distinguished by how the strips overlap (clockwise or anticlockwise) to form the first square.

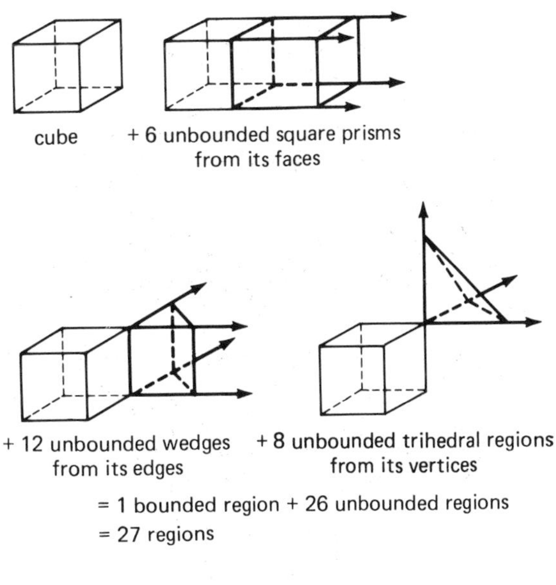

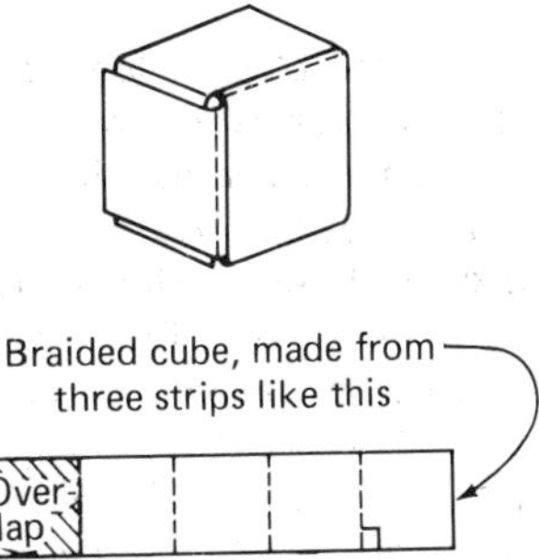

Fig. 8-9. Drawings showing one bounded region and 6 + 12 + 8 = 26 unbounded regions produced by the extended face planes of the cube.

To demonstrate another combinatorial aspect of these braided models (and to reinforce the deep connections that polyhedra share with combinatorial mathematics) let us turn to a question that I first heard about from John E. (Jack) Wetzel. A few years ago he and Jeanne W. Kerr coauthored a paper[10] which considered the question: If the face planes of a Platonic solid are extended, into how many regions (both bounded and unbounded) will space be partitioned?

Jack visited the University of Santa Clara to give a talk on the problem and to work with Gerald Alexanderson on natural generalizations of the question.[11] Jack borrowed one of my dodecahedra as a "prop," and we discussed the question he planned to talk about. After looking at some models, I realized that surfaces of my braided models can be interpreted so that the answer to the original question involving the unbounded regions of the cube, octahedron, and dodecahedron can simply be read off.[12]

Let us see how this is done, looking first at a tetrahedron. We see, from Fig. 8-8, that there is just one bounded region and 4 + 6 + 4 = 14 unbounded regions produced by the extended face planes. Of course, this is a very special case, since the tetrahedron is the only Platonic solid having faces that do not occur in pairs lying in parallel planes. In fact, it is this feature of the other Platonic solids that enables us to use the braided models for counting the unbounded regions produced by the extended face planes of that solid. By studying the illustrations in Fig. 8-9 you may begin to see how the braided cube in the figure relates to the unbounded regions for the extended face planes of the cube.

The face planes of the regular octahedron, when extended, form a slightly more complicated division of space.[13] First, eight tetrahedra appear on the octahedron's faces; these, together with the original octahedron, form Kepler's "stella octangula." Then the unbounded regions are formed. A new feature is that not all the unbounded regions grow straight out of a vertex, edge, or face of the original octahedron; some grow out of vertices, edges, or faces of the tetrahedra that occur on the faces of the original octahedron. The complete count is 9 bounded regions plus 50 unbounded regions for a total of 59 regions.

Now consider the braided diagonal cube of Fig. 8-7. The small square hole shown in the center of each face represents an unbounded polyhedral region; the portion of the diagonal from the center of each face to the vertex of

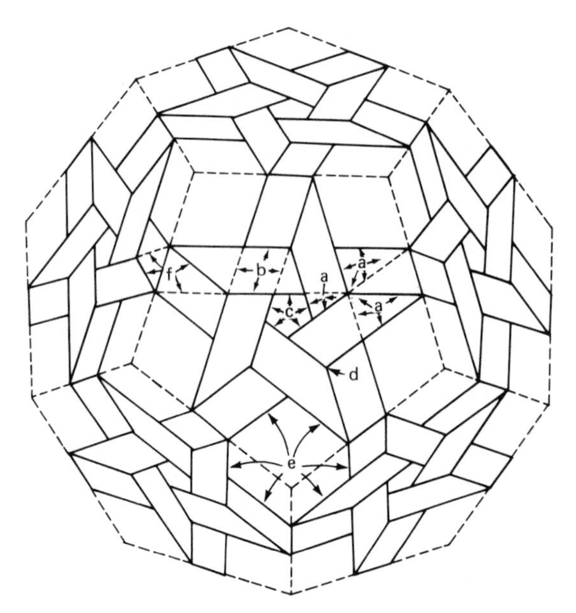

Fig. 8-10. Extended planes identified by edges of the 10 identical strips woven about the "ghost" of a dodecahedron intersect inside this model to form the regular icosahedron. The 362 unbounded regions formed by those 20 planes are (a) 15 × 12 trapezoid-like regions with four infinite faces, (b) 5 × 12 parallelogram-like regions with four infinite faces, (c) 1 × 12 polyhedral regions with five infinite faces, (d) 5 × 12 trihedral regions, (e) 20 polyhedral regions with six infinite faces from the vertices of the ghost, and (f) 30 regions with four infinite faces from the edges of the ghost.

the cube represents an unbounded wedgelike region, and so on. Now let us return to the question posed by Kerr and Wetzel, looking at the surface of the "golden dodecahedron," so named because on the folded tape from which it is made the ratio of the long fold lines to the short fold lines is the golden ratio. The dodecahedron in Fig. 8-1 is constructed so that the short fold lines become edges on the completed polyhedron; the dodecahedron in Fig. 8-2 is constructed so that the long fold lines become edges on the completed polyhedron. If tape of the same width is used to construct the great stellated dodecahedron and the golden dodecahedron, then each of these models has exactly sixty isosceles triangles visible on its surface and the great stellated dodecahedron fits inside the golden dodecahedron with the vertices of its pentagrams coinciding with the vertices of the golden dodecahedron. With the models in hand, it is easy to see that each of the six pairs of parallel planes from the original dodecahedron intersects the other five pairs of parallel planes on the surface of the golden dodecahedron. No more bounded regions can be produced, and the unbounded regions may be read off the golden dodecahedron.

So we have answered the question about the regions created by extended face planes for all of the Platonic solids except the icosahedron. Do you suppose there is a braided model that will help to answer the question in that case? It is a lot harder to draw the finite regions in this case;[14] and fifty-nine stellations give a very large number of finite regions!

Let us be a little less ambitious and try to find a model that delineates the unbounded regions created by the extended face planes of the icosahedron. Since the icosahedron has twenty faces, which occur in sets of parallel planes, the problem is to find ten straight strips that can be braided together to give a completed figure with the symmetry of the icosahedron. Then we could just count up the number of different types of region on the surface of the braided model.

There it is, in Fig. 8-10! The strips go around the "ghost of a dodecahedron." We can compute the number of strips on this model. There are three visible sections on every strip crossing the face closest to our view, so there must be six sections in every strip. But since each of the 12 faces of the ghost host is crossed by five sections, there are in all 60 sections; hence there must be exactly ten strips. This is the model we need. In fact, the planes determined by the top and bottom of these ten strips interpenetrate each other in the center of this model to form the faces of a little icosahedron. Just as with the other braided models, these planes cannot cross each other again to form any bounded regions, so this model can be used to count up the number of unbounded regions created by the extended face planes of the icosahedron.

I have discussed just a few examples of what studying polyhedra can lead to within mathematics. But people other than mathematicians and scientists are also fascinated by polyhedra and by the symmetries of real-life objects. In Kuala Lumpur they play a game with the Sepak Tackraw ball. As I understand

Fig. 8-11. Temari balls by Kazuko Yamamoto, Los Gatos, California.

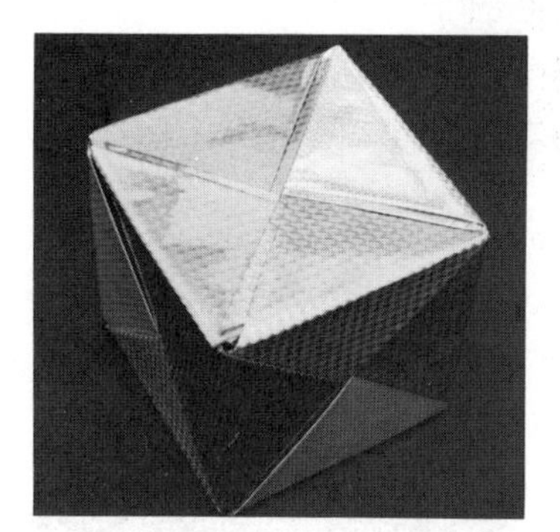

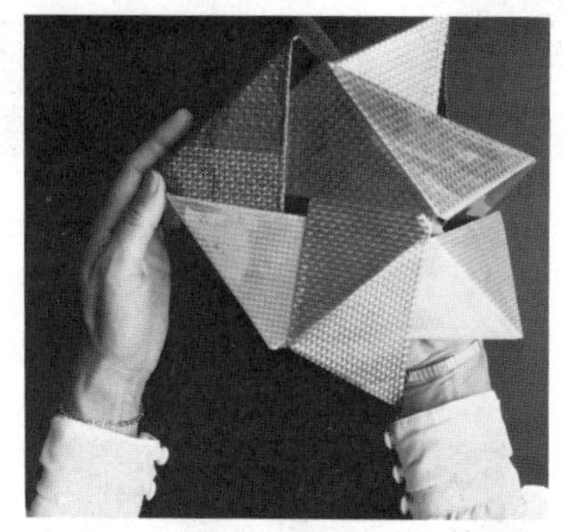

Fig. 8-12. The braided diagonal cube coming apart.

Fig. 8-13. The braided icosahedron coming apart.

it, the ball is hit only with your feet or your head, and that makes the game a very exciting spectator sport. This object has precisely the same symmetry as the golden dodecahedron.

There are many other examples of regular symmetries in nature and in man-made objects,[15] but I would like to show you just one more. Figure 8-11 illustrates some Temari balls. Temari balls have been made in China and Japan for over a thousand years. The most authentic ones have a center composed of rice hulls about which some material is wrapped. Threads are wound round it until the whole becomes spherical in shape. Then guidelines are laid out on the surface and designs are sewn on. Many of these designs have cyclic, icosahedral, or octahedral symmetry.

And, now, as I promised, I will take apart some of the braided models. (See Figs. 8-12–8-14.)

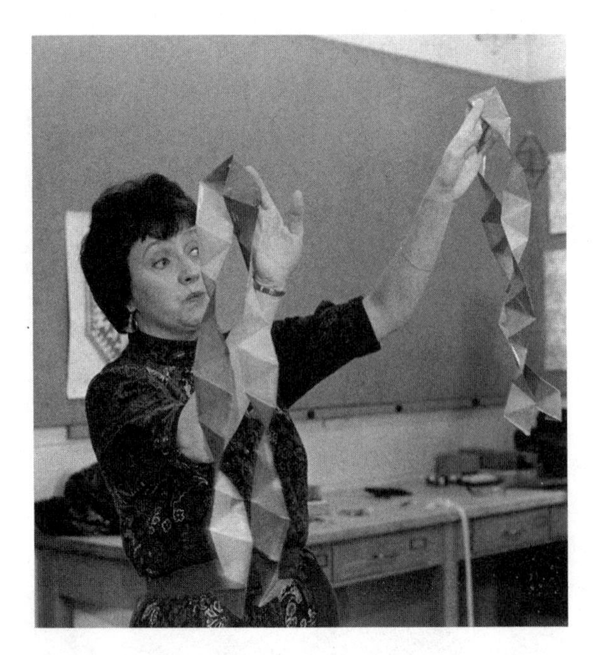

Fig. 8-14. Jean Pedersen demonstrating her "proof by destruction" in a workshop at the Shaping Space Conference. Photograph by Stan Sherer.

a

b

c

Fig. 8-15. (a) Seven braided models. On the left are the pentagonal bipyramid, the trigonal bipyramid and the ring of rotating tetrahedra (see Chapter 21). (b) The triangular bipyramid (in the left hand) coming apart into 19 triangles, and the pentagonal bipyramid coming apart into 31 triangles. (c) The ring of rotating tetrahedra comes apart. (See Chapter 2.)

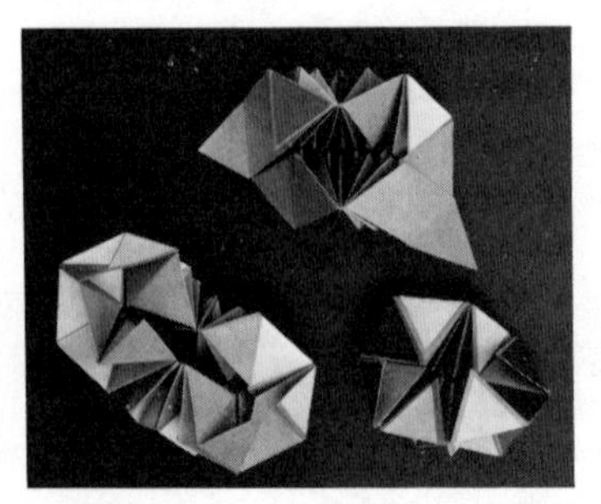

Fig. 8-16. Collapsing three polar collapsoids.

Figure 8-15a shows seven additional models. The pentagonal bipyramid is made[16] from a straight strip of 31 equilateral triangles. And the triangular bipyramid is made from a straight strip of 19 triangles (Fig. 8-15b). The rotating ring of 10 tetrahedra[17] comes apart into two long strips of equilateral triangles (Fig. 8-15c).

The other models in Fig. 8-15a, which I call "the collapsoids," resulted from my impatience in building polyhedra. When I wanted to know more about a rhombic dodecahedron, I thought it would be easier to build something "like" it in which each face was replaced with a pyramidlike cell made of four equilateral triangles. When the model was almost finished, my son picked it up, played with it, and announced: "Look, Mom! It folds up!" That intrigued me. I named these models "collapsoids" and I made others that were "like" some of the other regular rhombic polyhedra, also known as *zonohedra*.[18] Figure 8-16 shows how the 12-, 20-, and 30-celled polar collapsoids fold up. The same idea works for other zonohedra.

It is surely difficult to believe that anyone who has been exposed to the delights of constructing polyhedra and studying them could ask, "Why should we study polyhedra?" I have these models in my office and students come in and beg to know how to make them. They never ask, "What are they good for?" They know! And we know too. As George Pólya has said, "In the plane you have polygons—and, if plane geometry is interesting,—then solid geometry is more interesting!"[19]

Jennifer's Puzzle

Jennifer's Instructions

Instructions: Try It!

1. You get all of the little strips of five triangles each (there should be eight) and braid them into four tetrahedra.
2. Then you get the four strips of seven triangles each and braid an octahedron (that is an eight-faced polyhedron).
3. Now you take the two big strips of five triangles each and braid a large tetrahedron as before, but in this one you put the four little tetrahedra and the octahedron.

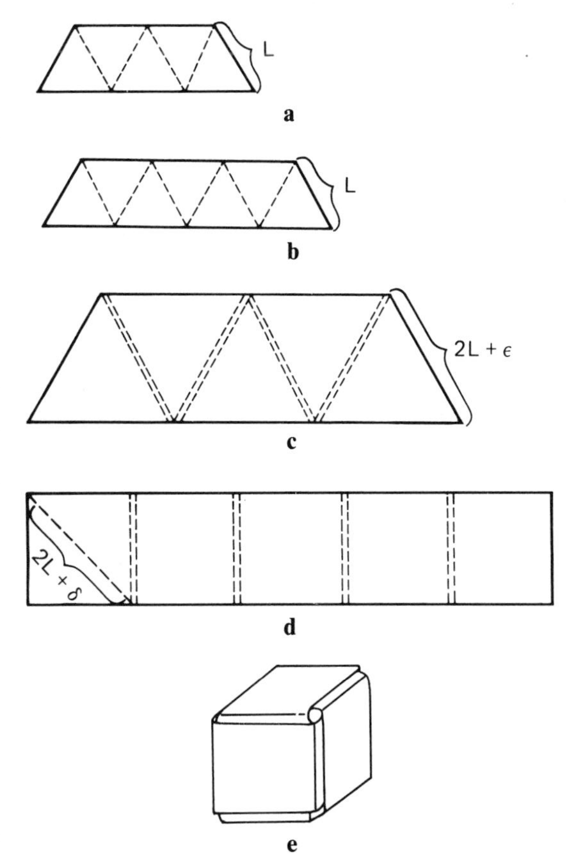

Fig. 8-17. Eight strips like (a) and four strips like (b) are required; these may be of lightweight paper. Two strips like (c) and three strips like (d) should be cut from heavier material. The lengths ε and δ are determined experimentally.

4. Finally, take the three strips of five squares each and braid a cube into which you put the large tetrahedron.

GOOD LUCK!

Jennifer Pederson
9th Grade Geometry Project
Castillero Junior High School

How to Make the Puzzle Pieces

Seventeen strips of paper are required. The choice of material for the strips shown in Figs. 8-17a and 8-17b is not important, so long as it has enough bulk and crispness to hold a good fold. The puzzle will be visually more interesting if different colors are used. For strips shown in Figs. 8-17c and 8-17d the material must be as substantial as lightweight card-

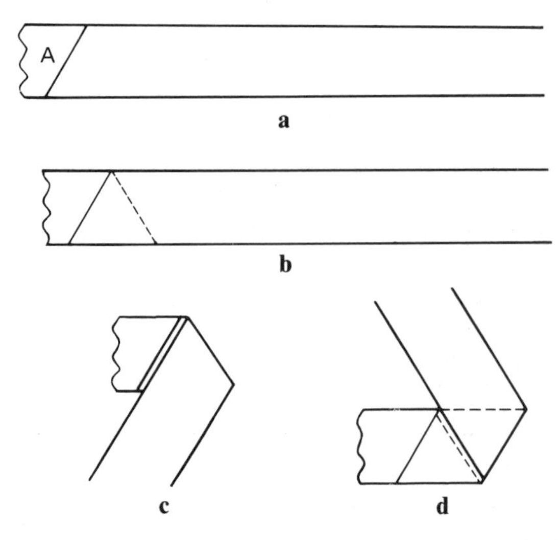

Fig. 8-18. Making the valley folds. (a) First draw a 60° line. (b) and (c) Form a crease by folding the paper with top edge aligned along the 60° line. (d) Succeeding folds are made by aligning an edge with a previously formed crease, *being sure to avoid covering any part of the crease produced by the previous fold.*

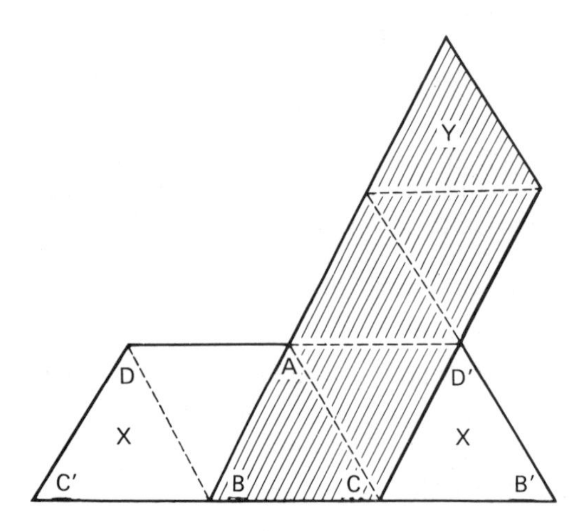

Fig. 8-19. Constructing a tetrahedron from two strips of paper.

board. The thicker the material, the greater the problems with folds.

Each pattern strip should be scored so that all dotted lines are valley folds. To fold the triangles, start with a strip longer than the finished piece and draw a line as shown in Fig. 8-18. Then fold the paper so that edge A lies along the 60° line. Along the fold is a crease whose width depends of the thickness of the paper; do not cover this crease when making the next fold. Continue making folds until the required number of triangles has been made.

The small tetrahedra and the octahedron must be braided and fitted together before the strips shown in Fig. 8-17c are made. The length $2L + \varepsilon$ is determined by measuring the length required so that the configuration of small polyhedra will fit into the completed large tetrahedron. The crease will be wider when heavier material is used; this width may be determined experimentally.

Strip 8-17c can be easily constructed by using a pattern triangle (cut from lightweight paper) to draw the fold lines, leaving an appropriate space between successive triangles for each crease. Score the crease lines firmly so that the strip will fold easily. Fold each strip so that the score lines will be on the inside of the completed model.

The three strips (8-17d) for the cube are constructed in the same way. The length $2L + \delta$ of the diagonal of a pattern square is determined by measuring the edge of the completed big tetrahedron. The allowance for the crease should be *twice* as wide as with the big tetrahedron, since the strips of the cube wrap around each other when the model is constructed.

Assembling the Polyhedra

Tetrahedron. Use two strips like the ones in Figs. 8-17a or 8-17c. On a flat table top, lay one strip over the other strip as shown in Fig. 8-19; the two strips should stay in contact during the assembly. Each fold line should be a valley fold as viewed from above. Think of triangle ABC as the base of the tetrahedron being formed; for the moment, ABC remains on the table. Fold the bottom strip into a tetrahedron by lifting the two triangles marked X and overlapping them so that C′ meets C, B′ meets B and D′ meets D. This produces a tetrahedron with three triangles sticking out from one edge. Complete the model by carefully picking up the whole configuration, holding the overlapping triangles X in position, wrapping the protruding strip around two faces of the tetrahedron, and finally tucking the Y triangle into the open slot along edge BC.

Octahedron. Use four strips like 8-17b. Begin with a pair of overlapping strips held together with a paper clip as indicated in Fig. 8-20a. Fold these strips into a double pyramid by placing triangle a_1 under triangle A_1, triangle a_2 under triangle A_2 and triangle b under triangle B. The overlapping triangles b and B are secured with another paper clip so that the configuration looks like Fig. 8-20b. Repeat this process with the other two strips.

Next place one pair of "braided" strips over the other pair of braided strips as shown in Fig. 8-20c, making certain that the flaps with the paper clips are oriented exactly as shown. Complete the octahedron by following (in order) the three steps indicated in 8-20c. An octahedron is formed by step 1; then step 2 places the flap with a paper clip against a face of the octahedron. In step 3 you should tuck the flap inside the model.

When you become adept at this process, you will be able to slip the paper clips off as you perform the last three steps. This is only an aesthetic consideration, since the clips would otherwise be concealed inside the completed model.

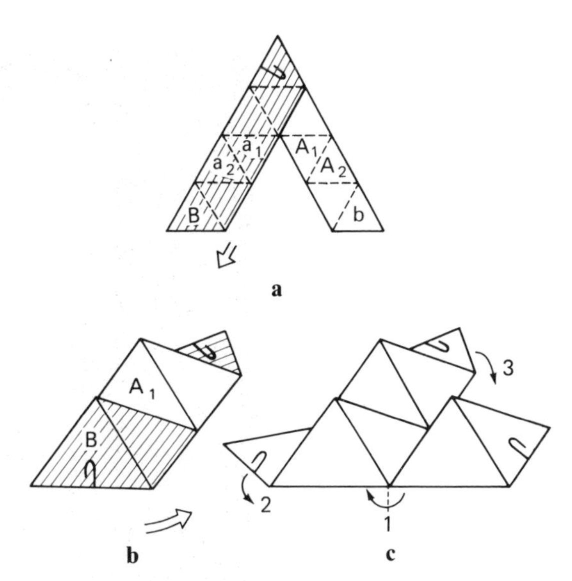

Fig. 8-20. Constructing an octahedron from two strips of paper.

Cube. Use three strips like the one in Fig. 8-17d. Fold one strip and use a paper clip to fasten together the end squares. Fold a second strip around the outside of the "cube" so that one square covers the clipped squares and the two end squares cover one of the square holes. Secure the end squares of the second strip with a paper clip, giving the configuration shown in Fig. 8-21a. Be sure that the overlapping squares of the second strip do not cover any squares from the first strip, and that the first paper clip is covered.

Slide the third strip underneath the top square so that two squares stick out on both the right- and the left-hand sides of the cube. Tuck the end squares of this third strip inside the model through slits along the bottom of the cube (see Fig. 8-21b).

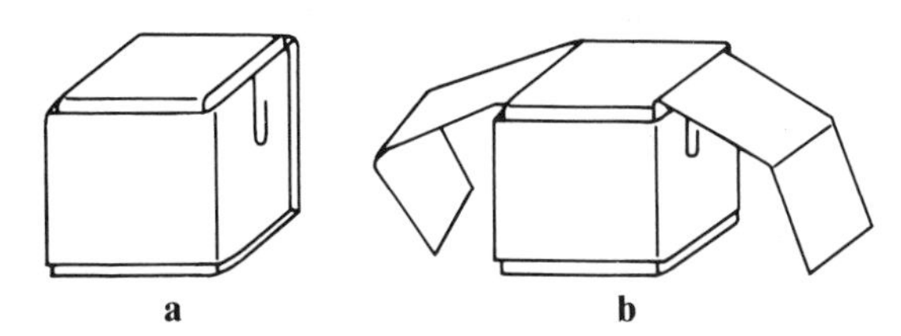

Fig. 8-21. Constructing a cube from three strips of paper.

Putting the Puzzle Together

The four small tetrahedra and the octahedron fit together (see Fig. 8-22) to form a tetrahedral cluster. The large tetrahedral model is braided around this arrangement of little polyhedra.

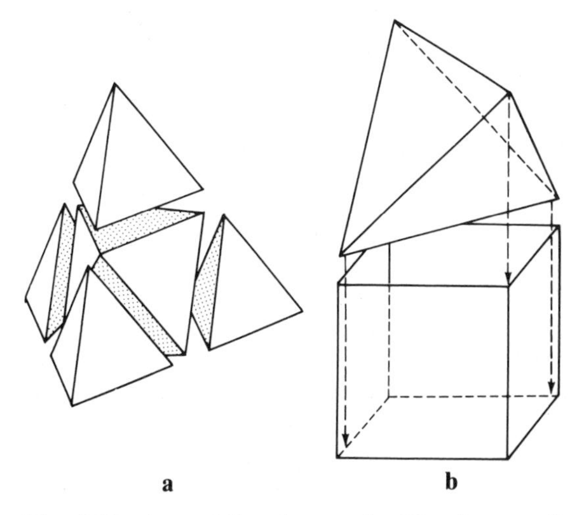

Fig. 8-22. Assembling the puzzle. The four small tetrahedra fit face-to-face against the octahedron (a); the large tetrahedron is made to fit around this configuration. The large tetrahedron then slips into the cube (b).

Fig. 8-23. H. S. M. Coxeter at Jean Pedersen's lecture. Photograph by Stan Sherer.

The completed cube can be opened by turning it upside-down, pulling (pull firmly, since this square is attached inside the model with a paper clip) on the strip that covers the top face, and folding back the flaps that were the last to be tucked inside. The large tetrahedron can then be inserted into the cube. The paper clips are no longer needed when the tetrahedron is inside. With some construction materials, friction alone holds the cube together even without the tetrahedron inside.

Variations and Modifications

If the cube is constructed with three strips, each a different color, the cube will have opposite faces with the same color. You can use the same strips to braid a cube in which pairs of adjacent faces are the same color. In this construction, each strip goes over two strips and then under two as it travels around the cube.

If the inner polyhedra turn out to be slightly too large for the puzzle to fit together, there is a way to salvage the project. If both edges of the strips are trimmed slightly, the resulting model has truncated vertices and can be accommodated in a space that would otherwise be too small.

Notes

Acknowledgements. I should like to thank Peter Hilton for many helpful discussions, and Chris Pedersen for the photographs in Figs. 8-1, 8-2, 8-11, 8-12, 8-13, 8-15, and 8-16.

[1] Martin Gardner's "Mathematical Games," *Scientific American* (September 1971):204–12, was the first written account of how to braid the Platonic solids from straight strips. The dodecahedron can be constructed without paper clips if strips with eleven trapezoids are used, rather than the seven suggested in this article. It goes together very nicely if each of three pairs of strips consists of two mirror-image strips. See also Martin Gardner, *Wheels, Life and Other Mathematical Amusements* (San Francisco: W. H. Freeman, 1983), pp. 106–114. For more detailed instructions for constructing the Platonic solids shown in Fig. 8-1, see Jean J. Pedersen, "Plaited Platonic Puzzles," *The Two-Year College Mathematics Journal 4*, no. 3 (Fall 1973):22–37, and Jean J. Pedersen, "Platonic

Solids from Strips and Clips," *The Australian Mathematics Journal 30,* no. 4 (Fall 1974):130–33.

[2] This is the result if you make them from straight, cylindrical strips of paper and if you require that all strips on a given model are either identical to, or the mirror image of, one another.

[3] H. S. M. Coxeter, *Regular Polytopes* (New York: Macmillan Mathematics Paperbacks, 1963). Peter Hilton and Jean Pedersen, "Descartes, Euler, Poincaré, Pólya and Polyhedra," *L'Enseignement Mathématique* (1982):327–43. Jean Pedersen and George Pólya, "On Problems with Solutions Attainable in More Than One Way," *The Two-Year College Mathematics Journal 15,* no. 3 (June 1984):218–28. Jean J. Pedersen, "The Challenge of Classifying Polyhedra," *The Two-Year College Mathematics Journal 11,* no. 3 (June 1980):162–73. George Pólya, *Mathematics and Plausible Reasoning,* vol. 1 (Princeton, N.J.; Princeton University Press, 1973), pp. 35–38; *Mathematical Discovery* (New York: John Wiley and Sons, 1981), combined edition, vol. II, pp. 149–156.

[4] See Coxeter, *Regular Polytopes.* Also see Jean Pedersen, "Combinatorics, Group Theory and Geometric Models," *Cahiers de Topologie et Géométrie Différentielle 22,* no. 4 (1981):407–28, and Jean Pedersen, "Geometry: The Unity of Theory and Practice," *The Mathematical Intelligencer 5,* no. 4 (1983):37–49.

[5] Hilton and Pedersen, "Descartes, Euler, Poincaré, Pólya and Polyhedra."

[6] Ibid.

[7] Topologists use the term "polyhedron" in a very general sense which includes the manifolds described above, their generalizations to higher dimensions, and all polygons.

[8] Peter Hilton and Jean Pedersen, "Regular Polygons, Star Polygons, and Number Theory," *Mitteilungen, Mathematisches Seminar, Universität Giessen,* vol. II, (in honor of H. S. M. Coxeter) (1984):217–44.

[9] Jean J. Pedersen, "Jennifer's Puzzle," *Matimyás Matematika 4,* no. 4 (October 1980):10–18, and Jean J. Pedersen, "The Magic of Reality: an Analogy between the Star of David and the Stella Octangula," *Matimyás Matematika* (1982).

[10] J. W. Kerr and J. E. Wetzel, "Platonic Divisions of Space," *Mathematics Magazine 51* (1978):229–34.

[11] G. L. Alexanderson and John E. Wetzel, "Dissections of a Plane Oval," *American Mathematics Monthly 84* (1977):224–29; "Dissections of a Tetrahedron," *Journal of Combinatorial Theory 11* (1971):58–66; "Simple Partitions of Space," *Mathematics Magazine 51* (1978):220–25; "Dissections of a Simplex," *Bulletin of the American Mathematical Society 79* (1973):170–71; "Divisions of Space by Parallels," *Transactions of the American Mathematical Society 291* (1985):363–77; "Arrangements of Planes in Space," *Discrete Mathematics 34* (1981):219–40; "A Simplicial 3-Arrangement of 21 Planes," *Discrete Mathematics 60* (1986):67–73.

[12] Jean J. Pedersen, "Visualizing Parallel Divisions of Space," *The Mathematical Gazette 62* (1978):250–62.

[13] To aid visualization of three-dimensional objects on a two-dimensional page, try the following: (a) make a reproduction of the illustrations on a copying machine; (b) use that copy to make a transparency for use on an overhead projector; and (c) cut apart the figures and overlay the various configurations so that the corresponding parts of the successive figures coincide. This is particularly effective for showing illustrations in the classroom. If particular planes are colored consistently, it will help to clarify the meaning of the illustrations.

[14] H. S. M. Coxeter, P. Duval, H. T. Flather, and J. F. Petrie, "The Fifty-Nine Icosahedra," *University of Toronto Studies Math/Science,* no. 6 (1938).

[15] Jean J. Pedersen, "Some Isonemal Fabrics on Polyhedral Surfaces," in *The Geometric Vein: The Coxeter Festschrift* (New York: Springer-Verlag, 1981), pp. 99–122. Jean Pedersen, "Geometry: The Unity of Theory and Practice," *The Mathematical Intelligencer 5,* no. 4 (1983):37–49.

[16] Peter Hilton and Jean Pedersen, *Build Your Own Polyhedra* (Menlo Park: Addison-Wesley, 1987). (This book contains the necessary detailed instructions for building all of the polyhedra mentioned in this chapter.)

[17] Jean J. Pedersen, "Braided Rotating Rings," *The Mathematical Gazette 62* (1978):15–18. Reprinted in *The Oregon Mathematics Teacher* (March 1979):13–15.

[18] Jean J. Pedersen, "Collapsoids," *The Mathematical Gazette 59* (1975):81–94.

[19] Jean Pedersen, "Pop-up Polyhedra," *California Mathematics 8,* no. 1 (April 1983):37–41.

Part III
Roles of Polyhedra in Science

9

Form, Function, and Functioning

GEORGE FLECK

Polyhedra are objects worthy of study and admiration in their own right. They have been inspirations for mathematicians, artists, and architects, and have also served as models for abstract notions about the biological and physical world. The sophistication of such modeling has evolved over the centuries, influencing both physical and mathematical theories. In studying polyhedra we see over and over again ways in which theory is inspired by nature, and ways in which science is inspired by theory.

The Polyhedron Kingdom lies within the realm of mathematics, and polyhedron theory deals with precise ways of talking about polyhedra, ways which seem comfortable to mathematicians (see Part IV). Some of polyhedron theory treats properties of space. Much of polyhedron theory has developed within the minds of mathematicians as old problems have suggested new ones. This theory sometimes appears connected only tenuously with the natural world.

Science and engineering are also fields of investigation in which abstract theory is formulated within people's minds, but the connections with the real world seem—to many nonmathematicians at any rate—both more necessary and more various than in mathematics. We shall look at some contemporary investigations in botany, microbiology, robotics, and chemistry (each discussed at the Shaping Space Conference) in which polyhedra play central roles, investigations which illustrate ways that the geometry formulated by mathematicians is related to the geometry used by scientists and engineers. This symbiotic relationship is dynamic and mutually enriching.

Does Form Explain Function? Science Looks to Geometry for Models

Geometry has been considered a fundamental source of insight into the nature of the universe since the time of Pythagoras (who died about 500 B.C.). We shall note how geometric ideas, and polyhedra especially, have been employed by some of the most creative and influential contributors to the development of natural philosophy (as natural science used to be called) and the contemporary sciences, providing models for atoms, viruses, robots . . . even the solar system.

Plato's Ideas

Plato (427–347 B.C.) argued in his *Timaeus*[1] that fire, earth, air, water, and the quintessence are the *kinds* from which all in the universe is compounded. "It is plain I presume," wrote Plato, "that fire and earth and water and air are solid bodies." He argued that the form in which fire has come to exist is tetrahedral, that the form of earth is cubic, of air octahedral, of water icosahedral, and of the quintessence dodecahedral.[2] Plato then related these forms to the properties, functions, and transformations of all matter, constructing an inclusive natural history based on what were later to be called the Platonic solids. Thus Plato modeled a richly featured, diverse, and constantly changing world in terms of a small number of geometric solids whose features derived only from numbers, lines, and triangles. Plato made use of the polyhedron theory of his day, and it is likely that his use inspired the variations later introduced into the theory by such persons as Archimedes (287?–212 B.C.).

Johannes Kepler (1571–1630) in his *Mysterium Cosmographicum*[3] modeled the solar system of planets in terms of that same set of Platonic solids. (Kepler's representation of this model is shown in Fig. 10-32.) Though manifestly incorrect, Kepler's model is detailed, quantitative, and provocative. Most importantly, the model is visually stimulating, giving a pictorial vocabulary for discussing ideas that might otherwise be too abstract for easy communication. The model captured the imaginations of many persons beyond those with special expertise in quantitative astronomy (see Chapters 3 and 4).

Spheres and Polyhedra as Models for Matter

It is appealing to think of the material world as composed of very small building units. With little information about the nature of those building blocks, early theorists speculated quite freely about them. We noted that Plato thought of them as the regular solids. Since earliest days, some theorists about the nature of matter have explicitly described the shapes of such building units as spheres, and others have described them as polyhedra. Sometimes particular units were chosen for convenience, without regard for a perfect correspondence between theory and reality, though surely some investigators intended their models to be faithful to the natural world.

Another issue which has been of concern for two millennia in various guises is whether matter must fill all space, or whether there can be interstitial voids between the building units. Plato seems vague about whether he thought there can be empty space, but Aristotle (384–322 B.C.) rejected empty space as inconsistent with his theory of motion. However, Epicurus (342–270 B.C.) and Lucretius[4] (99–50 B.C.) argued that motion is impossible unless there is empty space between particles of matter.

If there is no void, spheres alone cannot be the building units of condensed matter; two spheres can come no closer together than to touch at one point, and consequently some space in any packing of spheres is empty. Some combinations of polyhedra fill space completely; certain of those combinations have been known since antiquity. But a satisfactory theory of matter must be able to account for change. Thus if the units do fully fill space without void, they must somehow be able to deform and transform to permit both motion and chemical change. We shall see that detailed modeling of such transformations is an important part of current scientific research.

Spheres and Whirlpools as Models for Atoms and Molecules

Modern chemistry dates from the late eighteenth century. One of the first modern chemists was John Dalton (1766–1844) who, as early as 1810, constructed physical models which he could hold in his hands, modeling atoms and compounds of atoms with spheres joined with connecting rods. Dalton used these models as teaching tools, but we do not know how faithful to reality he believed these spheres to be. There is no evidence that he believed there to be any relationship between the shapes of his models and the shapes of what we now call molecules.

Ball-and-stick models became popular with many chemists during the last half of the nineteenth century. Chemists were rapidly acquiring structural information about molecules from the laboratory, and as the century closed their models were increasingly intended to portray the three-dimensional geometry of molecules. August Wilhelm Hofmann (1818–92) used a collection of elaborate croquet-ball models to illustrate his 1865 lectures at the Royal Institution in London (see Fig. 9-1). In his models, the centers of the croquet balls were coplanar; Hofmann seems to have used his models to represent only connectivity of atoms, not their three-dimensional geometry. Benjamin Collins Brodie (1817–80), the controversial Oxford chemist, strongly urged his colleagues during the 1860s and 1870s to avoid use of ball-and-stick models, warning that the models depicted much more detail about molecules than was warranted by experimental data. His warnings drew a mixed response. The majority of physical chemists in the late nineteenth century explicitly rejected geomet-

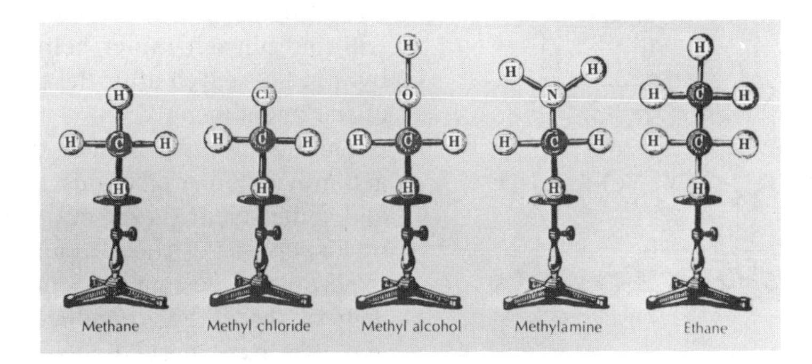

Fig. 9-1. Models of molecules, made from croquet balls, used to illustrate a lecture in 1865. After A. W. Hofmann, "On the Combining Power of Atoms," *Proceedings of the Royal Institution of Great Britain 4* (1865):401–30.

ric ideas about molecules, but most organic chemists of the same period found ball-and-stick models to be increasingly useful. The balls depicted atoms, the sticks were bonds, and the shapes of the models were usually thought to represent the shapes of real molecules. By the turn of the century it was an almost universal article of faith in organic chemistry that all possible molecules can be modeled faithfully with balls and sticks.

Many late-nineteenth-century scientists modeled the molecules of gases by billiard-ball spheres, but most of these theoreticians considered the billiard balls to be models of only certain properties of the gases. For example, the balls were good models of idealized temperature-pressure-volume behavior, but not of the chemical reactivity of the components of the gases.

The arguments of Epicurus and Lucretius about the void became in the nineteenth century arguments about the existence (and the properties) of various types of aether, and this controversy extended into the twentieth century. The billiard-ball model of a gas seems to require the notion of empty space between particles of a gas. Yet from the beginning of the nineteenth century proponents of the wave theory of light argued persuasively for a plenum (the subtle fluid which they called the *aether*) to transmit those waves, and by 1880 the luminiferous aether had become dogma.

The nature of the space within ordinary matter was widely and enthusiastically debated. As the nineteenth century closed, the luminiferous aether was joined by a whole collection of electromagnetic and dielectric aethers invented to explain phenomena such as radio waves, magnetic waves, and even gravitational waves. Speculations about the properties of aether produced a theory of vortex atoms which attempted to combine features of the continuous and discontinuous theories of matter and space. William Rankine (1820–1872) proposed[5] a theory of molecular vortices in 1849, and Hermann von Helmholtz (1821–1894) derived[6] mathematical expressions which show that in a frictionless, isotropic fluid of uniform density, vortices once formed would retain their identity forever. Both Lord Kelvin[7] (1824–1907) and Peter Tait (1831–1901) developed these ideas about whirlpools in the aether further. In his investigations of vortices, Tait combined mathematical theory with physical models. It has been said[8] that a lecture demonstration of smoke rings by Tait (to illustrate Helmholtz vortex motion) early in 1867 gave Kelvin the idea of the vortex atom. Tait described[9] an apparatus to produce smoke rings, telling about various ways those rings could model properties of atoms. Kelvin's theory, in turn, led Tait to extend his investigations on the analytic geometry of knots,[10] Tait believing that a mathematical theory of intertwining and knotting of vortices was necessary for understanding vortex atoms.

James Clerk Maxwell (1831–1879) proposed a gear-and-idle-wheel mechanical model of vortices in aether (Fig. 9-2), remarking that his model "serves to bring out the actual mechanical connections between the known electromagnetic phenomena; so that I venture to say that any one who understands the provisional

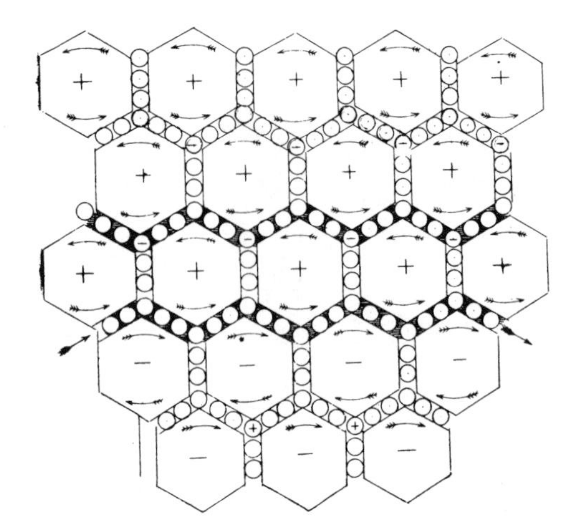

Fig. 9-2. Maxwell's mechanical model for the aether, using small idle wheels to permit all vortices to revolve in the same direction. The idle wheels represent electrical particles. Reprinted from *Philosophical Magazine,* series 4, *21,* (1861): Plate V, Figure 2.

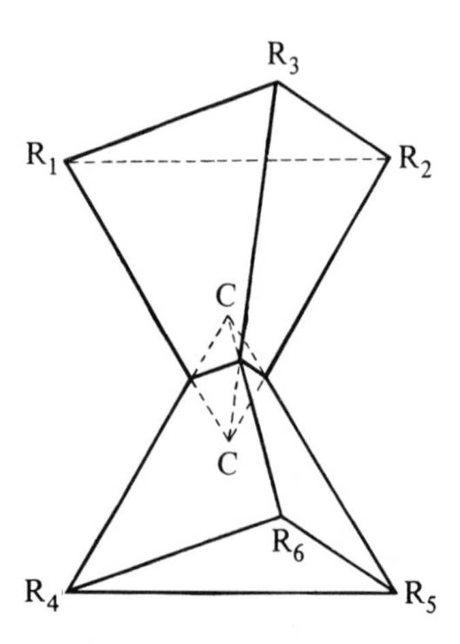

Fig. 9-3. Two interpenetrating tetrahedra presented by J. H. van 't Hoff as a model for a molecule with a carbon–carbon single bond and six different groups $R_1 \ldots R_6$ bonded to the carbons. Each carbon atom is at the center of one tetrahedron and at the vertex of the other. van 't Hoff suggested that such models could be constructed from hard rubber tubes as edges, hollow hard rubber balls as vertex connectors, and sealing wax to bond the parts together. J. H. van 't Hoff, *The Arrangement of Atoms in Space,* 2nd ed. (London: Longmans, Green, 1898), p. 54.

and temporary character of this hypothesis, will find himself rather helped than hindered by it in his search after the true interpretation of the phenomena."[11]

The quantitative models developed for the aether yielded predictions for its properties, and some scientists set out to test the models by experimental measurements. A series of experiments designed to measure predicted drift of the aether, conducted between 1880 and 1930 by Albert Michelson, Edward Morley, and Dayton Miller,[12] were widely interpreted as demonstrating that there is no aether. Without a detectable aether to swirl around, the vortex-atom theory died, but the questions that it attempted to answer remain with us. Listeners to conversations of present-day chemists about physical interpretations of chemical quantum theory have reason to conclude that the issue of whether there are voids in matter has not been resolved.

Polyhedra as Models for Atoms, Molecules, and Viruses

One of the most productive ideas of modern chemistry has been the model of an atom as a polyhedron. This model has been central to structural chemistry since the last decades of the nineteenth century when it was introduced into the mainstream of European chemistry independently by Joseph Le Bel (1847–1930)[13] and Jacobus Henricus van 't Hoff (1852–1911).[14] Plato had considered the ultimate units of matter to be polyhedra, but Le Bel and van 't Hoff extended this idea by showing how organic substances could be modeled by joining polyhedra in systematic ways to form molecules of great variety and complexity. The notion of three-dimensional molecular geometry was popularized by van 't Hoff, who considered carbon atoms to be situated at the centers of tetrahedra. He encouraged chemists to construct cardboard tetrahedra to examine the various geometrically possible arrangements of atoms.[15] The balls of the Daltonian models become the vertices of tetrahedra. But the polyhedral model did more than simulate the environment around single carbon atoms. A carbon–carbon single bond was modeled by two tetrahedra interpenetrating at vertices (Fig. 9-3), a carbon–carbon double bond was modeled by two tetrahedra interpenetrating along edges (Fig. 9-4), and a carbon–carbon triple bond was modeled by two tetrahedra sharing a common face (Fig. 9-5). These ideas were extended to include a wide range of molecules (see Chapter 10), and "the tetrahedral atom" became a central unifying concept of organic chemistry. Joined tetrahedra are still used by chemists for visualizing molecular form; see the representations by Hargittai and Hargittai (Fig. 10-10) and by Pauling and Hayward (Fig. 9-6). Indeed, a major contemporary journal of organic chemistry is titled *Tetrahedron* (see Fig. 9-7).

Polyhedral models are widely used also in inorganic chemistry. The spatial theory of molecular structure, based on polyhedra, was readily adaptable to the compounds of many elements. Nitrogen, depending on its oxidation state, could be modeled with tetrahedra or cubes. Alfred Werner (1866–1919) used octahedra to model metal complexes.[16] In 1902 Gilbert Newton Lewis (1875–1946) found it useful in teaching general chemistry to model atoms with cubes,[17] and later developed the cubic model in a more formal manner.[18] This polyhedral model of the atom placed an electron at each vertex of the cube and provided a context for discussing the role of electrons in chemical bonding. The Lewis model became known as the *octet* theory for chemical bonding, and the eight dots at the vertices became known as Lewis dots.

More recently, in the 1930s and later, Linus Pauling developed, utilized, and popularized a theory of coordinated polyhedra to predict structures for crystals. To visualize the consequences of this theory, he often built elaborate models. An example is Pauling's model for the structure of the mineral sodalite shown in Fig. 9-8.

Polyhedra play a significant role in contemporary research. Very recently, a simple and successful polyhedral model for molecules—the valence shell electron pair repulsion (VSEPR) model—has been developed. As well as a guide for chemical researchers, the VSEPR model has become an important pedagogical tool for teaching about molecular structure. The geometric problem posed by the VESPR model is well-known to geometers as "The Problem of Tammes." VSEPR theory is described by Hargittai and Hargittai in Chapter 10. An indication of the perceived importance of polyhedral chemical models is that a major journal of inorganic chemistry is titled *Polyhedron* (see Fig. 9-7). The structures of certain molecules reinforce that perception; especially interesting is the structure (see Fig. 9-9) of the 60-carbon cluster molecule named Buckminsterfullerene![19]

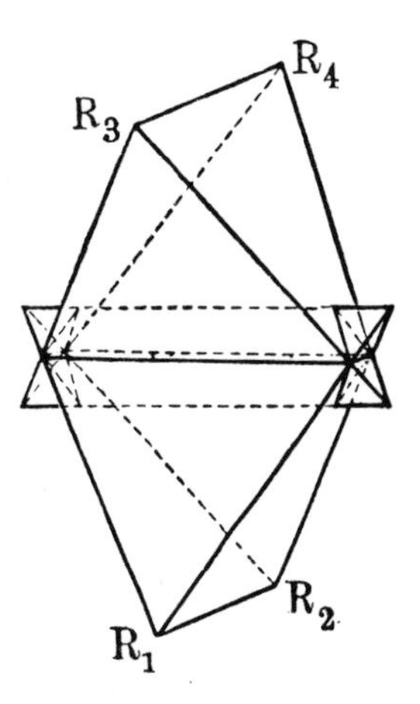

Fig. 9-4. A tetrahedral model of a compound with a carbon–carbon double bond. J. H. van 't Hoff, *The Arrangement of Atoms in Space,* 2nd ed. (London: Longmans, Green, 1898), p. 97.

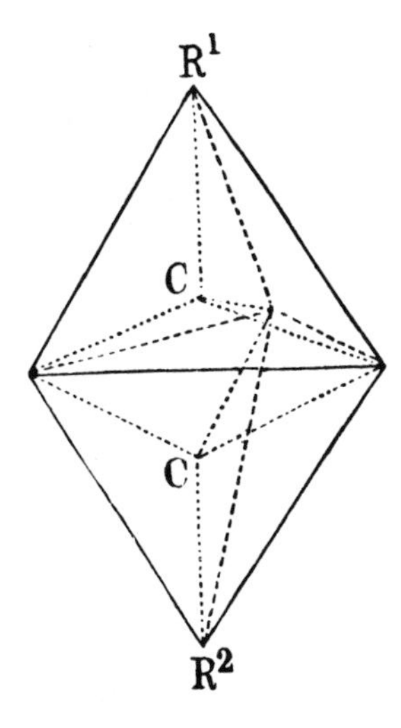

Fig. 9-5. A tetrahedral model of the compound R_1—C≡C—R_2 with three equivalent bonds joining the carbon atoms. J. H. van 't Hoff, *The Arrangement of Atoms in Space,* 2nd ed. (London: Longmans, Green, 1898), p. 104.

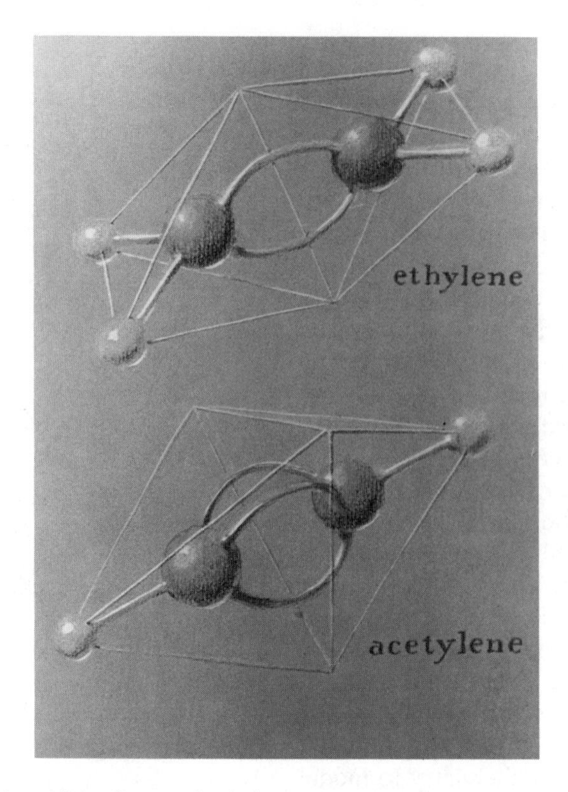

Fig. 9-6. Artistic conception of the forms of ethylene and acetylene molecules. From Linus Pauling and Roger Hayward, *The Architecture of Molecules*. W. H. Freeman and Company. Copyright © 1964.

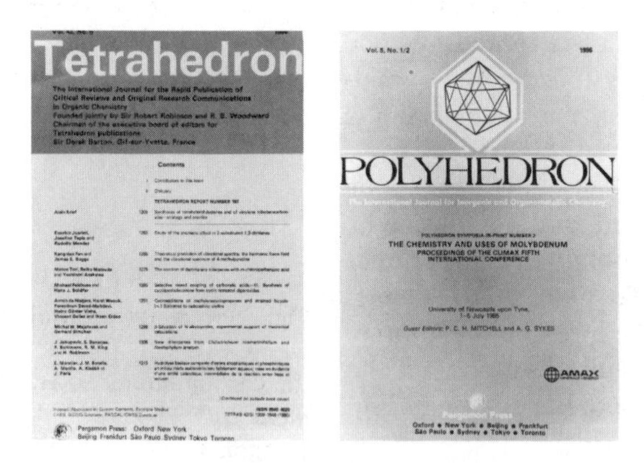

Fig. 9-7. International journals of chemistry have the titles *Tetrahedron* and *Polyhedron*. Reproduced by permission of the publisher, Pergamon Press.

Fig. 9-8. A model of the mineral sodalite, utilizing coordinated polyhedra. Reprinted from Linus Pauling, *The Nature of the Chemical Bond and the Structure of Molecules and Crystals,* 2nd ed. Copyright 1939, 1940, by Cornell University. Used by permission of the publisher, Cornell University Press.

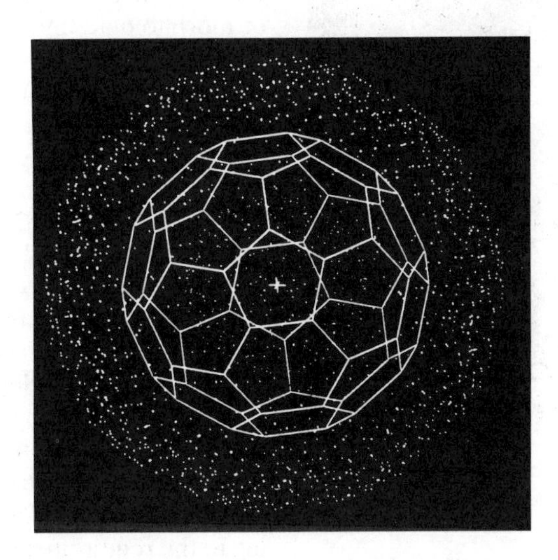

Fig. 9-9. A computer-generated depiction of the truncated icosahedral structure suggested for the $C_{60}La$ molecule. Used by permission of John C. Spurlino and Florentine A. Quiocho, Department of Biochemistry, Rice University.

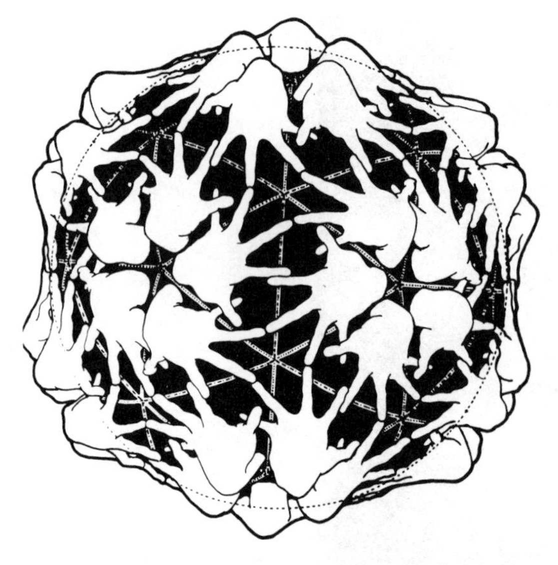

Fig. 9-10. A drawing by D. L. D. Caspar illustrating strict equivalence in a shell with icosahedral symmetry constructed from sixty identical left-handed units. The three classes of connections in this surface lattice are represented by the specific bonding relations: thumb-to-pinkie = pentamer bond; ring finger-to-middle finger = trimer bond; and index finger-to-index finger = dimer bond. Any two of these classes of bonds would hold the structure together. The triangles drawn under the hands define equivalent subdivisions defined by the three- and fivefold axes at their intersections.

Fig. 9-11. A geodesic dome built on a plan of a T = 12 icosahedral surface lattice.

The polyhedral models, which have proved so valuable in discussing atoms and molecules, have also been useful in investigating structures larger than single molecules. At the Shaping Space Conference, Donald L. D. Casper spoke[20] about his three decades' long fascination with the coincidence between the Platonic polyhedra and the structures of viruses. It began in 1956 when Crick and Watson[21] suggested that the design of "spherical" viruses, built from large numbers of identical protein subunits, would be based on the symmetry of the Platonic polyhedra. That same year Caspar obtained experimental evidence in Cambridge for an icosahedral structure of one of the small isometric plant viruses. Icosahedral symmetry requires sixty identical parts, and at that time the symmetry of icosahedral viruses was thought to be a consequence of specific bonding among identical units. Such a structure is illustrated in Fig. 9-10.

Caspar and his colleague Aaron Klug argued that the essential idea in the design of viral structures is that they "build themselves," that the design is embodied in the specific bonding properties of the parts.[22] Given the bonding rules, the units combine to form the structure automatically. The problem faced by the investigators was that most of the icosahedral viruses are not built of sixty identical subunits; indeed, by 1962 electron microscopy had revealed regular surface arrays of morphologic units that were neither multiples nor submultiples of 60. Their question was then twofold: Why icosahedral symmetry? What are the design possibilities for such icosahedrally symmetric structures?

The clue to the answer formulated by Caspar and Klug in 1962 was based on an icosageodesic dome (see Fig. 9-11) which, were it a complete sphere, would be divided into 720 truncated triangular facets grouped to form 12 pentamers and 110 hexamers. If a hand (such as one of those in Fig. 9-10) were placed in each of these triangles, one could imagine (with a certain amount of flexibility either in bonding or in the structure units themselves) that identical units could be connected according to the requirements of bond specificity, but allowing for some departure from strict equivalency. The term that Crick and Watson proposed for this was "quasi-equivalence." The units would be deformed in slightly different ways in symmetrically distinct but quasi-equivalent positions.

The design for such structures can be described by the ways that the plane hexagonal net can be folded into polyhedra. These designs can be enumerated completely and non-redundantly by the triangulation numbers

$$T = (h^2 + hk + k^2),$$

which designate the number of symmetrically distinct but quasi-equivalent situations for the $60T$ units in a design. The indices h and k can be any positive integers; one may be zero. These designs for indices 0, 1, and 2 (that is, for $T = 1$, $T = 3$, and $T = 7$) have been recognized for a number of icosahedral viruses. Some are built on the $T = 1$ plan. The $T = 3$ plan is also very common.

These surface lattices can be represented as icosadeltahedra, polyhedra consisting of $20T$ equilateral triangular facets. Capsid models can be built of $60T$ identical subunits grouped to form 12 pentamers and $10(T - 1)$ hexamers with quasi-equivalent bonding in the T symmetrically distinct environments. It is informative to build both rigid and flexible models of these structures.

A number of small tumor viruses are built on the $T = 7$ plan. But, to everyone's great surprise, a radical departure from the idea of quasi-equivalence was revealed with the discovery that the $T = 7$ icosahedral polyoma virus capsid is built of 72 pentamers[23] instead of the predicted 12 pentamers and 60 hexamers. Bonding specificity apparently is not conserved in this structure. Caspar described a polyhedral model of the polyoma capsid constructed from 72 pentagons connected using three more equivalent types of contacts which correspond to switching of bonding specificity. He said that this idea was so incompatible with the expectations of his theory that the referees who reviewed the paper said the model was not suitable for publication. "The theory was so good. Why throw away a good theory with experiments that haven't been thoroughly tested?"

However, further experiments have confirmed the structure, a design which appears inscrutable in geometric terms but which obviously has biological logic. Caspar commented: "The theories we have formulated in the past have given a very good explanation of why icosahedral viruses are icosahedral. Now we don't know!"

Modeling Condensed Matter

Polyhedra have been used not only in modeling molecules but also in modeling the space-filling qualities of whatever it is that forms the condensed states of matter. In crystals, polyhedra have been considered to be the units that repeat in three dimensions to fill space. As we have seen, a continum model (a model in which there is no empty space) for crystals of a pure substance (an element or a compound) cannot be achieved with spheres. But there are only a few polyhedra that fill space by periodic repetitions of themselves. There are even greater difficulties in modeling arbitrary mixtures of different substances with polyhedra, since most collections of *different* polyhedra do not fill space.

Theories about structure of the solid state have long been involved with the question of how polyhedra can be packed to fill three-dimensional space. Bricklayers have known about the packing of parallelopipeds since antiquity, but they have generally not been concerned about theories of the structure of matter. Aristotle (refuting Plato) asserted that, of the regular solids, both the cube and the tetrahedron fill space; he was wrong, however. Kepler considered the shapes of space-filling polyhedra which would be obtained if closely packed spheres were uniformly compressed; we shall see that this strategy for investigating the shaping has been fruitful in recent years. The quite complicated general question of how space can be filled by repetition of identical polyhedra has not yet been solved.

One experimental way to prepare space-filling collections of different polyhedra is to use the method of Kepler and compress a collection of plasticene balls.[24] Another way is to start with a packing of spherical objects, and increase the size of the spheres without increasing the volume of the collection; Stephen Hales (1677–1761) used this strategy in studying the shapes of peas which were swollen in a closed container.[25] It appears that the unfinished task of describing general collections belongs largely to the mathematicians, but their intuitions have benefited from these physical experiments.

Clearly the solid state cannot be described by a single geometric model. Some solids are almost perfectly ordered in a simple manner at the atomic scale, and others are almost completely random. Most solids have structures between these two extremes. Only the perfect crystalline structures—which do not really exist—are well understood. The structure of the solid state has been a subject of inquiry in which the interaction between geometry and the physical sciences has been particularly fruitful. Lord Kelvin took a very empirical approach in investigating "the division of space with minimum partitional area"[26] Extending experimental methods used by Plateau,[27] he observed and manipulated intersecting soap films in imaginative ways, and his paper is an instructive example of how a physical model can guide a geometric investigation, and how mathematics can permit generalization from a few simple physical observations.[28] Lord Kelvin concluded that space could reasonably be divided into modified truncated octahedra with 14 faces (eight hexagons and six squares); he called these polyhedra *tetrakaidecahedra,* and argued that they would pack to fill space with minimum partition area. Kelvin's shapes are not classical polyhedra; the edges are not straight lines and the faces are not plane surfaces.

Static space-filling units probably cannot model simultaneously both the space-filling qualities (such as the quasi-crystalline regions sometimes found near apparently chaotic regions in liquids, or the orientation of polar solvent molecules near solute ions) and the dynamic aspects (such as fluid flow, or diffusion of solvent and solutes) of liquid solutions. It would be interesting to try to model such systems with dynamically transforming polyhedra. Meanwhile, sphere-packing models continue to be useful.

Packing of Spheres of Various Sorts

Twentieth-century models of atoms, with a very small nucleus surrounded by wavelike electrons, have spherical symmetry (probably instantaneous spherical symmetry, but certainly at least time-averaged spherical symmetry) in isolation. The sphere is an excellent model for isolated atoms and has been developed by many investigators as a model for atoms in molecules, and for molecules in liquids and solids.

When pressed, few chemists insist that a collection of spheres with interstices is a realistic model for combinations of atoms, whether crystalline arrays or discrete molecules. They say that their ball-and-stick models are merely conveniences, that their ball-and-stick drawings are artistic conventions, and that their tables of covalent and ionic radii are just conventional ways of presenting data concisely. Yet it would be naive to believe that the ubiquitous presence of spherical models for two centuries has had no influence on the concepts held by chemists.

Much effort has been spent examining the consequences of spherical models for atoms, calculating atomic-radii for these spheres, and discussing how spheres of various sizes can pack together. William Barlow (1845–1934) observed[29] that there are two "closest-packed" ways of arranging identical spheres in space, one with cubic symmetry and the other with hexagonal symmetry. He was convinced that a thorough understanding of crystalline solids would necessarily involve both a geometrical theory of space groups (to which he contributed) and a complementary mechanical theory of crystal structures (toward which he worked for several decades). Lord Kelvin examined the problem of close-packing spheres with oriented binding sites.[30] Pauling developed in detail[31] the model of atoms-as-spheres, using what he called "covalent radii" to correlate distances within molecules and crystals among quite diverse compounds of particular elements.

Spherical models for atoms have been useful, even though atoms within an environment of "touching" nearest-neighbor atoms are not spherical, and even though the electrons in such environments fill space without interstices. All this has stimulated mathematical studies of sphere packings in spaces of three and higher dimensions. These studies, in turn, have surprising and important applications in coding theory.

Polyhedra as Models for Plant Structures

Ralph Erickson spoke[32] at the Shaping Space Conference about botanical research in which polyhedra have been used as models for plant cells. These models suggested novel experiments, involving such unlikely materials as lead spheres and soap bubbles. The experiments in turn have stimulated mathematical speculation.

Parenchymal tissues such as are found in a plant stem have cells that pack closely together. The cells are not classical polyhedra; there probably are no plane faces nor any straight edges in these cells. Erickson described experiments[33] in which James Marvin obtained single-cell geometric data about pith cells from the Joe-pye weed. Marvin then constructed paper polyhedral scale models of those cells. Previous investigators had taken Kelvin's tetrakaidecahedra[34] as a model for such space-filling cells, but the data of Matzke and co-workers showed that the real botanical world was more complex. A fundamental question is the extent to which the shapes of such cells are determined by geometry alone, and the extent to which the shapes are the results of other factors.

Erickson described studies in which Matzke and Marvin used lead shot and soap bubbles to model plant cells. In a series of Kepler-type experiments, Matzke[35] and Marvin[36] compressed collections of spherical lead shot with enough pressure to force the initially spherical pieces of lead into shapes that together fill space without interstices. As Kepler had shown centuries earlier, spheres transform into polyhedra when forced to be space filling. If uniform spheres were packed as a face-centered cubic array and compressed, rhombic 12-hedra resulted. If the lead shot were poured into the container "randomly," a distribution of irregular polyhedra (averaging about 14 faces) resulted. Matzke and Marvin undertook their studies in attempts to model cell shapes in plant tissue in terms of the polyhedra observed in the shot-deformation studies.

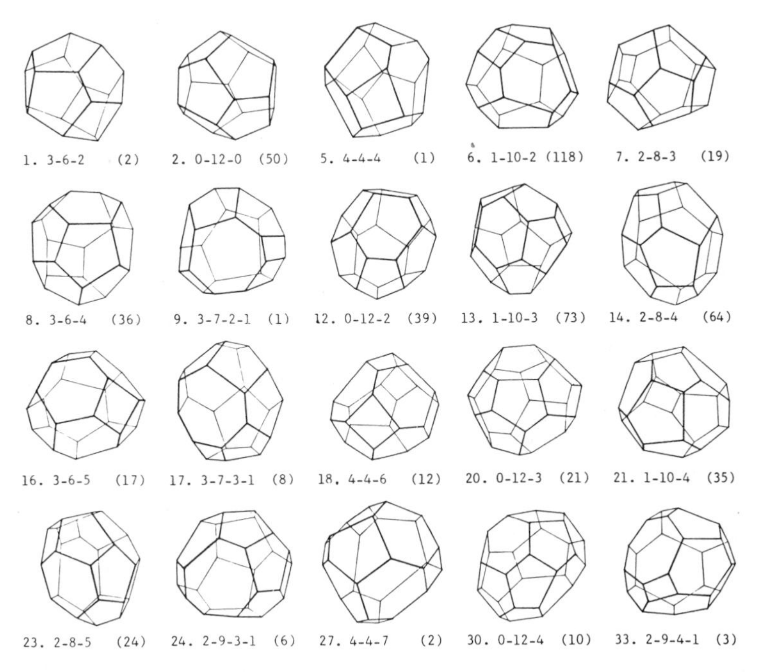

Fig. 9-12. Camera lucida drawings of representative soap bubbles from the center of a foam. Each drawing is specified by listing the numbers of rectangular, pentagonal, hexagonal (and heptagonal) faces. In parentheses is the frequency of occurrence of the polyhedron class in a group of 600 bubbles studied. From E. B. Matzke, "The Three-Dimensional Shape of Bubbles in Foam—An Analysis of the Rôle of Surface Forces in Three-Dimensional Cell Shape Determination," *American Journal of Botany 33* (1946):70, Figs. 27–46. Reprinted by permission.

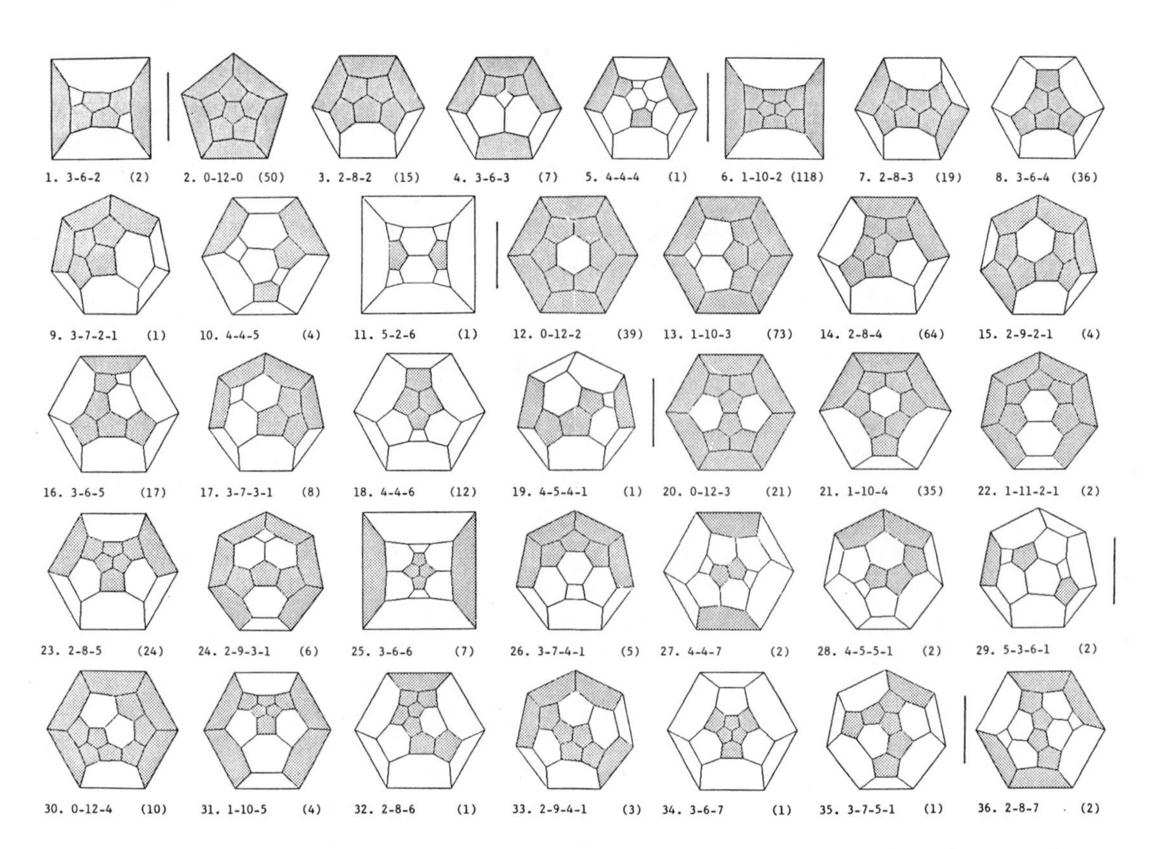

Fig. 9-13. Schlegel diagrams of central bubbles (some from Fig. 9-12) tabulated by Matzke. The numbering corresponds to the numbering in Fig. 9-12. Pentagons are shaded. Viewpoints are chosen to demonstrate symmetries when possible. Used by permission of Ralph O. Erickson.

Matzke[37] also used another model, reminiscent of the 1880s' work of Lord Kelvin. Several thousand soap bubbles were assembled in a transparent container, and the interior bubbles were examined microscopically (Fig. 9-12). These soap films partition space. We can distinguish between the *filling of space* by a collection of geometric shapes and the *partition of space* by surfaces that can be considered to be the faces of polyhedra. Matzke made detailed comparisons of these polyhedra, the polyhedra from the lead-shot experiments, and the polyhedra observed in plant cells.[38]

Erickson noted that accurate visualization of models is difficult, even when the models (such as polyhedra) are apparently tangible. A difficulty in using two-dimensional representations of stick models or solid models of polyhedra on the printed page is that all faces cannot be shown simultaneously. To understand a three-dimensional model, a person must pick up the physical object and turn it around. Erickson constructed paper models of Matzke's soap bubbles to aid visualization and he drew Shlegal diagrams[39,41] (Fig. 9-13) to aid classification. Shlegel diagrams distort shape, but they allow simultaneous viewing of all faces and of their connectivity. Many of these Shlegel diagrams have obvious symmetry. Consider number 0-12-2. It is highly symmetrical, with two hexagonal faces opposite each other, more or less as in an antiprism. But it is not an antiprism. The sides are pentagons. Symmetries are very prominent in these polyhedra; only 44 of the 600 bubbles are in the symmetry group that contains just the identity element.

Visualization becomes less of a problem when the investigators use several different schemes which appear to model the same geometry. Erickson constructed paper models so that he could look at them from all directions. He also used stick structures (representing the same polyhedra) built from semiflexible plastic tubing and four-arm connectors, the arms at the 109.5° tetrahedral angle.

All edge intersections of soap films are necessarily tetrahedral intersections. So by building models with these rigid connectors and semiflexible tubes, the model approximates the minimum-surface models. Lord Kelvin[40] went to some pains to point out that in the packing of soap bubbles the edges of the cells which he visualized were plane and curved, alternate edges curving in alternating ways. This very delicate structure of the Kelvin 14-hedron is only approximated by these skeletal models.

Kelvin's tetrakaidecahedron alone is not a sufficient model for either plant cells or soap bubbles. The tetrakaidecahedron has no pentagonal faces, whereas the majority of natural cells do have pentagonal faces. Matzke's data on 600 bubbles, probably the largest sample of such cells that exists in literature, reveal 36 polyhedral forms; some contain one heptagonal face. Erickson built them all both as stick models and also as cardboard models.

Figure 9-14 displays the fantastic differences in the frequency of the polyhedra. Some of them occurred only once in Matzke's 600 cells. Amazingly, a 13-sided polyhedron (1-10-2) occurred 118 times.

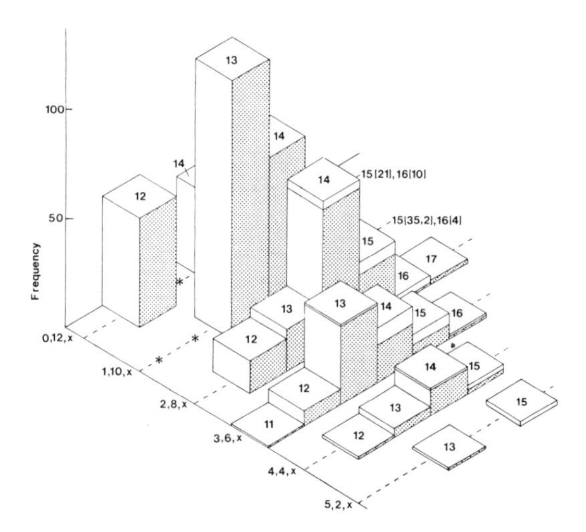

Fig. 9-14. Frequencies of polyhedral forms among 600 central bubbles. Each bar is labeled with the number of faces of the polyhedron. Unshaded portions of bars indicate frequencies of polyhedra having one heptagonal face. Asterisks indicate forms [(0-12-1), (1-10-0); (1-10-1)] which cannot be built. Five bars are hidden: (0-12-3); (0-12-4); (1-10-4); (1-11-2-1); and (1-10-5).

Does Form Explain Dynamic Functioning? Science Looks to Geometry for Mechanistic Models

Growth of a rigid plant stem, self-assembly of a virus, functioning of a robot arm: such dynamic processes are being simulated with dynamic polyhedral models which focus on transformations. Ralph Erickson, Donald Caspar, and Godfried Toussaint spoke at the Shaping Space Conference on aspects of their research in this frontier area in which the symbiotic relationships among theory, mechanical models, and the natural world are strikingly evident. We shall examine some of their work.

Plant Growth and Polyhedral Transformations

A rigid arrangement of parenchymal plant cells is a dynamic system with very interesting spatial properties. The plant cells divide as the tissue grows. If those cells are space-filling polyhedra, the polyhedra must be capable of local transformations which result in changing the number of packed polyhedra without weakening the plant structure. We would expect that division of a single cell would occur with minimal disruption of intersections and edges in adjacent cells.

An advantage of the skeletal, stick models discussed by Erickson for modeling transformation of plant cells is that these models can be manipulated and transformed rather easily. A face can be added by breaking a couple of connections and inserting three edges and two connectors. Another operation which is very helpful in exploring the possible transformational forms consists of disconnecting and then reconnecting an edge and its four neighboring edges. The result is promotion of square faces to pentagons and demotion of two hexagons to pentagons. Other 14-hedra can be found by carrying out similar manipulations on the edges. R. E. Williams has proposed[42] a model for a cell packing based on repetition of this neighbor-switching operation, creating cells with two square faces, eight pentagons, and four hexagons.

Erickson noted that a puckered surface can be traced through the packing of the truncated octahedra. It consists of hexagons and squares, going up and down like valleys and ridges. The squares, most interestingly, are oriented at 90° to each other along a path. Neighbor-switching operations within this packing can be used[43] to discuss dislocations in cellular structures as well as creation and annihilation of cells. He proposed that it should be possible to orient two puckered surfaces properly, connect them appropriately, and create a uniform packing of polyhedra which will fill space. With a different orientation and edge-connections, another space-filling packing of 14-hedra would be created. This can be done extensively, if not exhaustively. The convincing way to do that is with stick models; there are too many possibilities and transformations for cardboard models to be feasible for this task. Erickson said that his models covered half of the kitchen floor!

Polyhedral Models for Self-Assembly of Viruses

Caspar has used a variety of model-building strategies to explore both the geometry and the energetics of protein assemblies.[44] Caspar and Klug first illustrated[45] their idea of self-assembly with a dynamic model with wooden-peg subunits designed to assemble in the $T = 3$ icosahedral surface lattice. The structural unit, and various stages in assembly, are shown in Fig. 9-15.

This model shows that it is possible to design a single unit with bonds that can be switched by interaction with identical copies of itself to bond in the three different environments of the $T = 3$ lattice. Such a model is too simple to explain all features of the control mechanisms that were postulated for icosahedral virus self-assembly, but it does display some of the essential interaction properties of a structural unit designed to form an icosahedral capsid.

The model shown in Figs. 9-15 and 9-16 illustrates some of Caspar's further ideas about designing mechanical models to represent the dynamics of macromolecules. Such models should illustrate how the energy of interaction is distributed throughout the structure. One strategy has been to underdesign, so that the structure is not unintentionally overdetermined. The construction of a mechanical model is often a trial-and-error process. This is, in fact, Nature's way for adaption and evolutionary development. Analogies with human technology and behavior have indeed provided essential keys for understanding the operation of biological systems, and, conversely, analogizing Nature's methods is a natural way to make analogs of Nature's machines.

Caspar concluded by discussing another biological structure at a higher level of organization, an all-pentamer radiolarian skeleton in which there are 5-around-1 arrangements, 6-around-1 arrangements, and paired pentamers next to each other. The skeleton is a regular organized structure without obvious symmetry. He suggested that in nature where there is regularity, with structures built of identical parts, there are likely to be regular plans. Geometric considerations are always important in these plans, and sometimes they predominate. However, satisfactory a priori predictions about what in fact happens in nature cannot be made. The only way to find out is to look.

Robotics and Motions of Polyhedra

Godfried Toussaint spoke[46] about dynamic computational geometry, a new area in computer science which has evolved from work in graphics and visual design, inspired by robotics and problems of movement. A fundamental problem in robotics theory involves the ways in which a set of objects can be moved without collisions. Toussaint and others have been studying how sets of polygons in the plane can be translated without collisions.[47] Toussaint discussed some aspects of his work in generalizing such studies to three-dimensional movement of polyhedra.

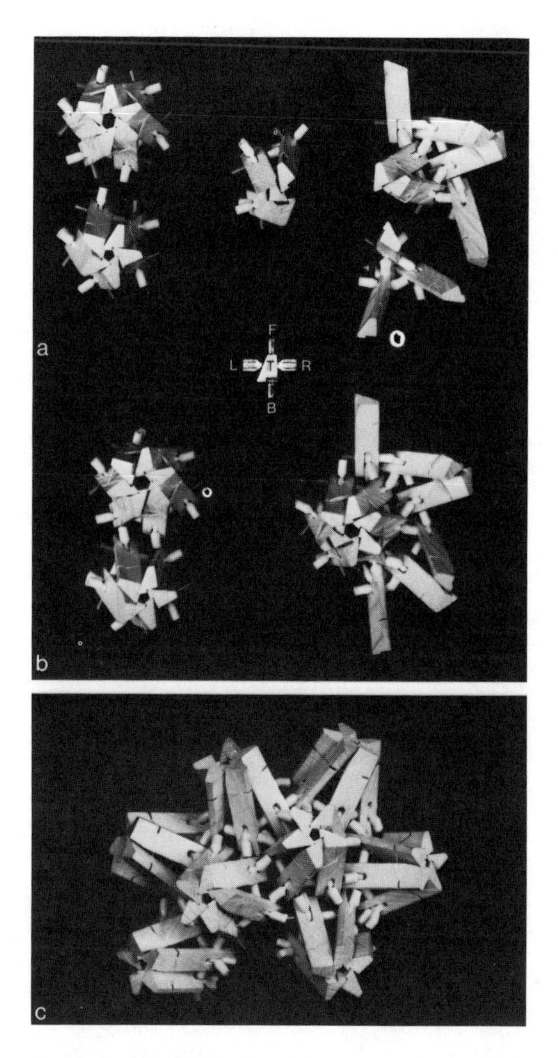

Fig. 9-15. Constituent parts of Caspar's self-assembly model. (a) On the left: a pentamer and hexamer; center, part of a hexamer or pentamer; right, a trimer and two trimers bonded together. (b) The same units with more bonds formed. (c) Misassembly of eight pentamers attempting to form a $T = 1$ shell. Models made by Charles Ingersoll, Sr. Photographs by Fred Clow.

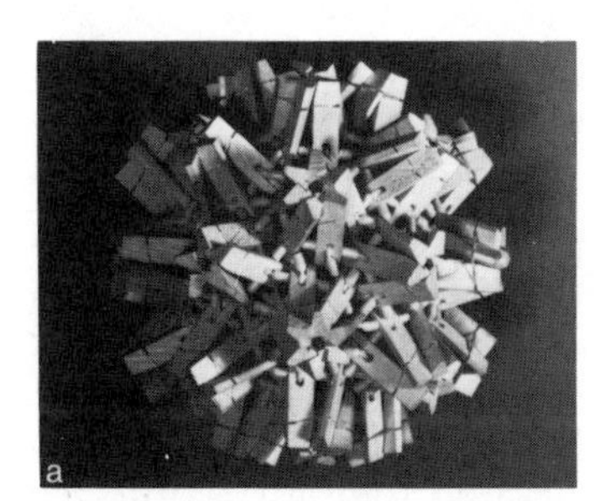

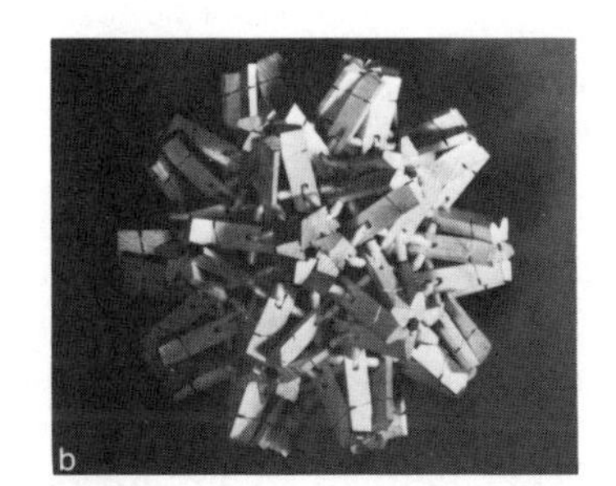

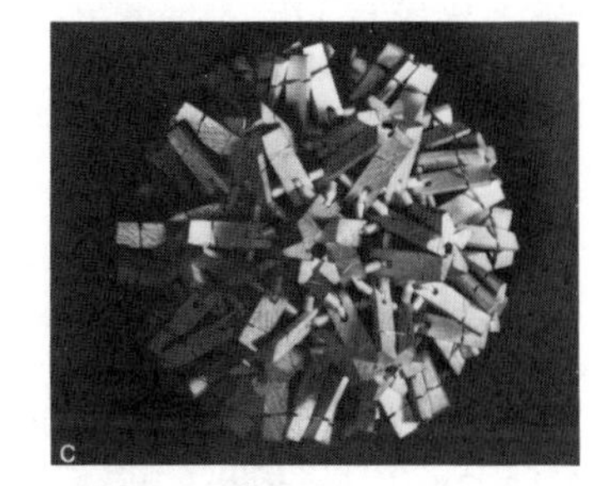

Fig. 9-16. Completed $T = 3$ self-assembly model viewed down: (a) twofold axis. (b) fivefold axis. (c) threefold axis. Models made by Charles Ingersoll, Sr. Photographs by Fred Clow.

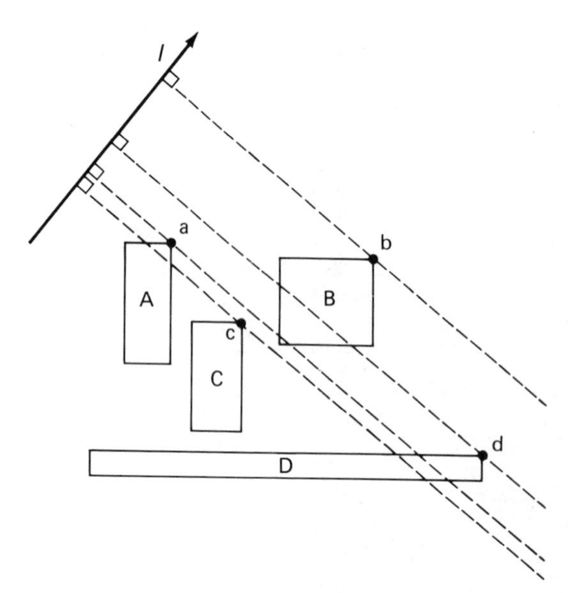

Fig. 9-17. An illustration of a failure of the *line-sweep heuristic* for a set of isothetic rectangles. Note that rectangle D cannot be translated before rectangle C in direction *l*, even though d occurs before c on line *l*.

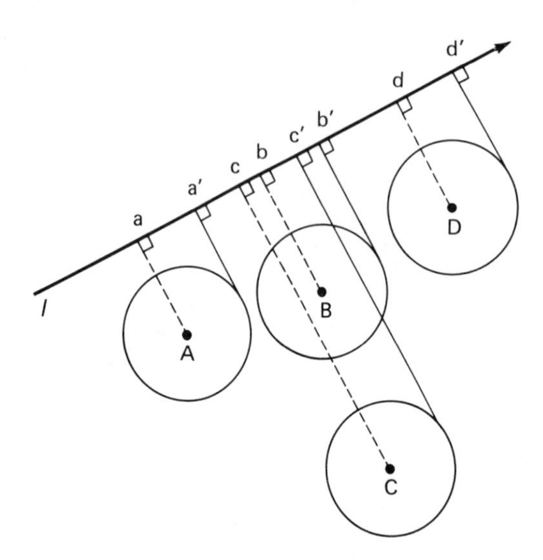

Fig. 9-18. When the objects are circles of the same size, their centers yield the same ordering as their support points. The line-sweep heuristic gives a valid ordering for this set.

He began with a problem: Can all members of a set of nonintersecting rectangles in the plane with their edges parallel to the *x* and *y* axes (such rectangles are called *isothetic*) be translated in the same direction by some common vector to a final destination, subject to the constraints that the rectangles are moved only one at a time, and that no collisions occur? (A collision occurs whenever the interiors of two rectangles intersect.)

The answer is yes.[48] There always exists (for any direction) at least one order in which the polygons can be moved to permit such a translation. This property holds for every finite set of convex polygons, and the ordering can be determined by computation. We say that convex polygons in the plane exhibit the translation-ordering property.

Efficient computation of translation ordering is important for a robot who is given the task of separating such polygons. A simple, intuitive algorithm which sometimes works is the *line-sweep heuristic* illustrated in Fig. 9-17. The vector *l* is the desired direction of translation. Lines perpendicular to *l* are constructed to intersect the rectangles at *support vertices* a, b, c, and d. According to the line-sweep heuristic, the translation ordering is the order of the projections of the support vertices on *l*. The algorithm fails in this particular case, since the method requires D to move second, even though it is blocked by C.

But is there a class of objects for which the line-sweep heuristic always works? It works if the objects are all circles of the same size, and with a slight modification (use of centers instead of support points) for any set of circles of arbitrary sizes (see Fig. 9-18).[49] In three dimensions, this method is called the *plane-sweep heuristic*; it works for sets of spheres and for some sets of isothetic polyhedra.

This problem can be generalized[50] for other types of polygons, and for motions other than simple translations. When the convexity constraint is relaxed, there results a class of problems concerning interlocking polygons. It becomes interesting to ask whether a collection of polygons is "movably separable" in a specified sense.

In three-dimensional space, four isothetic rectangular polyhedra can be arranged so that no translation ordering exists for some directions. Such an arrangement is shown in Fig. 9-19.[51] Some sets of convex polyhedra interlock in all directions, with no translation ordering in any direction. Such a configuration can be built from 12 long, flat, very-thin sticks. The first step is to construct a set of three interlaced sticks A, B, and C (a *triplet*) on the *xy* plane as shown in Fig. 9-20, with their lengths parallel to α, β, and γ directions. Define a "hole" at the center of the configuration, and three "overlap" regions where the sticks touch each other. Three more triplets are constructed (see Fig. 9-21) on planes perpendicular to the *xy* plane, planes individually parallel to the α, β, and γ directions. The key feature of the final configuration is that each new triplet added embraces an overlap region of two sticks in the original triplet; each overlap region of the original triplet lies in the central hole of an embracing triplet.

Inspection of Fig. 9-21 reveals that there is no direction in which more than two of the dozen sticks can be translated. Even though each polyhedral stick can be individually translated away from the configuration without disturbing the others, there exists no translation ordering in any direction.

Are there sets of polyhedra in which no member can be moved out without disturbing the others? If the constraint that the polyhedra be rectangular solids is relaxed, a configuration of twelve convex polyhedra can be constructed in which no polyhedron can be translated in any direction without disturbing the others.[52] K. A. Post found an example of six convex polyhedra that interlock in such a way that no one can be moved without disturbing the others.[53]

Toussaint discussed problems which arise in generalizing results from polygons in the plane to polyhedra in 3-space. Star-shaped polygons generalize in a straightforward way to star-shaped polyhedra. Any two star-shaped polygons can be separated with a single translation,[54] and this property also holds for star-shaped polyhedra in 3-space.[55] The situation is more complicated with *monotonic* figures. (A polygon is monotone if there exist two extreme vertices in some direction connected by two polygonal chains in which the vertices of the chain occur in the same order as their projections onto a line in that direction.) Any two monotone polygons can be separated with a single translation in at least one direction.[56]

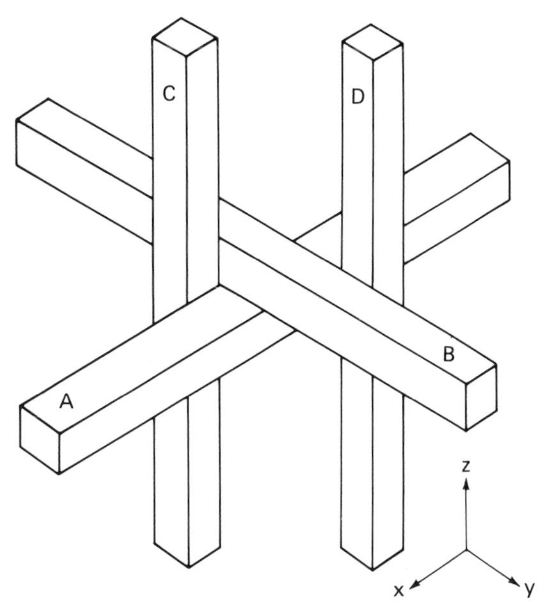

Fig. 9-19. A set of isothetic convex polyhedra that does not allow a translation ordering in the direction $x + y$. This example was discussed by Guibas and Yao.[51]

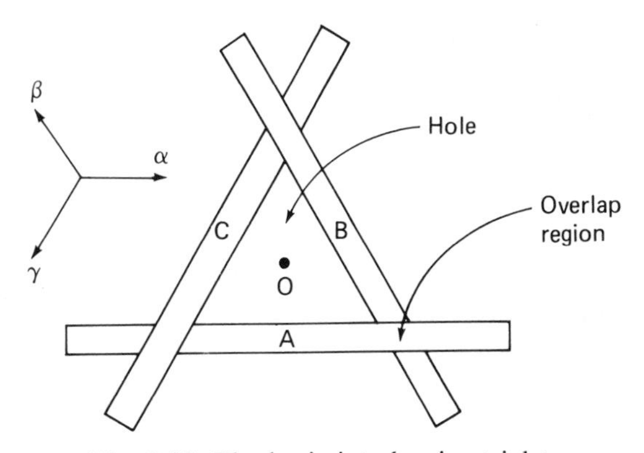

Fig. 9-20. The basic interleaving triplet.

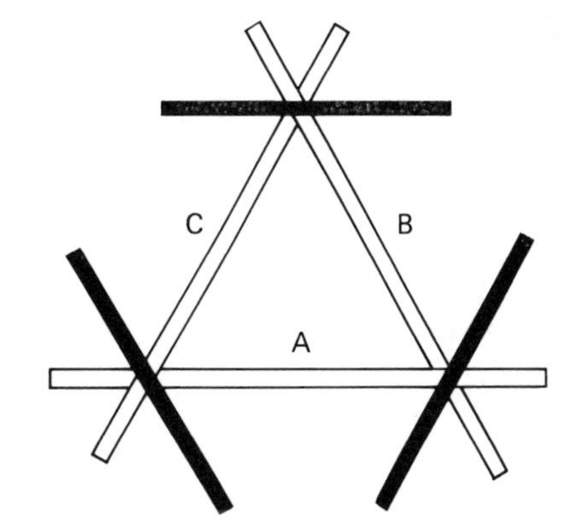

Fig. 9-21. Three more triplets added to the basic one.

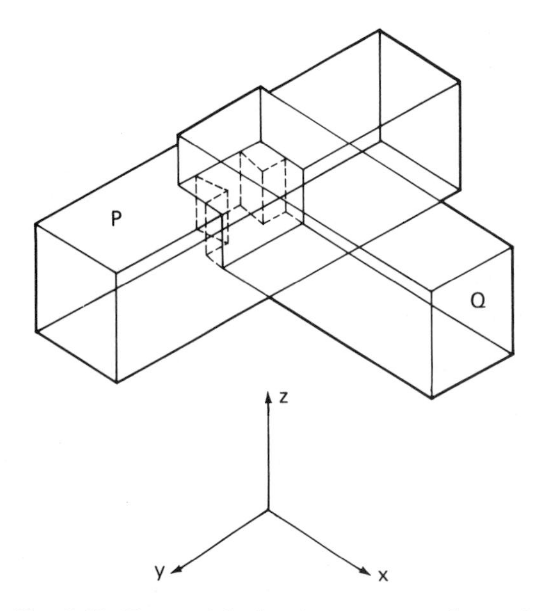

Fig. 9-22. Two polyhedra (components of an ari-kake joint) strongly monotonic with respect to PL(z). The only way to separate this pair is to translate either P or Q in the z direction.

The property of monotonicity does not generalize straightforwardly and uniquely to three dimensions. Toussaint defined a polyhedron as *weakly monotonic* in a direction if the intersection (with the polyhedron) of each plane perpendicular to that direction is a simple polygon, a line segment, or a point. These weakly monotonic polyhedra can be classified in terms of the properties of the polygons of intersections. It turns out that two polyhedra weakly monotonic with respect to a common direction can interlock under all motions. Further, a polyhedron is *strongly monotonic* with respect to a direction PL(l) if the polygons which result from the intersection of planes parallel to l are all monotonic in a direction orthogonal to l. Figure 9-22 illustrates two polyhedra, P and Q, which are strongly monotonic with respect to PL(z). It is known in Japanese carpentry as the ari-kake joint,[57] a dovetail which can be separated only by a translation in the z direction. Such polyhedra are always separable if they share a common direction of monotonicity. It is an open question whether they are necessarily separable when they do not share such a common direction.

Polyhedron Theory Accommodates Changing Expectations

What can we expect when time and change are included in polyhedron theory? What are the new questions, and what might be the form of the answers?

We have seen that static polyhedral units probably do not suffice as bases for chemical modeling of the *dynamic* aspects of solution structure. More needs to be known about the transformational properties of three-dimensional arrays of nonidentical polyhedra. Some important questions are: What local transformations can be achieved without disrupting long-range structure? In an array, can a rotation of one polyhedron be achieved by interactions only with its nearest neighbors? Under what conditions can a polyhedron migrate through a space-filling array of transformable polyhedra?

We have looked at skeletal models for transformations of the polyhedra which serve as models for plant cells in growing tissue. Their detachable, rigid tetrahedral connectors and flexible tubing model both the individual plant cells and the transformations of individual cells, as well as the packing of cells into large arrays and transformations within arrays. These arrays have long-range patterns which can be seen, for instance, as paths along puckered surfaces. Their short-range transformational possibilities are seen in the neighbor-switching operations which produce dislocation, creation, or annihilation of cells. Such tangible models are suggestive, but general conclusions seem elusive. In this fundamental research further collaboration between botanists and geometers will surely be fruitful.

Nineteenth-century scientists effectively used billiard-ball models to discuss gases, without necessarily abandoning belief in an aether that pervaded the space in which those billiard balls moved. Since no model of reality is complete, realists should notice that complementary models, although inconsistent in detail, may yield complementary information about the natural world. Thus a chemist, or a botanist, or a biophysicist may see in Toussaint's dynamic computational geometry applications beyond robotics to more general

Fig. 9-23. A mechanical model illustrating rigidity in a $T = 3$ icosahedral lattice, built of 90 identical pieces of $\frac{1}{32}$-inch-thick sheet aluminum bent to represent dimers connected by pentamer/hexamer bonds. The diameter along the threefold axis is 11.5 inches, and the model weighs 2 pounds. Its shape approximates a truncated Platonic icosahedron, and it supports without distortion a 32-pound plaster bust of Plato. Model made by Charles Ingersoll, Sr. Photograph by William Saunders.

problems about how objects move and how they become bound to one another. Biochemists are concerned about how complex, stable structures are assembled by bringing together (in the proper orientations) individual molecules which one might have expected to be moving randomly in solution. Geometers with various perspectives may see beyond these qualitative connections.

One central fact of biology is that living systems build structures in the midst of apparent molecular chaos. Biochemical explanations of such structure creation are, at best, incomplete and inadequate. Figure 9-23, in which a representation of Plato is held before us by a mechanical model of a chemical system, may well be an appropriate image for an appropriate strategy for future symbiotic progress in natural science, in engineering and in mathematics. It is clear that many fields of inquiry are enriched by both the results and the questions from other disciplines. It is also clear that the possibilities for learning from the work of others are vast.

Notes

[1] *Plato IX,* R. G. Bury, trans., Loeb Classical Library (Cambridge, Mass.: Harvard University Press, 1929), p. 127.

[2] Ibid., p. 133.

[3] Johannes Kepler, *Mysterium Cosmographicum (Tübingen, 1596; Frankfurt, 1621).* An English translation *Mysterium Cosmographicum: The Secret of the Universe* was prepared by A. M. Duncan (New York: Arbaris Books, 1981.)

[4] Titus Lucretius Carus, *De Rerum Natura:* book I, lines 330–575; book II, lines 85–308; see Cyril Bailey, trans., *Lucretius on the Nature of Things* (Oxford: Oxford University Press, 1910), pp. 38–45, 68–75.

[5] William John Macquorn Rankine, "On the Hypothesis of Molecular Vortices, or Centrifugal Theory of Elasticity, and Its Connexion with the Theory of Heat," *Philosophical Magazine,* ser. 4, *10* (1855):411.

[6] H. Helmholtz, "On Integrals of the Hydrodynamical Equations, Which Express Vortex-Motion," *Philosophical Magazine,* ser. 4, *33* suppl. (1867):485; "from Crelle's *Journal,* LV (1858), kindly communicated by Professor Tait."

[7] Sir William Thomson, "On Vortex Atoms," *Philosophical Magazine,* ser. 4, *34* (1867):15.

[8] Cargill Gilston Knott, *Life and Scientific Work of Peter Guthrie Tait* (Cambridge, England: Cambridge University Press, 1911), p. 68.

[9] P. G. Tait, *Lectures on Some Recent Advances in Physical Science* (London, 1876), p. 291.

[10] Peter Guthrie Tait, *Scientific Papers,* 2 vols. (Cambridge: Cambridge University Press, 1898), I, pp. 270–347; papers originally published 1867–85.

[11] J. C. Maxwell, "On Physical Lines of Force. Part II.—The Theory of Molecular Vortices applied to Electric Currents." *Philosophical Magazine,* ser. 4, *21* (1861):281–91.

[12] Loyd S. Swenson, Jr., "The Michelson-Morley-Miller Experiments before and after 1905," *Journal for the History of Astronomy 1* (1970):56–78.

[13] Joseph Le Bel, *Bulletin de la Société Chimique* [Paris], *22* (November 1874), 337–47.

[14] J. H. van 't Hoff, *Voorstel tot Uitbreiding der Structuur-Formules,* Utrecht (September 1874). J. H. van 't Hoff gives credit to earlier investigators: "In my 'Dix Années dans l'Histoire d'une Théorie' I mentioned Gaudin and his 'Architecture du Monde' (1873); then Meyerhoffer in his 'Stereochemie' added Paterno [*Giorn. di Scienze Naturali ed Econ.* vol. v., Palermo; *Gazz. Chim.* 1893, 35], who in 1869 proposed to explain isomeric bromethylenes by a tetrahedral grouping round carbon; and Rosenstiehl [*Bull. Soc. Chim. 11,* 393], who in the same year represented benzene by six tetrahedra; and now Eiloart, in his 'Guide to Stereochemistry,' goes back to Swedenborg's 'Prodromus Principiorum Rerum Naturalium sive Novorum Tentaminum Chymicam et Physicam Experimentalem geometrice explicandi' [Jan Osterwyk, Amsterdam, 1721]."—J. H. van 't Hoff, *The Arrangement of Atoms in Space,* 2nd ed. (London: Longmans, Green, 1898), p. 1.

[15] van 't Hoff, *Arrangement of Atoms in Space,* p. 8.

[16] A. Werner, "Beitrag zur Konstitution anorganischer Verbindungen," *Zeitschrift für anorganische Chemie 3* (1893):267–330.

[17] G. N. Lewis, *Valence and The Structure of Atoms and Molecules* (New York: The Chemical Catalog Company, 1923), p. 29.

[18] G. N. Lewis, "The Atom and the Molecule," *Journal of the American Chemical Society 38* (1916):762–85.

[19] H. W. Kroto, J. R. Heath, S. C. O'Brien, R. F. Curl, and R. E. Smalley, "C_{60}: Buckminsterfullerene," *Nature 318* (1985):162–63.

[20] Donald L. D. Caspar. "Dynamic Models of Icosahedral Viruses." Shaping Space Conference.

[21] F. H. C. Crick and J. D. Watson, "Structure of Small Viruses," *Nature 177* (1956):473–75.

[22] D. L. D. Caspar and A. Klug. "Physical Principles in the Construction of Regular Viruses." *Cold Spring Harbor Symposia on Quantitative Biology 27* (1962):1–24.

[23] I. Rayment, T. S. Baker, D. L. D. Caspar, and W. T. Murakami, "Polyoma Virus Capsid Structure at 22.5 Å Resolution," *Nature 295* (1982):110–15.

[24] See J. D. Bernal, "A Geometrical Approach to the Structure of Liquids, *Nature 183* (1959):141–47; "The Structure of Liquids," *Scientific American 203* (1960):124–34. See also J. L. Finney, "Random Packings and the Structure of Simple Liquids. I. The Geometry of Random Close Packing," *Proceedings of the Royal Society* [London], *sec. A, 319* (1970):479–93.

[25] Stephen Hales, *Vegetable Staticks,* London: W. & J. Innys, 1727. Reprint (Canton, Mass.: Watson, Neale, 1969).

[26] Sir William Thomson. "On the Division of Space with Minimum Partitional Area." *The London, Edinburgh, and Dublin Philosophical Magazine and Journal of Science,* ser. 5, *24* (1887):503–14.

[27] Joseph A. F. Plateau, *Statique expérimentale et théorique des liquides soumis aux seules forces moléculaires* (Paris: Gand, 1873).

[28] Two excellent contemporary monographs on this subject are C. Isenberg, *The Science of Soap Bubbles and Soap Films* (Tieto, 1978), and S. Hildebrandt and A. Tramba, *Mathematics and Optimal Form,* Scientific American Library (San Francisco: Freeman, 1985).

[29] W. Barlow, "Probable Nature of the Internal Symmetry of Crystals," *Nature 29* (1883–84):186–88, 205–7, 404.

[30] Sir William Thomson, "Molecular Constitution of Matter," *Proceedings of the Royal Society of Edinburgh 16* (1888–89):693–724.

[31] Linus Pauling, "Interatomic Distances and Their Relation to the Structure of Molecules and Crystals," ch. 5 in *The Nature of the Chemical Bond* (Ithaca, N.Y.: Cornell University Press, 1939).

[32] Ralph Erickson, "Polyhedral Cell Shapes," Shaping Space Conference. See also Ralph O. Erickson, "Polyhedral Cell Shapes," in Grzegorz Rozenberg and Arto Salomaa, eds, *The Book of L* (Berlin: Springer-Verlag, 1986), pp. 111–124.

[33] James W. Marvin. "Cell Shape Studies in the Pith of *Eupatorium purpureum,*" *American Journal of Botany 26* (1939):487–504. A survey of other similar work is given in Edwin B. Matzke and Regina M. Duffy, "The Three-Dimensional Shape of Interphase Cells within the Apical Meristem of *Anacharis densa,*" *American Journal of Botany 42* (1955):937–45.

[34] Thomson, "On the Division of Space with Minimum Partitional Area."

[35] Edwin B. Matzke, "Volume-Shape Relationships in Lead Shot and Their Bearing on Cell Shapes," *American Journal of Botany 26* (1939): 288–95.

[36] James W. Marvin, "The Shape of Compressed Lead Shot and its Relation to Cell Shape," *American Journal of Botany 26* (1939):280–88.

[37] E. B. Matzke, "The Three-dimensional Shape of Bubbles in Foam—An Analysis of the Rôle of Surface Forces in Three-dimensional Cell Shape Determination," *American Journal of Botany 33* (1946):58–80.

[38] In addition to the references in notes 33, 35, 36, and 37, see F. T. Lewis, "A Geometric Accounting for Diverse Shapes of 14-Hedral Cells: The Transition from Dodecahedra to Tetrakaidecahedra," *American Journal of Botany 30,* (1943):74–81, and F. T. Lewis, "The Analogous Shapes of Cells and Bubble," *Proceedings of the American Academy of Arts and Sciences, 77* (1949):147–86.

[39] M. Brückner, *Vielecke and Vielfläche* (Leipzig: Teubner, 1900).

[40] Thomson, "On the Division of Space with Minimum Partitional Area."

[41] H. S. M. Coxeter, *Introduction to Geometry,* (New York: John Wiley, 1961), p. 162.

[42] R. E. Williams, "Space-Filling Polyhedron: Its Relation to Aggregates of Bubbles, Plant Cells, and Metal Crystallites," *Science 161* (1968):276–77. See also R. E. Williams, *The Geometrical Foundation of Natural Structure,* 2nd ed. (New York: Dover Publications, 1979).

[43] J. E. Morrall and M. F. Ashby, "Dislocated Cellular Structures," *Acta Metallurgica 22* (1974):567–75.

[44] D. L. D. Caspar, "Movement and Self-Control in Protein Assemblies: Quasi-Equivalence Revisited," *Biophysical Journal 32* (1980):103–35.

[45] D. L. D. Caspar and A. Klug, "Physical Principles in the Construction of Regular Virus Particles," *Cold Spring Harbor Symposia in Quantitative Biology 27* (1962):1–24.

[46] Godfried T. Toussaint. "On Translating a Set of Polyhedra." Shaping Space Conference; McGill University School of Computer Science Technical Report No. SOCS-84.6, 1984. See also Godfried T. Toussaint, "Movable Separability of Sets," in G. T. Toussaint, ed., *Computational Geometry,* (Amsterdam: North-Holland, 1985), pp. 335–75.

[47] Godfried T. Toussaint, "Shortest Path Solves Translation Separability of Polygons," McGill University School of Computer Science Technical Report No. SOCS-85.27, 1985.

[48] Binay K. Bhattacharya and Godfried T. Toussaint, "A Linear Algorithm for Determining Translation Separability of Two Simple Polygons," McGill University School of Computer Science Technical Report No. SOCS-86.1, 1986.

[49] G. T. Toussaint, "On Translating a Set of Spheres," Technical Report SOCS-84.4, School of Computer Science, McGill University, March 1984.

[50] G. T. Toussaint, "The Complexity of Movement," *IEEE International Symposium on Information Theory,* St. Jovite, Canada, September 1983; J.-R. Sack and G. T. Toussaint, "Movability of Objects," *IEEE International Symposium on Information Theory,* St. Jovite, Canada, September 1983. G. T. Toussaint and J.-R. Sack, "Some New Results on Moving Polygons in the Plane," *Proceedings of the Robotic Intelligence and Productivity Conference,* Detroit: Wayne State University (1983) pp. 158–63.

[51] L. J. Guibas and Y. F. Yao. "On Translating a Set of Rectangles." *Proceedings of the 12th Annual ACM Symposium on Theory of Computing,* Association for Computing Machinery Symposium on Theory of Computing, *Conference Proceedings, 12* (1980), pp. 154–160.

[52] Robert Dawson, "On Removing a Ball Without Disturbing the Others," *Mathematics Magazine 57,* no. 1 (1984):27–30.

[53] K. A. Post, "Six Interlocking Cylinders with Respect to All Directions," unpublished paper, University of Eindhoven, The Netherlands, December 1983.

[54] Toussaint, "On Translating a Set of Spheres."

[55] Toussaint, "On Translating a Set of Polyhedra."

[56] Godfried T. Toussaint and Hossam A. El Gindy, "Separation of Two Monotone Polygons in Linear Time," *Robotica 2* (1984):215–20.

[57] Kiyosi Seike, *The Art of Japanese Joinery* (New York: John Weatherhill, 1977).

10

Polyhedral Molecular Geometries

István Hargittai and Magdolna Hargittai

Professor Coxeter writes: "the chief reason for studying regular polyhedra is still the same as in the times of the Pythagoreans, namely, that their symmetrical shapes appeal to one's artistic sense."[1] The success of modern molecular chemistry affirms the validity of this statement; there is no doubt that aesthetic appeal has contributed to the rapid development of what could be termed polyhedral chemistry.[2] The late Professor Muetterties movingly described his attraction to boron hydride chemistry, comparing it to Escher's devotion to periodic drawings:

> When I retrace my early attraction to boron hydride chemistry, Escher's poetic introspections strike a familiar note. As a student intrigued by early descriptions of the extraordinary hydrides, I had not the prescience to see the future synthesis developments nor did I have then a scientific appreciation of symmetry, symmetry operations, and group theory. Nevertheless, some inner force also seemed to drive me but in the direction of boron hydride chemistry. In my initial synthesis efforts, I was not the master of these molecules; they seemed to have destinies unperturbed by my then amateurish tactics. Later as the developments in polyhedral borane chemistry were evident on the horizon, I found my general outlook changed in a characteristic fashion. For example, my doodling, an inevitable activity of mine during meetings, changed from characters of nondescript form to polyhedra, fused polyhedra, and graphs.
>
> I (and others, my own discoveries were not unique nor were they the first) was profoundly impressed by the ubiquitous character of the three-center relationship in bonding (e.g., the boranes) and nonbonding situations. I found a singular uniformity in geometric relationships throughout organic, inorganic, and organometallic chemistry: The favored geometry in coordination compounds, boron hydrides, and metal clusters is the polyhedron that has all faces equilateral or near equilateral triangles.[3]

Molecular geometry describes the relative positions of atomic nuclei. Although positions may be given by position vectors or coordinates of all nuclei in the molecule, chemists usually give the positions by bond lengths, bond angles, and angles of internal rotation. This second way greatly facilitates the understanding and comparison of various structures. The most qualitative but nonetheless a very important feature of molecular geometry is the shape of the molecule. Polyhedra are especially useful in expressing molecular shapes for molecules with a certain amount of symmetry.

The molecules As_4 and CH_4 both have tetrahedral shapes (Fig. 10-1) and T_d symmetry, but there is an important difference in their structures. In As_4 all nuclei are at vertices of a regular tetrahedron and each edge of this tetrahedron is a chemical bond. Methane has a central carbon atom, with four chemical bonds directed from it to vertices of a tetrahedron

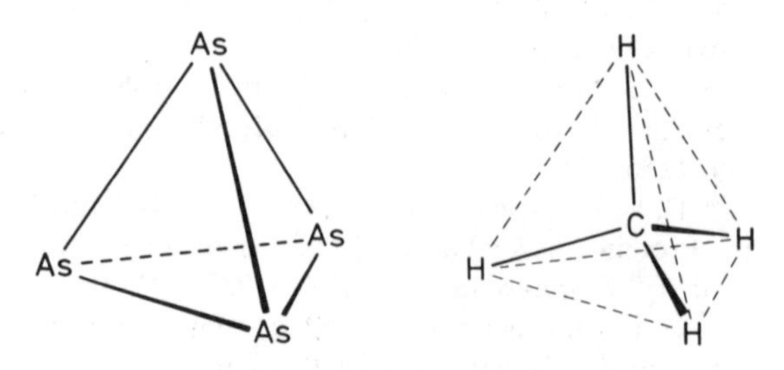

Fig. 10-1. The molecular shapes of As_4 and CH_4.

where the protons are located; no edge is a chemical bond. The As_4 and CH_4 molecules are clear-cut examples of the two distinctly different arrangements. Such distinctions are not always so unambiguous. An interesting example is zirconium borohydride, $Zr(BH_4)_4$. Two independent studies[4] describe its structure by the same polyhedral configuration, but give different interpretations (Fig. 10-2) of the bonding between the central zirconium atom and the four boron atoms at the vertices of a regular tetrahedron. In one interpretation,[5] there are four Zr—B bonds; according to the other,[6] each boron atom is linked to zirconium by three hydrogen bridges, and there is no direct Zr—B bond.

Real molecules ceaselessly perform intramolecular vibrations. In even small-amplitude vibrations, nuclear displacements amount to several percent of the internuclear separations; large-amplitude vibrations may permute atomic nuclei in a molecule. In describing a molecule by a highly symmetric polyhedron, we refer to the hypothetical motionless molecule. The importance and consequences of intramolecular motion in the polyhedral description of molecules are discussed in the final section.

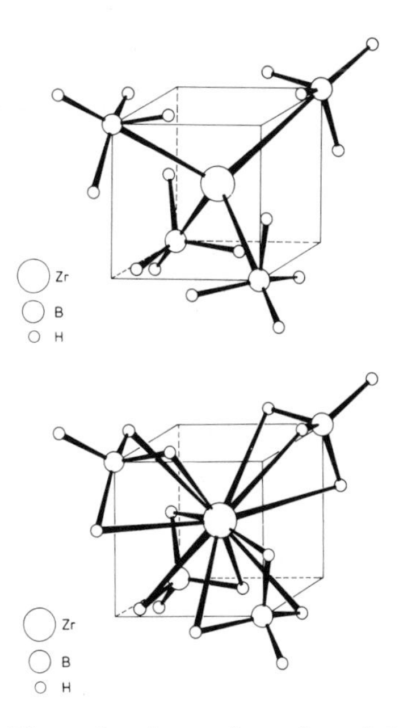

Fig. 10-2. The molecular configuration of zirconium borohydride, $Zr(BH_4)_4$, in two interpretations but described by the same polyhedral shapes.

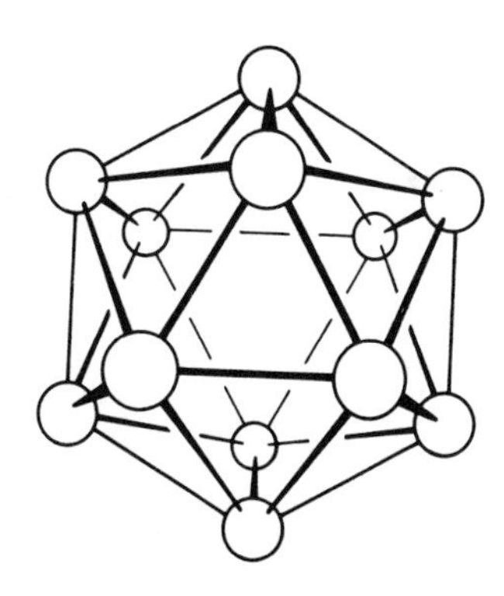

Fig. 10-3. The regular icosahedral configuration of the $B_{12}H_{12}^{2-}$ ion. Only the boron skeleton is shown.

Boron Hydride Cages

All faces of boron hydride polyhedra are equilateral or nearly equilateral triangles. Boron hydrides with a complete polyhedral shape are called *closo* boranes (Greek *closo:* 'closed'). One of the most symmetrical and most stable of the polyhedral boranes is the $B_{12}H_{12}^{2-}$ ion; its regular icosahedral configuration is shown in Fig. 10-3. Table 10-1 presents structural systematics of $B_nH_n^{2-}$ *closo* boranes and the related $C_2B_{n-2}H_n$ carboranes in which some boron sites are taken by carbon atoms.[7] The so-called quasi-*closo* boranes are derived from the *closo* boranes by replacing a framework atom with a pair of electrons.

Figure 10-4 shows the systematics of borane polyhedral fragments obtained by removing one or more polyhedral sites from *closo* boranes.[8] Since all faces of the polyhedral skeletons are triangular, they may be called deltahedra.[9] The derived deltahedral fragments are the tetrahedron, trigonal bipyramid, octahedron, pentagonal bipyramid, bisdisphenoid, symmetrically tricapped trigonal prism, bicapped square antiprism, octadecahedron, and icosahedron. Only the octadecahedron is not a convex polyhedron.

A *nido* (nestlike) boron hydride is derived from a *closo* borane by removal of one skeleton atom. An *arachno* (weblike) boron hydride is derived from a *closo* borane by removal of two adjacent skeletal atoms. In either case, if the starting *closo* borane is not a regular polyhedron, then the atom removed is the one at a vertex with the highest connec-

Table 10-1. Structural Systematics of $B_nH_n^{2-}$ *closo* Boranes and $C_2B_{n-2}H_n$ *closo* Carboranes after Muetterties[3]

Polyhedron and point group	Boranes	Dicarboranes
Tetrahedron, T_d	(B_4Cl_4)*	—
Trigonal bipyramid, D_{3h}	—	$C_2B_3H_5$
Octahedron, O_h	$B_6H_6^{2-}$	$C_2B_4H_6$
Pentagonal bipyramid, D_{5h}	$B_7H_7^{2-}$	$C_2B_5H_7$
Dodecahedron (triangulated), D_{2d}	$B_8H_8^{2-}$	$C_2B_6H_8$
Tricapped trigonal prism, D_{3h}	$B_9H_9^{2-}$	$C_2B_7H_9$
Bicapped square antiprism, D_{4d}	$B_{10}H_{10}^{2-}$	$C_2B_8H_{10}$
Octadecahedron, C_{2v}	$B_{11}H_{11}^{2-}$	$C_2B_9H_{11}$
Icosahedron, I_h	$B_{12}H_{12}^{2-}$	$C_2B_{10}H_{12}$

* No boron hydride

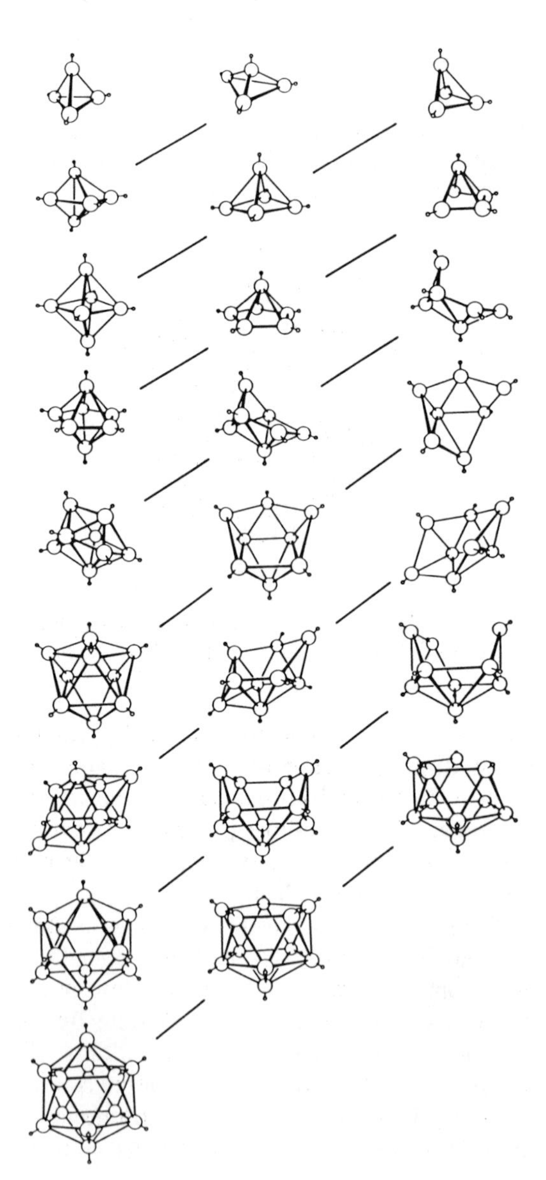

tivity. Complete *nido* and *arachno* structures are shown[3] together with starting boranes in Fig. 10-5.

Polycyclic Hydrocarbons

Fundamental polyhedral shapes are realized among polycyclic hydrocarbons where the edges are C—C bonds and there is no central atom. Such bond arrangements may be far from the energetically most advantageous,[10] and particular arrangements may be too unstable to exist. Yet the fundamental character of these shapes, their high symmetry, and their aesthetic appeal make them an attractive and challenging playground[11] for organic chemists. These substances also have practical importance as building blocks for such natural products as steroids, alkaloids, vitamins, carbohydrates, and antibiotics.

Tetrahedrane (Fig. 10-6a) is the simplest regular polycyclic hydrocarbon. The synthesis of this highly strained molecule may not be possible. Its derivative, tetra-*tert*-butyltetrahedrane (Fig. 10-6b), is amazingly stable,[12] perhaps because the substituents help "clasp" the molecule together. Cubane (Fig. 10-6c) has

Fig. 10-4. *Closo, nido,* and *arachno* boranes. The genetic relationships are indicated by diagonal lines. Reprinted with permission from Ralph W. Rudolph, "Boranes and Heteroboranes: A Paradigm for the Electron Requirements of Clusters?" *Accounts of Chemical Research, 9* (1976):446–52. Copyright 1976 American Chemical Society.

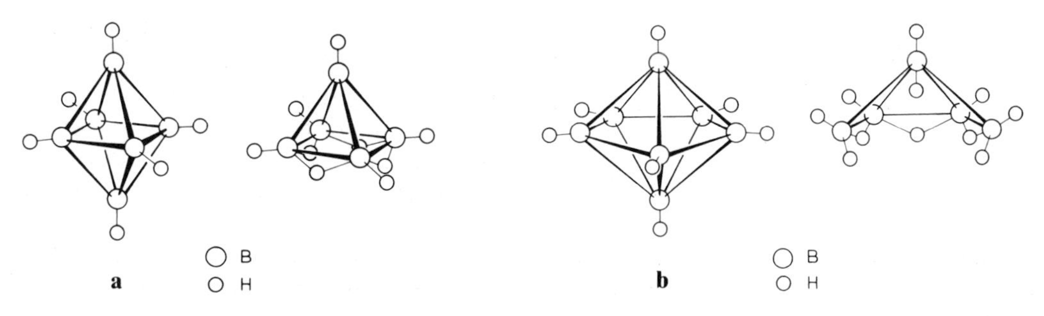

Fig. 10-5. Examples of *closo*/*nido* and *closo*/*arachno* structural relationships. (a) *Closo*-$B_6H_6^{2-}$ and *nido*-B_5H_9. (b) *Closo*-$B_7H_7^{2-}$ and *arachno*-B_5H_{11}.

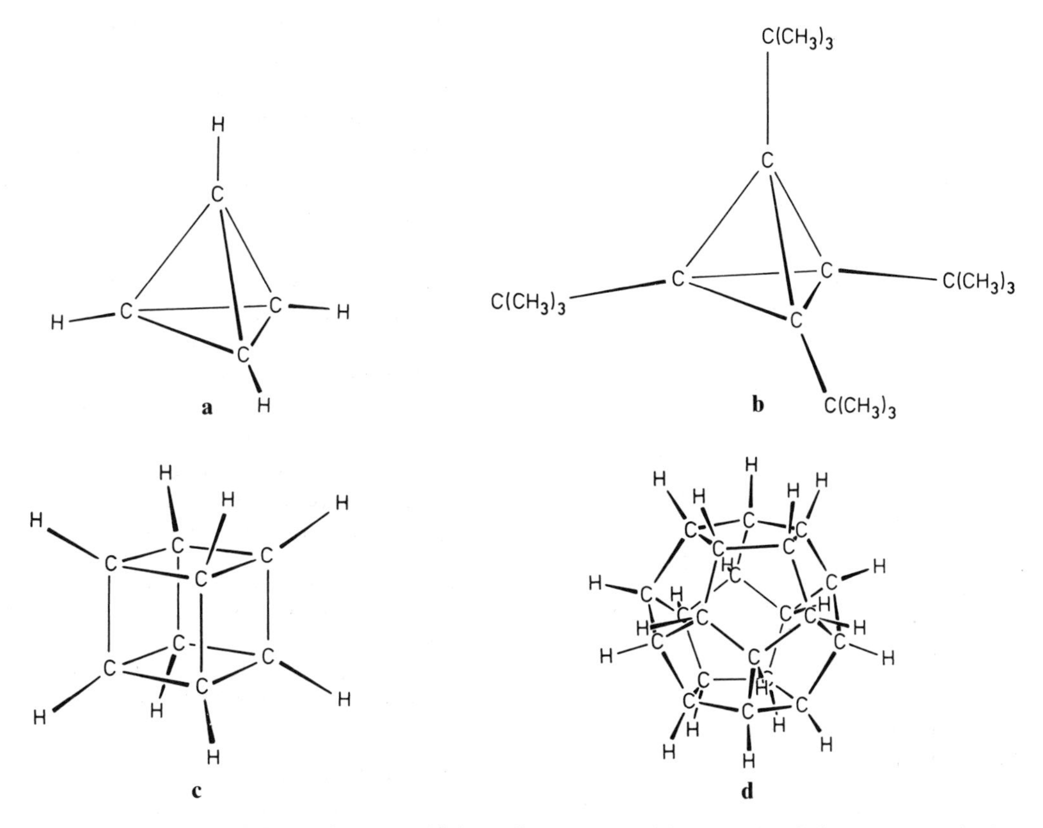

Fig. 10-6. (a) Tetrahedrane, $(CH)_4$. It has very high strain energy and has not (yet?) been prepared. (b) Tetra-*tert*-butyltetrahedrane, $\{C[C(CH_3)_3]\}_4$. (c) Cubane, $(CH)_8$. (d) Dodecahedrane, $(CH)_{20}$.

been known for some time.[13] Dodecahedrane (Fig. 10-6d), prepared only recently,[14] was predicted[15] two decades ago to have "almost ideal geometry . . . practically a miniature ball bearing!" Its carbanion was predicted to be stabilized by a "rolling charge" effect, delocalizing the extra electron over twenty equivalent carbon atoms.

In the $(CH)_n$ convex polyhedral hydrocarbons, each carbon atom is bonded to three other carbon atoms; the fourth bond is directed externally to a hydrogen atom. Around the all-carbon polyhedron is thus a similar polyhedron whose vertices are protons. The edges of the inner polyhedron are C—C bonds. Because four bonds would meet a carbon atom at the vertices of an octahedron, and five in an icosahedron, the enveloping-polyhedra structure is not possible for these Platonic solids. For similar reasons, only seven of the 14 Archimedean polyhedra can be considered in the $(CH)_n$ polyhedral series.

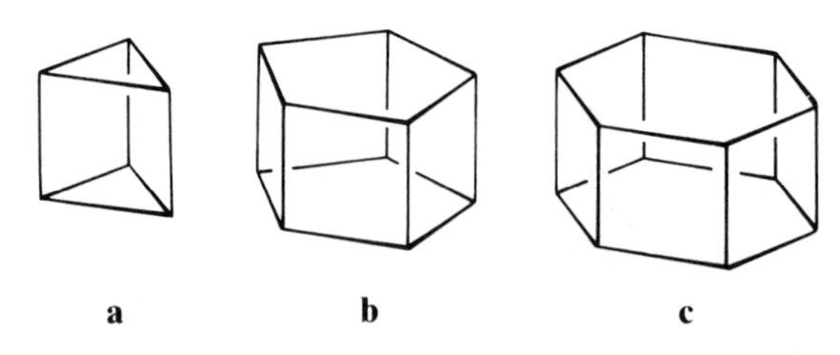

Fig. 10-7. (a) Triprismane, $(CH)_6$. (b) Pentaprismane, $(CH)_{10}$. (c) Hexaprismane, $(CH)_{12}$, not yet prepared.

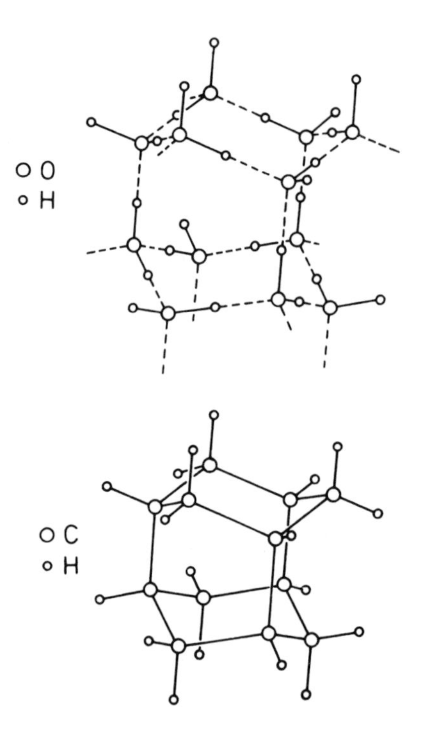

Fig. 10-8. Ice crystal structure (*top*) and the iceane hydrocarbon, $C_{12}H_{18}$ (*bottom*).

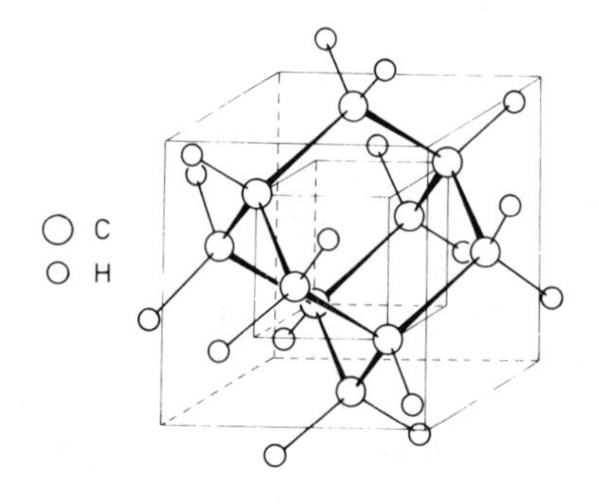

Fig. 10-9. Adamantane, $C_{10}H_{16}$ or $(CH)_4(CH_2)_6$.

Cubane may also be described as tetraprismane, composed of eight identical methine units arranged at the corners of a regular tetragonal prism with O_h symmetry, bound into two parallel four-membered rings conjoined by four four-membered rings. Triprismane, $(CH)_6$, has D_{3h} symmetry,[16,17] and pentaprismane, $(CH)_{10}$, has D_{5h} symmetry.[18] The quest for pentaprismane is a long story.[18] Hexaprismane, $(CH)_{12}$, the face-to-face dimer of benzene, has yet to be prepared. The molecular models are shown in Fig. 10-7.

Structural varieties become virtually endless if one reaches beyond the most symmetrical convex polyhedra. There are 5,291 isomeric tetracyclic structures of $C_{12}H_{18}$ hydrocarbons; only a few are stable.[19] One is iceane (Fig. 10-8), which may be visualized as two chair cyclohexanes connected by three axial bonds; it can also be described as three fused boat cyclohexanes. The name *iceane* was proposed by Fieser[20] almost a decade before its preparation. Considering water molecules in an ice crystal, he noticed three *vertical* hexagons with boat conformations. The emerging *horizontal* $(H_2O)_6$ units possess three equatorial hydrogen atoms and three equatorial hydrogen bonds available for horizontal building. He noted that this structure

> suggests the possible existence of a hydrocarbon of analogous conformation of the formula $C_{12}H_{18}$, which might be named "iceane." The model indicates a stable strain-free structure analogous to adamantane and twistane. "Iceane" thus presents a challenging target for synthesis.

The challenge was met.[21]

The adamantane molecule, $C_{10}H_{16}$, and the diamond crystal are closely related. Diamond has even been called the "infinite adamantylogue of adamantane."[22] The high symmetry of adamantane is emphasized when its structure is described[23] by four imaginary cubes packed one inside the other; two are shown in Fig. 10-9.

Structures with a Central Atom

Tetrahedral AX_4 molecules belong to the point group T_d. Successive substitution of the X ligands by B ligands leads to other tetrahedral configurations of the following symmetries:

AX_4	AX_3B	AX_2B_2	AXB_3	AB_4
T_d	C_{3v}	C_{2v}	C_{3v}	T_d

If each substitution introduces a new kind of ligand, then the resulting tetrahedral configurations will have the following symmetries:

AX_4	AX_3B	AX_2BC	$AXBCD$
T_d	C_{3v}	C_s	C_1

Important structures may be derived by joining two tetrahedra or two octahedra at a common vertex, edge, or face (Fig. 10-10). Ethane ($H_3C—CH_3$), ethylene ($H_2C=CH_2$) and acetylene ($HC≡CH$) may be derived formally in such a way. Joined tetrahedra are even more obvious in some metal halide structures with halogen bridges.[24]

Complex formation has similar consequences in molecular shape and symmetry.[25] The $H_3N \cdot AlCl_3$ donor–acceptor complex[26] has a triangular antiprismatic shape with C_{3v} symmetry (Fig. 10-11). Complex formation can be viewed as completion of the tetrahedral bond configuration around the central atoms of the donor (NH_3, C_{3v}) and the acceptor ($AlCl_3$, D_{3h}). The structure of the mixed metal–halogen complex potassium tetrafluoroaluminate, $KAlF_4$, can be viewed as formed from KF and AlF_3, with completion of the aluminum tetrahedron. The tetrahedral tetrafluoroaluminate structural unit is relatively rigid, whereas the position of the potassium atom around the AlF_4 tetrahedron is rather loose. The most plausible models are shown in Fig. 10-12; the model with two halogen bridges best approximates the experimental data.[27] The $KAlF_4$ molecule is representative of a class of compounds of growing practical importance: the mixed halides have much higher volatility than the individual metal halides.

The prismatic cyclopentadienyl and benzene complexes of transition metals are reminiscent of the prismanes. Figure 10-13a shows ferrocene, $(C_5H_5)_2Fe$, for which both the barrier to rotation and the free-energy difference between the prismatic (eclipsed) and antiprismatic (staggered) conformations are very small.[28] Figure 10-13b presents a prismatic model with D_{6h} symmetry for dibenzene chromium, $(C_6H_6)_2Cr$. Molecules with multiple

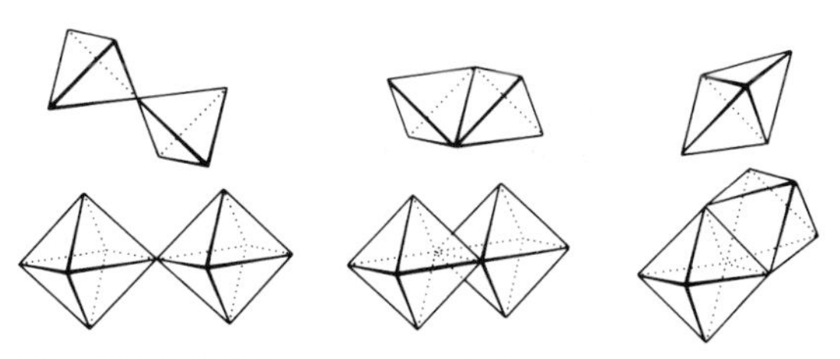

Fig. 10-10. Joining two tetrahedra (and two octahedra) at a common vertex, edge, or face.

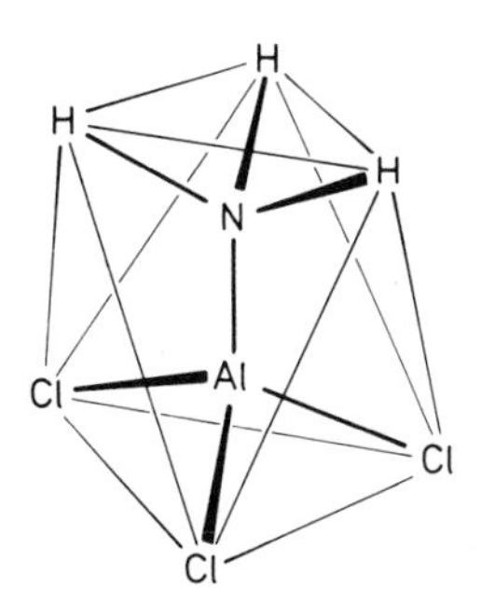

Fig. 10-11. The triangular antiprismatic shape of the $H_3N \cdot AlCl_3$ donor–acceptor complex.

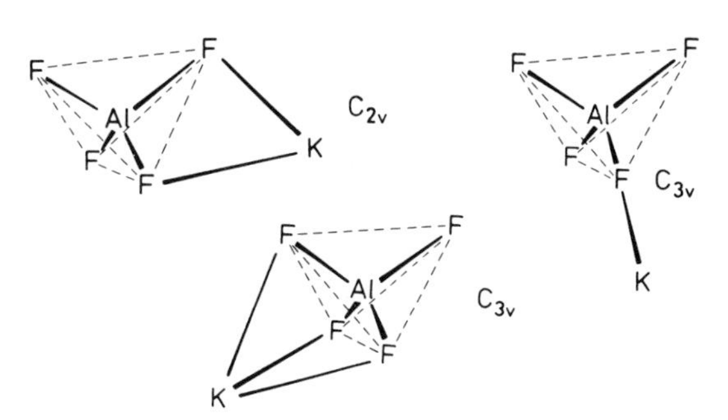

Fig. 10-12. Alternative models of the $KAlF_4$ molecule.

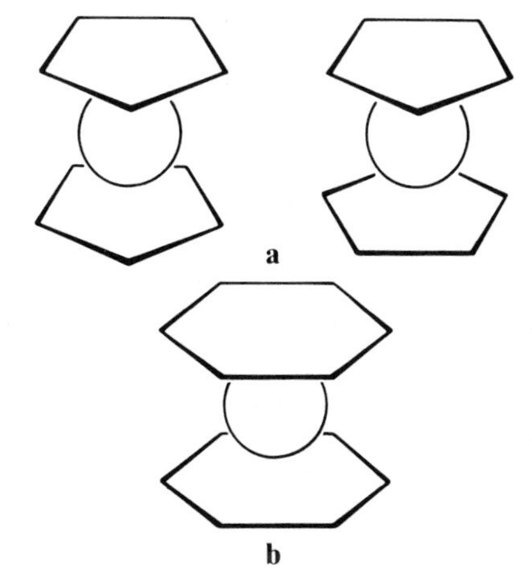

Fig. 10-13. (a) Prismatic (D_{5h}) and antiprismatic (D_{5d}) models of ferrocene. (b) Prismatic model (D_{6h}) of dibenzene chromium.

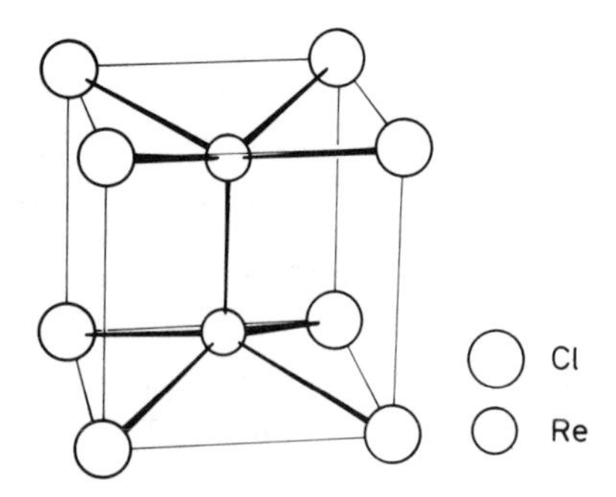

Fig. 10-14. The square prismatic model of the $[Re_2Cl_8]^{2-}$ ion.

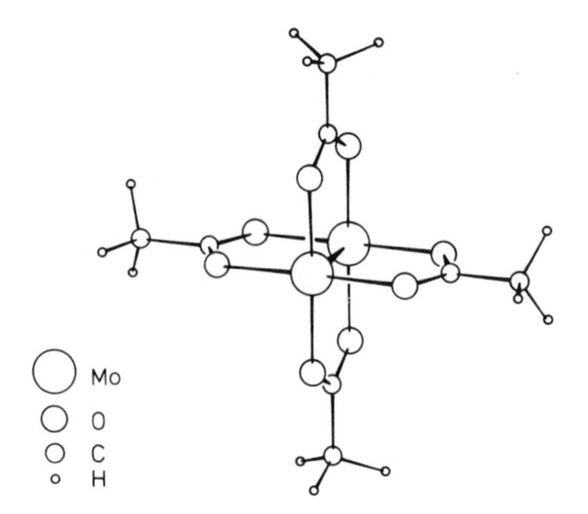

Fig. 10-15. The paddlelike structure of the anhydrous dimolybdenum tetra-acetate, $Mo_2(O_2CCH_3)_4$.

Fig. 10-16. (a) [2.2.2.2]-paddlane, $C_{10}H_{16}$, not yet prepared. (b) [1.1.1]-propellane, C_5H_6.

bonds between metal atoms often have structures with beautiful and highly symmetric polyhedral shapes.[29] One example is the square prismatic $[Re_2Cl_8]^{2-}$ ion[30] which played an important role in the discovery of metal–metal multiple bonds (Fig. 10-14). Another is the paddlelike structure[31] of dimolybdenum tetra-acetate, $Mo_2(O_2CCH_3)_4$ (Fig. 10-15).

There are hydrocarbons called *paddlanes* for their similarity to the shape of riverboat paddles.[32] The most symmetrical, highly strained [2.2.2.2]-paddlane (Fig. 10-16a) has not yet been prepared. The most unusual parent hydrocarbon known is the related [1.1.1]-propellane[33] (Fig. 10-16b) in which interactions between bridgehead carbons have been interpreted[34] by three-center, two-electron orbitals. The hydrocarbon skeleton seems to be electron deficient, while extra electron density is on the outside of the skeleton.

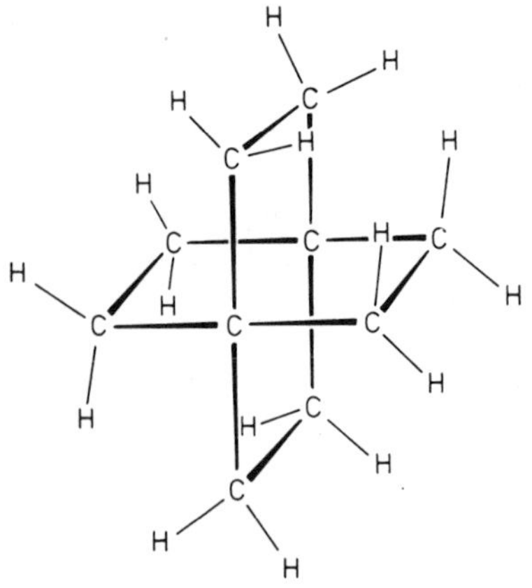

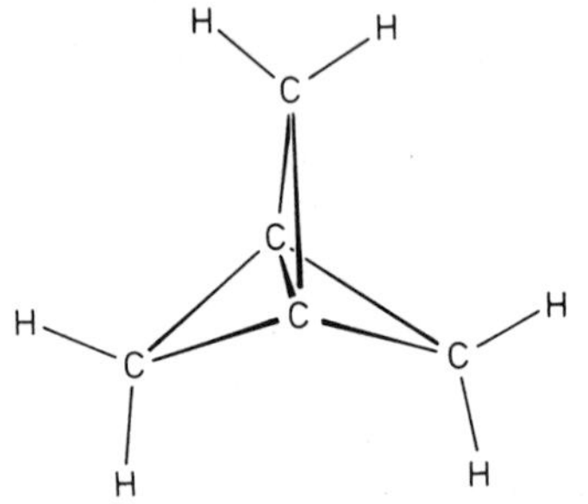

Regularities in Nonbonded Distances

There is no chemical bond between bridgehead carbons of [1.1.1]-propellane, even though the atoms are in a pseudobonding situation with proper bonding geometry. A reverse situation is seen in the ONF_3 molecule (Fig. 10-17), an essentially regular tetrahedron formed by three fluorines and one oxygen, each bonded to the central nitrogen atom. The nonbonded F···F and F···O distances are equal within experimental error.[35]

Certain intramolecular 1,3 separations (the 1,3 label referring to two atoms each bonded to a third) are constant throughout a series of related molecules. The 1,3 distance may remain constant even though bond distances and bond angles in the rest of the molecule change considerably. A controversy between two

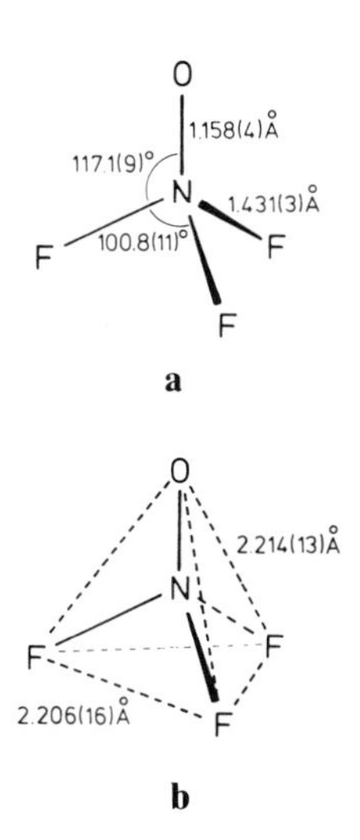

Fig. 10-17. The molecular geometry of ONF_3. (a) Bond lengths and bond angles. (b) Nonbonded distances.

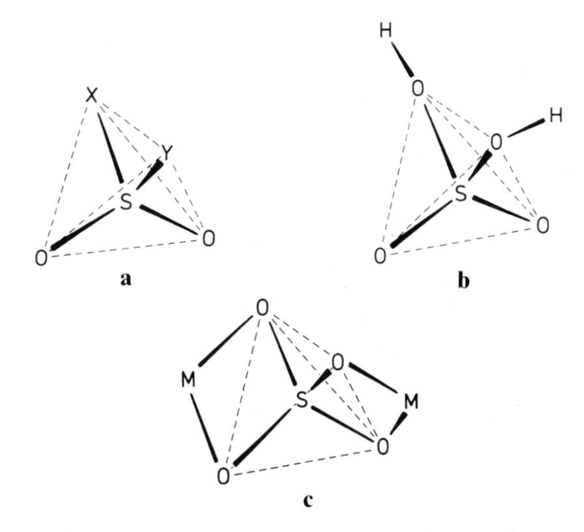

Fig. 10-18. The molecular geometry of (a) sulfones, XSO_2Y, (b) sulfuric acid, H_2SO_4, and (c) metal sulfates, M_2SO_4.

structure determinations of tetrafluoro-1,3-dithietane,

```
      S
    /   \
F2C       CF2
    \   /
      S
```

was settled by considering the F···F nonbonded distances.[36] The mean of the F···F 1,3 distances in 40 molecules containing a CF_3 group was found to be 2.162 Å, with a standard deviation of 0.008 Å!

The O···O nonbonded distances in XSO_2Y sulfones are remarkably constant at 2.48 Å in a series of compounds[37] in which the S=O bond lengths vary by 0.05 Å and the O=S=O bond angles by 5°. Geometric variations in the sulfone series can be visualized (Fig. 10-18a) as if the oxygen ligands were firmly attached to two vertices of the ligand tetrahedron around the sulfur atom, and this central atom were moving along the bisector of the O=S=O angle, depending on the X and Y ligands.[38]

The oxygen atoms in a sulfuric acid molecule, $(HO)SO_2(OH)$, form a nearly regular tetrahedron around the sulfur (Fig. 10-18b). The largest difference between the O···O distances is 0.07 Å, even though the SO bond distances differ by 0.15 Å, and the OSO bond angles[39] by 20°. Structures of alkali sulfate molecules were written in old textbooks as

```
Na—O    O
    \  //
     S
    /  \\
Na—O    O
```

In fact, the SO_4 groups in such molecules are nearly regular tetrahedral,[40] and the metal atoms are located on axes perpendicular to the edges of the tetrahedra; this structure is bicyclic (Fig. 10-18c). Sulfate and tetrafluoroaluminate structures are markedly similar; each has a well-defined tetrahedral nucleus around which atoms occupy relatively loose positions.

The VSEPR Model

Why is a methane molecule tetrahedral, whereas xenon tetrafluoride is planar? Why is ammonia pyramidal rather than planar? Why is water bent, rather than linear?

A simple and successful model,[41] designed to answer just such questions about molecules with a central atom, is based on the following postulate: *The geometry of the molecule is determined by the repulsions among the electron*

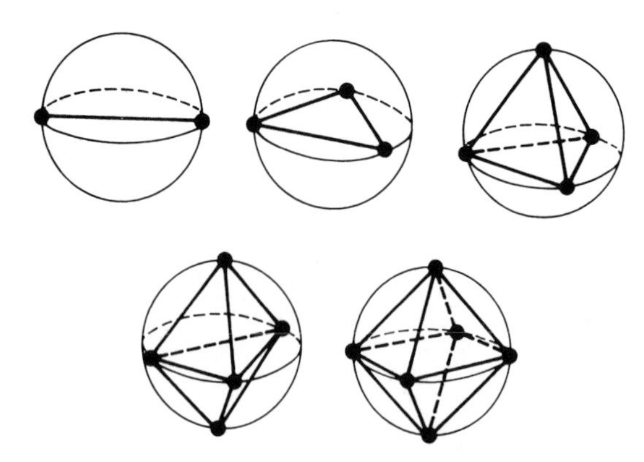

Fig. 10-19. Points-on-a-sphere configurations.

Fig. 10-20. Shapes of groups of balloons.

Fig. 10-21. Walnut clusters drawn by the artist Ferenc Lantos.

pairs in the valence shell of its central atom. We shall illustrate the utility of this valence shell electron pair repulsion (VSEPR) model, showing the importance of the polyhedral description of molecular structure.

If the electron distribution around a central atom has spherical symmetry, then all the electron pairs in its valence shell will be equidistant from the nucleus. Distances among the electron pairs will be maximized in the following arrangements:

Number of Electron Pairs	Arrangement
2	Linear
3	Equilateral triangle
4	Regular tetrahedron
5	Trigonal bipyramid
6	Octahedron

The arrangements are shown in Fig. 10-19 where electron pairs are represented by points on a sphere. For four or more electron pairs, the arrangements are polyhedral. Of the three polyhedra in Fig. 10-19, two are regular. The trigonal bipyramid is not a strongly unique solution; the square pyramidal configuration for five electron pairs is only slightly less advantageous. The space requirement and mutual repulsion of electron pairs are nicely simulated by balloons[42]; a natural simulation is provided by nut clusters on walnut trees.[43] (see Figs. 10-20 and 10-21.)

To predict the bond configuration around a central atom, the number of valence shell electron pairs must be known. The formula of a binary compound AX_n may be given as AX_nE_m, E denoting a lone pair of electrons. For methane (CH_4 or CH_4E_0), ammonia (NH_3 or NH_3E_1), water (OH_2 or OH_2E_2) and hydrogen fluoride (FH or FH_1E_3), $n + m = 4$; accordingly, each has a tetrahedral electron-pair configuration. The corresponding nuclear configurations are tetrahedral, pyramidal, bent, and linear. The ideal tetrahedral bond angles of 109°28′ occur only when all electron pairs are equivalent. A double bond or a lone electron pair requires more space than a single bond, repelling neighboring electron pairs more strongly. A bond to a strongly electronegative ligand such as fluorine has less electron density and repels electron pairs more weakly than a bond to a less electronegative ligand such as hydrogen.

Do differences in electron-pair repulsions influence the symmetry of a molecule? The AX_4, AX_3E, and AX_2E_2 molecules have T_d, C_{3v}, and C_{2v} symmetries, regardless of the ligands. For trigonal bipyramidal systems where $n + m = 5$, however, the nature of the ligands may be decisive in determining the symmetry. The axial and equatorial positions in the D_{3h} trigonal bipyramidal configuration are not equivalent. While the PF_5 molecule as an AX_5E_0 system has D_{3h} symmetry, it is not trivial to predict the symmetry of the SF_4 molecule as an AX_4E_1 system. The question is: In which of the two possible positions will the lone electron pair occur? The lone pair has the larger space requirement, and the equatorial position is more spacious than the axial; thus the lone-pair position is equatorial, and the SF_4 structure has C_{2v} symmetry. For the same reason, lone pairs are equatorial in ClF_3 (AX_3E_2) and XeF_2 (AX_2E_3). Double bonds require more space than single bonds, and behave in the VSEPR model similarly to lone pairs (Fig. 10-22).

Consider octahedral arrangements in which a central atom has six electron pairs in its valence shell. The symmetry is unambiguously O_h for AX_6; an example is SF_6. The IF_5 molecule (AX_5E_1) is a tetragonal pyramid; the electron pair may be at any of the six equivalent sites. When there are two lone pairs, they occupy positions maximally distant; thus XeF_4 (AX_4E_2) is square planar, D_{4h} (Fig. 10-23). Difficulties encountered with five-electron-pair valence shells are intensified for the case of seven electron pairs. Seven vertices cannot describe a regular polyhedron; the number of nonisomorphic polyhedra with seven vertices is large, but no one is relatively very stable. One of the early successes of VSEPR model was that it correctly predicted the nonoctahedral structure of XeF_6, as it is indeed a seven-electron-pair case (AX_6E_1).

Complete geometrical characterization of the valence shell configuration for a molecule with more than one lone pair requires more than specification of the bond angles. Sometimes, though by far not always, the angles made by lone pairs may be attainable from experimental bond angles. For example, the E—P—F angle of PF_3 can be calculated from the F—P—F angle by virtue of the C_{3v} symmetry. On the other hand, the E—S—E and

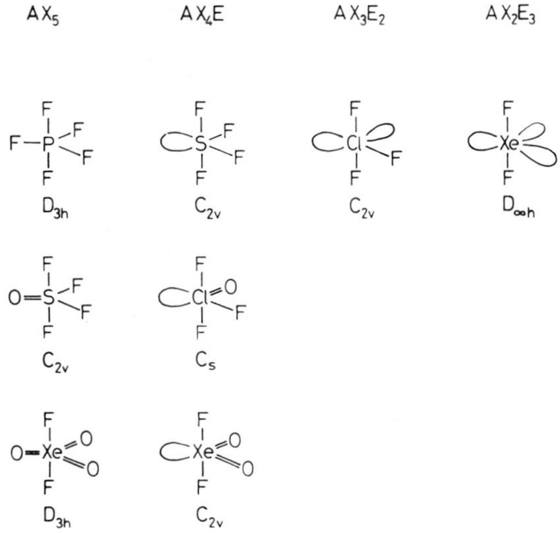

Fig. 10-22. Molecules with trigonal bipyramidal and related configurations.

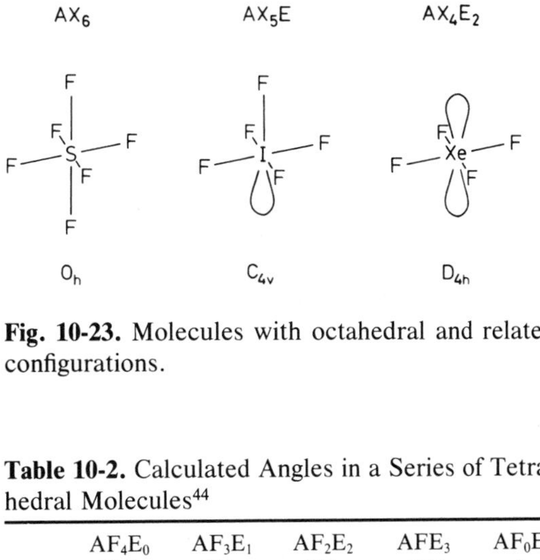

Fig. 10-23. Molecules with octahedral and related configurations.

Table 10-2. Calculated Angles in a Series of Tetrahedral Molecules[44]

	AF_4E_0 SiF_4	AF_3E_1 PF_3	AF_2E_2 SF_2	AFE_3 ClF	AF_0E_4 Ar
FAF	109.5°*	96.9°	98.1°	—	—
FAE	—	120.2°	104.3°	101.6°	—
EAE	—	—	135.8°	116.1°	109.5°*

* By virtue of T_d symmetry

Fig. 10-24. Matisse, Henri, *Dance (first version)*. (1909, early) Oil on canvas. 8′6 1/2″ × 12′9 1/2″. Collection, The Museum of Modern Art, New York. Gift of Nelson A. Rockefeller in honor of Alfred H. Barr, Jr.

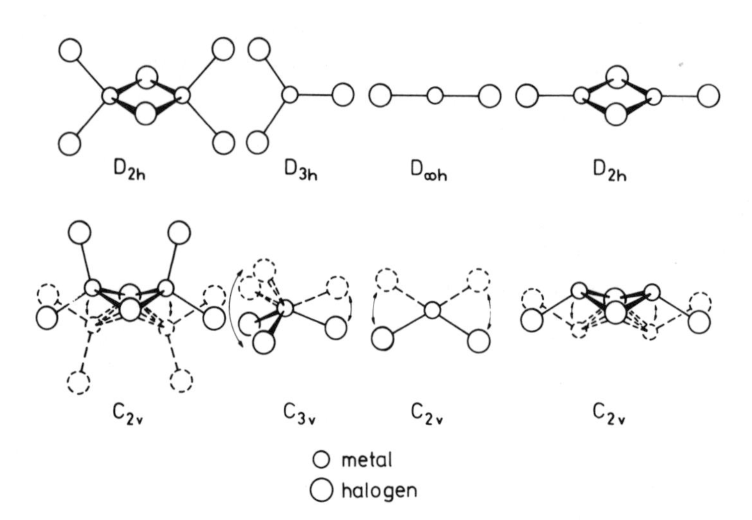

Fig. 10-25. Equilibrium versus average structures of metal halide molecules with low-frequency, large-amplitude deformation vibrations.

E—S—F angles of the C_{2v} SF_2 molecule cannot be calculated from the F—S—F bond angle.

Even when angles between lone pairs can be calculated from experimental data or deduced from quantum mechanical calculations, they are often ignored. Proper application of the VSEPR model should direct at least as much attention to angles of lone pairs and their variations as to bond angles.[44] As an example of the consistency of variations in all angles of a series of tetrahedral molecules, Table 10-2 presents a set of results from quantum chemical calculations.[45]

Consequences of Intramolecular Motion

Imagine watching the dynamic dance shown in Matisse's *Dance* (Fig. 10-24). As choreographed, one dancer jumps out of the plane of the other four. As soon as this dancer returns into the plane of the others, it is the role of the next to jump, and so on. The exchange of roles from one dancer to another throughout the five-membered troupe is so quick that a photograph with slow shutter speed would give a blurred picture; only a short exposure can identify a well-defined configuration of dancers. Matisse's *Dance* simulates the *pseudorotation* of the cyclopentane molecule, $(CH_2)_5$, which has a special degree of freedom in which the out-of-plane carbon atom exchanges roles with one of the four in-plane atoms. The process is equivalent to a permutation of two carbon atoms (and their hydrogen ligands), and is also equivalent to a rotation by $2\pi/5$ about the axis perpendicular to the plane of the four coplanar carbons.[46]

It is an extreme approach to disregard intramolecular motion. The motionless state, although hypothetical, is well defined: it is the equilibrium structure, the structure with minimum potential energy, a structure that emerges from quantum chemical calculations. Yet real molecules are never motionless, and experimentalists study real molecules. As with Matisse's *Dance*, the relationship between the lifetime of a configuration and the time scale of the investigating technique has crucial importance.[47]

Large-amplitude, low-frequency intramolecular vibrations may lower the molecular symmetry of the average structure versus the higher symmetry equilibrium structure. Some examples from metal halide molecules are shown in Fig. 10-25, although it is not yet un-

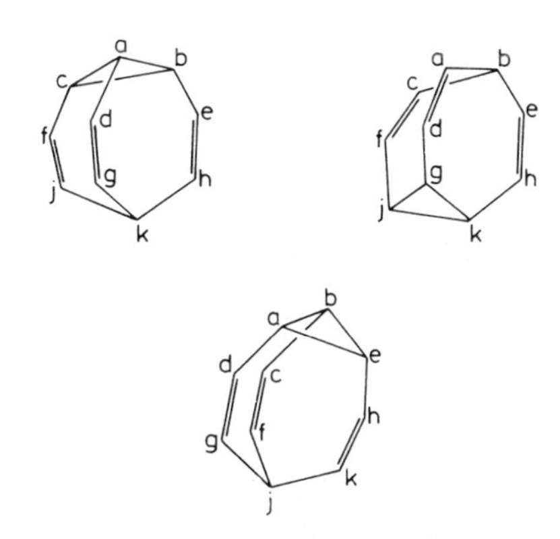

Fig. 10-26. Interconversion of nuclear positions in bullvalene.

Fig. 10-27. Interconversion of nuclear positions in hypostrophene.

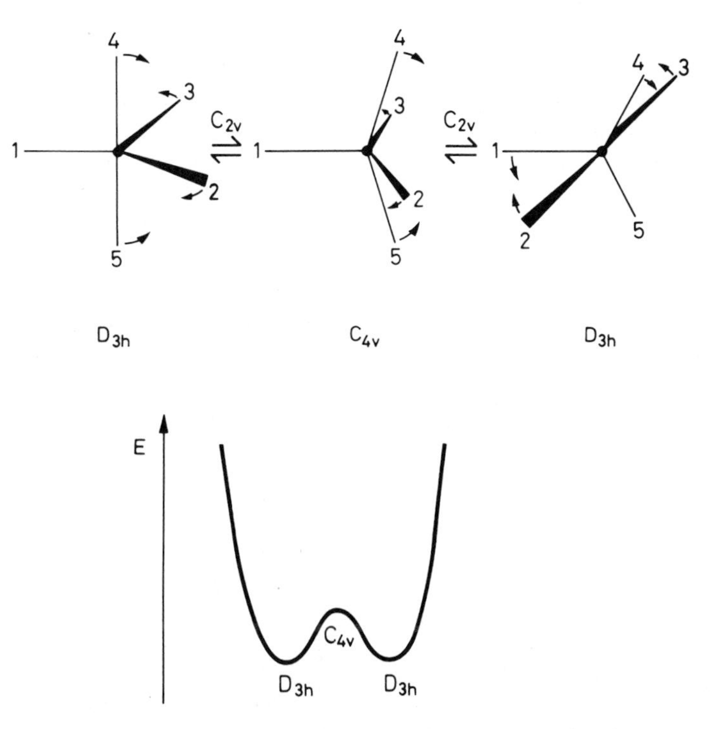

Fig. 10-28. Berry-pseudorotation of PF_5-type molecules.

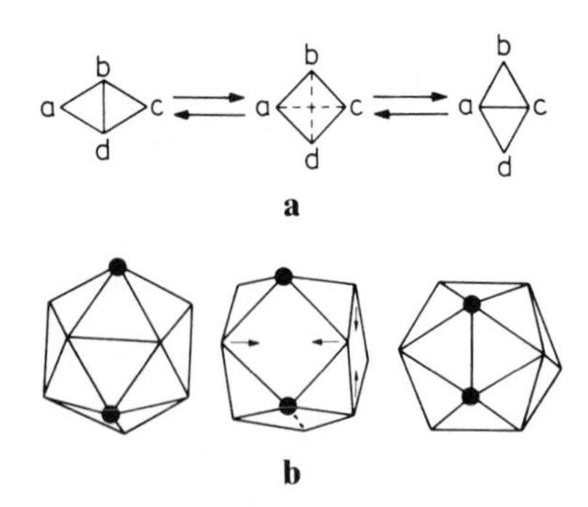

Fig. 10-29. (a) The Lipscomb model of the rearrangements in polyhedral boranes, and (b) an example of icosahedron/cuboctahedron/icosahedron rearrangement.

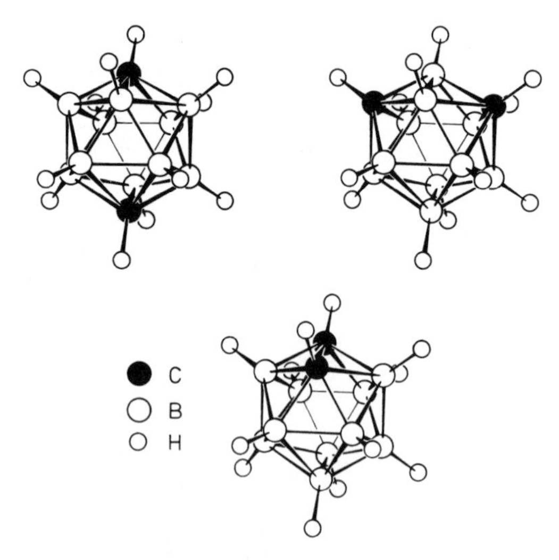

Fig. 10-30. Ortho-, meta-, and para-dicarba-*closo*-dodecaboranes. Whereas the ortho isomer easily transforms into the meta, the para isomer is obtained only under more drastic conditions and only in small amounts.

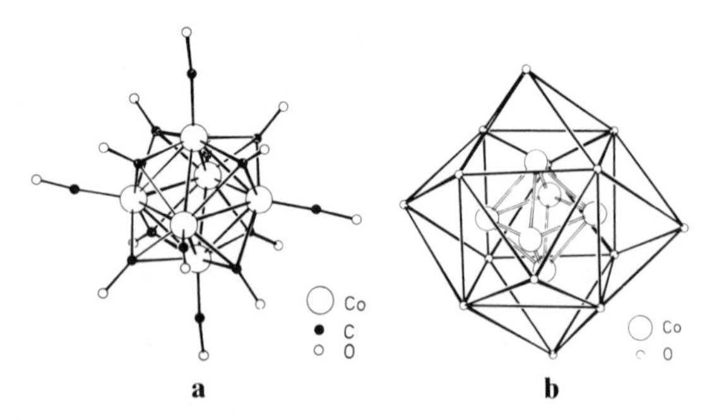

Fig. 10-31. The structure of $[Co_6(CO)_{14}]^{4-}$in two representations. (a) The octahedron of the cobalt cluster possesses six terminal and eight triply bridging carbonyl groups. (b) An omnicapped cube of the carbonyl oxygens envelopes the cobalt octahedron.

ambiguously determined whether those are indeed the equilibrium structures. Today it is only supposed that the experimentally determined structures occur because of averaging over all molecular vibrations.[48] No permutations of the nuclei are involved in these intramolecular vibrations on the time scale considered.

Rapid interconversion of nuclei takes place in a molecule of bullvalene, $(CH)_{10}$, under very mild conditions (Fig. 10-26). Bonds are made and broken, but nuclei shift only slightly. Four different kinds of carbon positions interconvert simultaneously.[49] Hypostrophene (Greek *hypostrophe:* 'turning about', 'recurrence') is a $(CH)_{10}$ hydrocarbon whose trivial name was chosen to reflect its behavior.[50] The molecule ceaselessly undergoes intramolecular rearrangements indicated in Fig. 10-27. The atoms have a complete time-average equivalence, yet hypostrophene could not be turned into pentaprismane.

Permutational isomerism among inorganic trigonal bipyramidal structures was discov–ered by Berry.[51] Although the D_{3h} trigonal bipyramid and the C_{4v} tetragonal pyramid have very different symmetries, they are easily interconverted by bending vibrations (Fig. 10-28). Permutations in an AX_5 molecule (e.g., PF_5) are easy to visualize as the two axial ligands replacing two equatorial ones, while the third equatorial ligand becomes axial in the transitional tetragonal pyramidal structure. Rearrangements quickly follow one another, with no position being unique. The C_{4v} form originates from a D_{3h} structure and yields another D_{3h} form. A similar pathway was established[52] for the $(CH_3)_2NPF_4$ molecule in which the dimethylamine group is permanently locked into an equatorial position whereas the fluorines exchange in pairs all the time. The PF_5 rearrangement also well describes the permutation of nuclei in five-atom polyhedral boranes.[53] In one mechanism[54] for rearrangements of polyhedral boranes, two common triangular faces are stretched to a square face. This intermediate may revert to the original polyhedron with no net change, or may turn into a structure isomeric with the original (Fig. 10.29). This mechanism is illustrated by interconversion of the *ortho* and *meta* isomers of dicarba-*closo*-dodecaborane[55] (Fig. 10-30); the

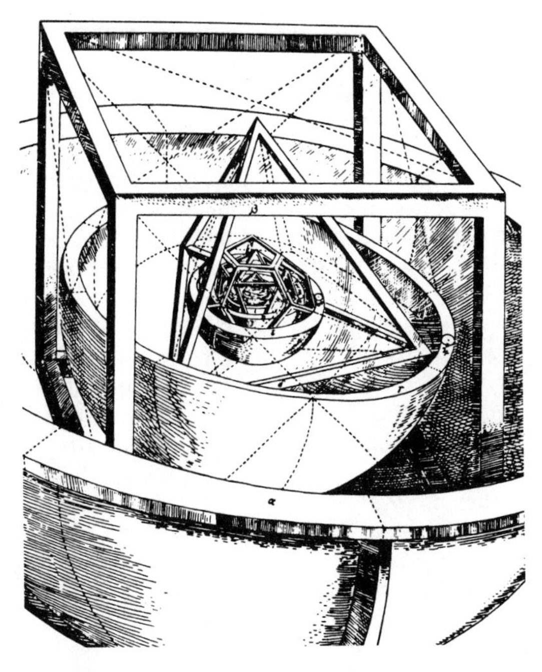

Fig. 10-32. The planetary system of Johannes Kepler, detail. From Kepler's *Mysterium Cosmographicum* (1595).

para isomer is obtained under more drastic conditions and only in small amounts. A similar model has been proposed[56] for carbonyl-scrambling in $Co_4(CO)_{12}$, $Rh_4(CO)_{12}$, and $Ir_4(CO)_{12}$.

Rapid interconversion among different modes of carbonyl coordination is possible, even in the solid state,[57] in transition-metal carbonyl molecules of the form $M_m(CO)_n$. The usually small m-atom metal cluster polyhedron is enveloped by another polyhedron whose vertices are occupied by carbonyl oxygens.[58] An attractive example is $[Co_6(CO)_{14}]^{4-}$ in which the octahedral metal cluster has six terminal and eight triply bridging carbonyl groups (Fig. 10-31a). This structure may also be represented[59] by an omnicapped cube enveloping an octahedron (Fig. 10-31b). These models are a reminder of another model in which polyhedra envelop other polyhedra; that model is Kepler's planetary system[60] (Fig. 10-32).

Notes

Acknowledgments: This chapter was written when we were at the University of Connecticut (István Hargittai as Visiting Professor of Physics (1983–84) and Visiting Professor of Chemistry (1984–85) and Magdolna Hargittai as Visiting Scientist at the Institute of Materials Science), both on leave from the Hungarian Academy of Sciences, Budapest. We express our appreciation to the University of Connecticut and our colleagues there for hospitality, and to Professor Arthur Greenberg of the New Jersey Institute of Technology for many useful references on polycyclic hydrocarbons and for his comments on the manuscript.

[1] H. S. M. Coxeter, "Preface to the First Edition," in *Regular Polytopes* 3rd ed. (New York: Dover Publications, 1973).

[2] A more extensive discussion of the shapes and symmetries of molecules and crystals can be found in I. Hargittai and M. Hargittai, *Symmetry through the Eyes of a Chemist* (Weinheim: VCH Verlagsgesellschaft, 1986).

[3] Earl L. Muetterties, ed., *Boron Hydride Chemistry* (New York: Academic Press, 1975).

[4] V. P. Spiridonov and G. I. Mamaeva, "Issledovanie molekuli borgidrida tsirkoniya metodom gazovoi elektronografii," *Zhurnal Strukturnoi Khimii, 10* (1969):133–35. Vernon Plato and Kenneth Hedberg, "An Electron-Diffraction Investigation of Zirconium Tetraborohydride, $Zr(BH_4)_4$," *Inorganic Chemistry 10* (1970):590–94.

[5] Spiridonov and Mamaeva, "Issledovanie mo-

lekuli borgidrida tsirkoniya metodom gazovoi elektronografii."

[6] Plato and Hedberg, "An Electron-Diffraction Investigation of Zirconium Tetraborohydride, $Zr(BH_4)_4$."

[7] Muetterties, *Boron Hydride Chemistry.*

[8] Robert E. Williams, "Carboranes and Boranes; Polyhedra and Polyhedral Fragments," *Inorganic Chemistry 10* (1971):210–14. Ralph W. Rudolph, "Boranes and Heteroboranes: A Paradigm for the Electron Requirements of Clusters?" *Accounts of Chemical Research 9* (1976):446–52.

[9] R. W. Rudolph and W. R. Pretzer, "Hückel-Type Rules and the Systematization of Borane and Heteroborane Chemistry," *Inorganic Chemistry 11* (1972):1974–78.

[10] A. Greenberg and J. F. Liebman, *Strained Organic Molecules* (New York: Academic Press, 1978).

[11] See ibid. and Lloyd N. Ferguson, "Alicyclic Chemistry: The Playground for Organic Chemists," *Journal of Chemical Education 46,* (1969):404–12.

[12] Günther Maier, Stephan Pfriem, Ulrich Shäfer, and Rudolf Matush, "Tetra-*tert*-butyltetrahedrane," *Angewandte Chemie, International Edition in English 17* (1978):520–21.

[13] Philip E. Eaton and Thomas J. Cole, "Cubane," *Journal of the American Chemical Society 86* (1964):3157–58.

[14] Robert J. Ternansky, Douglas W. Balogh, and Leo A. Paquette, "Dodecahedrane," *Journal of the American Chemical Society 104* (1982):4503–04.

[15] H. P. Schultz, "Topological Organic Chemistry. Polyhedranes and Prismanes," *Journal of Organic Chemistry 30* (1965):1361–64.

[16] Thomas J. Katz and Nancy Acton, "Synthesis of Prismane," *Journal of the American Chemical Society 95* (1973): 2738–39.

[17] Nicholas J. Turro, V. Ramamurthy, and Thomas J. Katz, "Energy Storage and Release. Direct and Sensitized Photoreactions of Dewar Benzene and Prismane," *Nouveau Journal de Chimie 1* (1977):363–65.

[18] Philip E. Eaton, Yat Sun Or, and Stephen J. Branca, "Pentaprismane," *Journal of the American Chemical Society 103* (1981):2134–36.

[19] Dan Farcasiu, Erik Wiskott, Eiji Osawa, Wilfried Thielecke, Edward M. Engler, Joel Slutsky, Paul v. R. Schleyer, and Gerald J. Kent, "Ethanoadamantane. The Most Stable $C_{12}H_{18}$ Isomer," *Journal of the American Chemical Society 96* (1974):4669–71.

[20] Louis F. Fieser, "Extensions in the Use of Plastic Tetrahedral Models," *Journal of Chemical Education 42* (1965):408–12.

[21] Chris A. Cupas and Leonard Hodakowski, "Iceane," *Journal of the American Chemical Society 96* (1974):4668–69.

[22] Chris Cupas, P. von R. Schleyer, and David J. Trecker, "Congressane," *Journal of the American Chemical Society 87* (1965):917–18. Tamara M. Gund, Eiji Osawa, Van Zandt Williams, and P. v. R. Schleyer, "Diamantane. I. Preparation of Diamantane. Physical and Spectral Properties." *Journal of Organic Chemistry 39* (1974):2979–87.

[23] Here it is assumed that the two different kinds of C—H bonds in adamantane have equal length. See I. Hargittai and K. Hedberg, in S. J. Cyvin, ed., *Molecular Structures and Vibrations* (Amsterdam: Elsevier, 1972).

[24] Magdolna Hargittai, "Az alumínium-klorid más fém-kloridokkal alkotott komplexeinek szerkezetéröl I. Kondenzált fázisú rendszerek," *Kémiai Közlemények 50* (1978):371–94; Magdolna Hargittai, "Az alumínium-klorid más fém-kloridokkal alkotott komplexeinek szerkezetéröl II. Alkáli-[kloroaluminátok]. Gáz-halmazállapotú komplexek," *Kémiai Közlemények 50* (1978):489–514.

[25] M. Hargittai and I. Hargittai, *The Molecular Geometries of Coordination Compounds in the Vapour Phase* (Budapest: Akadémiai Kiadó; Amsterdam: Elsevier, 1977).

[26] Magdolna Hargittai, István Hargittai, Victor P. Spiridonov, Michel Pelissier, and Jean F. Labarre, "Electron Diffraction Study and CNDO/2 Calculations on the Complex of Aluminum Trichloride with Ammonia," *Journal of Molecular Structure 24* (1975):27–39.

[27] E. Vajda, I. Hargittai, and J. Tremmel, "Electron Diffraction Investigation of the Vapour Phase Molecular Structure of Potassium Tetrafluoro Aluminate," *Inorganic Chimica Acta 25* (1977): L143–L145.

[28] A. Haaland and J. E. Nilsson, "The Determination of Barriers to Internal Rotation by Means of Electron Diffraction. Ferrocene and Ruthenocene," *Acta Chemica Scandinavica 22* (1968):2653–70.

[29] F. A. Cotton and R. A. Walton, *Multiple Bonds between Metal Atoms* (New York: Wiley-Interscience, 1982).

[30] F. A. Cotton and C. B. Harris, "The Crystal and Molecular Structure of Dipotassium Octachlorodirhenate(III) Dihydrate, $K_2[Re_2Cl_8]\cdot 2H_2O$," *Inorganic Chemistry 4* (1965):330–33; V. G. Kuznetsov and P. A. Koz'min, "A Study of the Structure of (PyH)$HReCl_4$," *Zhurnal Strukturnoi Khimii 4* (1963): 55–62.

[31] M. H. Kelley and M. Fink, "The Molecular Structure of Dimolybdenum Tetra-acetate," *Journal of Chemical Physics 76* (1982):1407–16.

[32] E. H. Hahn, H. Bohm, and D. Ginsburg, "The Synthesis of Paddlanes: Compounds in Which Quaternary Bridgehead Carbons Are Joined by Four Chains," *Tetrahedron Letters* (1973):507–10.

[33] Kenneth B. Wiberg and Frederick H. Walker, "[1.1.1] Propellane," *Journal of the American Chemical Society 104* (1982):5239–40.

[34] James E. Jackson and Leland C. Allen, "The C_1—C_3 Bond in [1.1.1] Propellane," *Journal of the American Chemical Society 106* (1984):591–99.

[35] Vernon Plato, William D. Hartford, and Kenneth Hedberg, "Electron-Diffraction Investigation of the Molecular Structure of Trifluoramine Oxide, F_3NO," *Journal of Chemical Physics 53* (1970):3488–94.

[36] István Hargittai, "On the Size of the Tetrafluoro-1,3-dithietane Molecule," *Journal of Molecular Structure 54* (1979):287–88.

[37] István Hargittai, "Group Electronegativities: as Empirically Estimated from Geometrical and Vibrational Data on Sulphones," *Zeitschrift für Naturforschung,* part A, *34* (1979):755–60.

[38] I. Hargittai, *The Structure of Volatile Sulphur Compounds* (Budapest: Akadémiai Kiadó; Dordrecht: Reidel, 1985).

[39] Robert L. Kuczkowski, R. D. Suenram, and Frank J. Lovas, "Microwave Spectrum, Structure and Dipole Moment of Sulfuric Acid," *Journal of the American Chemical Society 103* (1981):2561–66.

[40] K. P. Petrov, V. V. Ugarov, and N. G. Rambidi, "Elektronograficheskoe issledovanie stroeniya molekuli Tl_2SO_4," *Zhurnal Strukturnoi Khimii 21* (1980):159–61.

[41] R. J. Gillespie, *Molecular Geometry* (New York: Van Nostrand Reinhold, 1972).

[42] H. R. Jones and R. B. Bentley, "Electron-Pair Repulsions, A Mechanical Analogy," *Proceedings of the Chemical Society* [London]. (1961):438–440.

[43] Gavril Niac and Cornel Florea, "Walnut Models of Simple Molecules," *Journal of Chemical Education 57* (1980):429.

[44] I. Hargittai, "Trigonal-Bipyramidal Molecular Structures and the VSEPR Model," *Inorganic Chemistry 21* (1982):4334–35.

[45] Ann Schmiedekamp, D. W. J. Cruickshank, Steen Skaarup, Péter Pulay, István Hargittai, and James E. Boggs, "Investigation of the Basis of the Valence Shell Electron Pair Repulsion Model by Ab Initio Calculation of Geometry Variations in a Series of Tetrahedral and Related Molecules," *Journal of the American Chemical Society 101* (1979):2002–10; P. Scharfenberg, L. Harsányi, and I. Hargittai, unpublished calculations, 1984.

[46] R. S. Berry, "The New Experimental Challenges to Theorists," in R. G. Woolley, ed., *Quantum Dynamics of Molecules* (New York: Plenum Press, 1980).

[47] E. L. Muetterties, "Stereochemically Nonrigid Structures," *Inorganic Chemistry 4* (1965):769–71.

[48] Magdolna Hargittai, János Tremmel, and István Hargittai, "Molecular Structure of Dimeric Iron Trichloride in the Vapour Phase as Determined by Electron Diffraction," *Journal of the Chemical Society* [London] *Dalton Transactions,* (1980):87–89; Magdolna Hargittai, István Hargittai, and János Tremmel, "Molecular Structure of Monomeric Manganese(II) Bromide with Evidence on the Structure of the Dimer from Electron Diffraction," *Chemical Physics Letters 83* (1981):207–10; M. Hargittai and I. Hargittai, "On the Linearity of Iron Dichloride," *Journal of Molecular Spectroscopy 108* (1984):155–59.

[49] W. v. E. Doering and W. R. Roth, "A Rapidly Reversible Degenerate Cope Rearrangement. Bicyclo[5.1.0]octa-2, 5-diene," *Tetrahedron 19* (1963):715–737; G. Schroder, "Preparation and Properties of Tricyclo[3, 3, 2, $0^{4,6}$]deca-2, 7, 9-triene (Bullvalene)," *Angewandte Chemie, International Edition in English 2* (1963):481–82; Martin Saunders, "Measurement of the Rate of Rearrangement of Bullvalene," *Tetrahedron Letters* (1963):1699–1702.

[50] J. S. McKennis, Lazaro Brener, J. S. Ward, and R. Pettit, "The Degenerate Cope Rearrangement in Hypostrophene, a Novel $C_{10}H_{10}$ Hydrocarbon," *Journal of the American Chemical Society 93* (1971):4957–58.

[51] R. Stephen Berry, "Correlation of Rates of Intramolecular Tunneling Processes, with Application to Some Group V Compounds," *Journal of Chemical Physics 32* (1960): 933–38.

[52] George M. Whitesides and H. Lee Mitchell, "Pseudorotation in $(CH_3)_2NPF_4$," *Journal of the American Chemical Society 91* (1969):5384–86.

[53] Earl L. Muetterties and Walter H. Knoth, *Polyhedral Boranes* (New York: Marcel Dekker, 1968).

[54] W. N. Lipscomb, "Framework Rearrangement in Boranes and Carboranes," *Science 153* (1966): 373–78.

[55] R. K. Bohn and M. D. Bohn, "The Molecular Structures of 1,2-, 1,7-, and 1,12-Dicarba-*closo*-dodecaborane(12), $B_{10}C_2H_{12}$," *Inorganic Chemistry 10* (1971):350–55.

[56] Brian F. G. Johnson and Robert E. Benfield, "Structures of Binary Carbonyls and Related Compounds. Part 1. A New Approach to Fluxional Behaviour," *Journal of the Chemical Society* [London] *Dalton Transactions* (1978):1554–68.

[57] Brian E. Hanson, Mark J. Sullivan, and Robert

J. Davis, "Direct Evidence for Bridge-Terminal Carbonyl Exchange in Solid Dicobalt Octacarbonyl by Variable Temperature Magic Angle Spinning ^{13}C NMR Spectroscopy," *Journal of the American Chemical Society 106* (1984):251–53.

[58] Robert E. Benfield and Brian F. G. Johnson, "The Structures and Fluxional Behaviour of the Binary Carbonyls; A New Approach. Part 2. Cluster Carbonyls $M_m(CO)_n$ (n = 12, 13, 14, 15, or 16)," *Journal of the Chemical Society* [London] *Dalton Transactions* (1980):1743–67.

[59] Ibid.

[60] Johannes Kepler, *Mysterium Cosmographicum,* 1595.

Part IV
Theory of Polyhedra

11

Introduction to Polyhedron Theory

MARJORIE SENECHAL

This part of the book is concerned with the mathematical theory of polyhedra at the research level. The broad spectrum of current research presented at the Shaping Space Conference is reproduced here essentially in its original form. By way of assistance this chapter provides a brief introduction to some relevant aspects of polyhedron theory; after that you are on your own. The part concludes with a collection of unsolved problems, solicited by Douglas Dunham (Chapter 18). We hope that these chapters and problems will be of interest both to specialists and to nonspecialists who are interested in learning about current work in this field.

What Is a Polyhedron?

You have met enough polyhedra by now to be able to guess that this question will not be easy to answer. In fact, there is not one answer, but many. The problem in choosing a definition was explained many years ago by Robert Frost in his poem "Mending Wall": "Before I'd build a wall, I'd want to know what I was walling in, and what I was walling out."

The delightful book *Proofs and Refutations*, written in the form of a classroom discussion, is required reading for all polyhedron enthusiasts.[1] In it the author, Imre Lakatos, shows how a careful examination of the implications of a definition forces us to construct our walls very carefully. Consider, for example, the following excerpt:

GAMMA: A polyhedron is a solid whose surface consists of polygonal faces . . .

DELTA: Your definition is incorrect. A polyhedron must be a *surface*: it has faces, edges, vertices, it can be deformed, stretched out on a blackboard, and has nothing to do with the concept of "solid." *A polyhedron is a surface consisting of a system of polygons.*

TEACHER: For the moment let us accept Delta's definition. Can you refute our conjecture (Euler's formula, $F - E + V = 2$, which the class is trying to prove) now if by polyhedron we mean a surface?

ALPHA: Certainly. Take two tetrahedra which have an edge in common. Or, take two tetrahedra which have a vertex in common. Both these twins are connected, both constitute one single surface. And, you may check that for both $V - E + F = 3$.

DELTA: I admire your perverted imagination, but of course I did not mean that *any* system of polygons is a polyhedron. By polyhedron I meant a *system of polygons arranged in such a way* that (1) *exactly two polygons meet at every edge* and (2) *it is possible to get from the inside of any polygon to the inside of any other polygon by a route which never crosses any edge at a vertex*. Your first twins will be excluded by the first criterion in my definition, your second twins by the second criterion.

Delta believed that Alpha's twin tetrahedra are not polyhedra, but "monsters" which can and must be barred by a proper definition. Monster-barring, argued Lakatos, is often the reason that complicated, abstract definitions like Delta's second one appear in mathematics. (Unfortunately they are usually presented to the student in a take-it-or-leave-it way, with no explanation of how or why anyone would ever come up with them.)

So, if you wish to define "polyhedron," you should think about the kinds of objects that you are willing to accept as polyhedra. For example, you might (or might not) want to include

- Star polyhedra
- Toroidal polyhedra
- Infinite polyhedra
- Polyhedra whose faces are skew polygons
- Either pair of Delta's tetrahedral twins
- A finite capped cylinder (it has three faces, two edges, and no vertices!)

Of course, you might want to include some of these in some cases, and exclude them in others, depending on what properties you are studying.

A widely accepted definition of "polyhedron" is that given[2] by Branko Grünbaum; it is based on the following definition of a polygon:

Definition 11.1. A finite *polygon* is the figure formed by a finite sequence of vertices in E^3, $V_1, V_2, \ldots, V_n$, together with the edges, $[V_i, V_{i+1}]$, $i = 1, 2, \ldots, n-1$, and $[V_n, V_1]$. (An infinite polygon is defined in a suitably analogous way.)

Definition 11.2. A *polyhedron P* is any family of polygons (called faces of P) that has the following properties: (i) Each edge of one of the faces is an edge of just one other face. (ii) The family of polygons is connected; that is, for any two edges E and E' of P there exists a chain $E = E_0, P_1, E_1, P_2, E_2, \ldots, P_n, E_n = E'$, of edges and faces of P, in which each P_i is incident with E_{i-1} and with E_i. (iii) Each compact[3] set meets only finitely many faces.

(How does this definition compare with Delta's second one?)

Notice that Definition 11.1 does not require the vertices of a polygon to be coplanar. Thus the polyhedra permitted by Definition 11.2 may have skew nonplanar "faces"—and there may be infinitely many of them! Definition 11.2 is broad enough to include most of the polyhedra you have met in this book, and then some. On the other hand, it may be too broad for some purposes: for example, we may want to restrict ourselves to convex polyhedra. It might not be broad enough for other purposes, however. Do we really want to exclude the twin tetrahedra which share only a vertex, for example? They have been used to interpret molecular structures. Other important applications call for polyhedra with movable parts. Will this definition be able to accommodate the demands of scientists for a broader theory? (See Chapter 9.)

Why a Theory of Polyhedra?

Why don't we simply take them as we find them, admiring them for their beauty and the wonderful things that they represent? Of course, one could do this, but then our understanding would be extremely limited. In this section we will discuss some of the reasons, besides intellectual curiosity and aesthetic pleasure, why mathematicians are engaged in polyhedral research.

Many practical problems involve polyhedra built to specification, either by nature or by man. To understand these structures, and to be able to create new ones, we must know what the specifications are, why they are necessary, and what sorts of objects are characterized by them. For example, if we want to design bridges and buildings that stay up, we must study the form and dynamics of trusses and braces, and this leads to questions about polyhedral stability. Problems like these have been a focus of research in the Structural Topology Research Group in Montreal for many years. Other problems motivated by the needs of science and technology are discussed in Chapter 9.

Moreover, the need for a theory of polyhedra arises almost spontaneously when we try to make clear to colleagues and students what we are talking about. For example, if someone asks what we mean by the word "polyhedron" we might begin, like Gamma and Delta, by specifying certain characteristics we think all polyhedra have in common; these are usually characteristics of the polyhedra that we already know. Having listed them, it is then natural to wonder, like Alpha, whether all objects which have these characteristics are necessarily things that we want to call polyhedra. Or, we might ask whether there are new, as yet undiscovered, structures which also have these properties.

This line of questioning often leads us to generalize familiar concepts. For example, Definition 11.1 implies that there are many kinds of regular polygons in addition to the convex and star plane polygons: it implicitly permits regular prismatic and antiprismatic[4] polygons, and regular infinite polygons whose edges lie on straight lines, or zigzag, or are helical. By a careful analysis of the possible regular polyhedra with such polygons as faces, Grünbaum found that, in addition to the regular convex and star polyhedra, there are six additional families: the infinite regular plane tessellations, the Petrie–Coxeter polyhedra, a class of nine regular polyhedra with finite skew polygons as faces, infinite regular polyhedra with finite skew polygons as faces, regular polyhedra whose faces are infinite zigzag polygons, and regular polyhedra whose faces are infinite helical polygons! In this way our understanding of regular three-dimensional polyhedra has been greatly enriched.

Polyhedra can be generalized in other ways. One of these is to investigate their analogues—"polytopes"—in higher dimensions. The *regular convex* polytopes in n dimensions were discovered by Ludwig Schläfli in the early 1850s. In spaces of five or more dimensions, there are only three of them: the higher dimensional analogues of the tetrahedron, the cube, and the octahedron. But in four-dimensional space, there are six regular convex polytopes: the three just mentioned, one with 120 dodecahedral cells (cells are three-dimensional "faces"), one with 600 tetrahedral cells, and one with 24 octahedral cells. These six polytopes are not easy for us to visualize, but there are various geometric and algebraic techniques which, together with computer graphics, can help a great deal. One of the most interesting of these is described by Thomas Banchoff in Chapter 16.

Or we may generalize the definition of regularity to apply to a broader class of polyhedra. In Chapter 14, Jörg Wills discusses such a generalization of the regular solids, called Platonohedra. Here the polyhedra are "equivelar"; that is, their faces are regular k-gons and q meet at each vertex, but the polyhedra can be toroidal or indeed of any genus.[5] The "symmetries" of these objects include some transformations which preserve their combinatorial structures but, strictly speaking, not their metrical properties.

Sometimes problems are generalized because the original problem is too hard. The ancient question: "Which polyhedra fill space?" is one of these. The difficulties are so severe that it makes good sense to begin with the more tractable one: "Which combinatorial types of polyhedra fill space?" Progress on this question has recently been made by Egon Schulte (Chapter 12).

Another impetus for the development of a theory of polyhedra is the need to clarify fundamental concepts. As we have seen, the history of polyhedra is long and it has many roots. On close examination, we sometimes find that well-entrenched definitions and classifications are not as clear as we once thought. Or, as new classes of polyhedra are discovered, we may find old characterizations inadequate. This confusion leads to new questions about what it is we are talking about, and these questions generate research. For example, as Grünbaum and Shephard point out in Chapter 13, there is no problem reconciling the several widely held (but distinct) concepts of duality as long as we are talking about convex polyhedra, but with more general types of polyhedra it is no longer even clear what "dual" is supposed to mean.

Finally, it often happens that in the course of investigating one problem, surprising and illuminating links are found with others. The equivalence of the seemingly unrelated concepts of convex polyhedra, Dirichlet tessellations, and "spider webs" discussed in Chapter 17 is an exciting example.[6]

Polyhedral Themes

In addition to the variety of motivations for studying polyhedra theory, several themes run throughout the following papers and problems.

Symmetry. The aesthetic link between symmetry, beauty, and perfection was undoubtedly the reason why the regular polyhedra were first noticed and singled out for attention. A great deal has already been said about symmetry in this book, so we will not review the basic concepts here. The following remarks, however, may be helpful to keep in mind while reading the following chapters.

Symmetry theory is not a museum piece, but a valuable tool in the study of polyhedra. We have just seen that symmetry often suggests interesting generalizations. It can also be a guide in searching for new kinds of polyhedra. The semiregular polyhedra discovered by Archimedes were (at that time) a new class of highly symmetrical polyhedra, and Archimedes probably used symmetry considerations to ensure that this would be so. In fact, the eleven that can be obtained from the Platonic "solids" by truncation are obtained by truncating symmetrically: first a vertex (or edge) of a solid is truncated in such a way that its contribution to the symmetry of the polyhedron is not destroyed, and then all the other vertices are truncated in exactly the same way. This ensures that the truncated polyhedron has all the symmetries of the original regular one.

A requirement of symmetry can also help us to restrict a problem to a reasonable size. Obviously all sorts of idiosyncratic constructions are admitted under Definition 11.2; we cannot hope to survey them all. By focusing on those that have some symmetry properties, however, we obtain a manageable class of objects. And because symmetry is a hierarchical concept, we can broaden our study later by omitting its restrictions one by one. Symmetry can help to organize and present complex information. Several authors (H. S. M. Coxeter in Chapter 3, Arthur L. Loeb in Chapter 6, and Barry Monson in Chapter 15) point out the effectiveness of symmetry arguments in characterizing the coordinates of certain polyhedra; in his short note Monson points out an interesting connection between this problem and the theory of numbers. (Indeed, the theory of polyhedra has connections with, and implications for, almost every branch of mathematics.)

Symmetry theory can also be used to study properties of polyhedra which are inadequately characterized by their geometry. For example, we have seen that a carbon atom is often represented as a regular tetrahedron, because it is 4-valent. But when an atom joins with other atoms in a molecule or crystal, its four bonds may no longer be equivalent. To incorporate this information into the geometry of the tetrahedron, we can color its vertices or faces in such a way that equivalence is properly indicated. This leads us to the concept of "color symmetry," which has been extensively studied both for polyhedra and for tessellations of the plane. The colored tilings of interlocking creatures designed by M. C. Escher are typical examples of patterns with color symmetry, but less orderly colorings are important too. Again we must decide what we want to wall in and what we want to wall out. In a lecture at the Shaping Space Conference, David Harker argued that there are many interesting colored polyhedra which do not satisfy even the least restrictive definitions that have been proposed. Clearly the theory is still evolving.[7]

Networks. By now you are very familiar with Euler's deceptively simple formula for the vertices, edges, and faces of a convex polyhedron, $V - E + F = 2$. This formula turns out to have a wealth of implications, even though it says nothing whatever about angles, edge lengths, or other metric properties of polyhedra, being concerned only with the networks formed by edges and vertices. For example, it implies that there are at most five regular polyhedra (in addition to the so-called digonal polyhedra and dihedra). It is very surprising that this ancient and famous result does not in fact depend on either the symmetry or metric properties of the polyhedra, but only on their combinatorial properties.

The numbers of faces, vertices, and (consequently) edges of a polyhedron constitute its *f-vector* (V, E, F). Some of the questions one might ask about *f*-vectors are: If V, E, F are integers which satisfy the Euler relation, are there any corresponding polyhedral networks of edges and vertices? For example, there is no polyhedron with *f*-vector (0, 4, 6), but there are two very different polyhedra with *f*-vector (8, 12, 6). (One is the cube; what is the other?) Thus the relation $V - E + F = 2$ is a necessary condition for the existence of a polyhedron with *f*-vector (V, E, F) but it is not always sufficient. As Margaret Bayer explains in Chapter 18, the additional conditions that must be satisfied by V, E, and F have been found for the three-dimensional case, but the analogous problem in higher dimensions remains unsolved.

Another important question is: If a polyhedron with f-vector (V, E, F) does exist, what kinds of faces does it have? (How many triangles, how many quadrilaterals, and so forth?) In other words, what is its *face sequence* $[f_k]$, $k = 3, 4, 5, 6, \ldots$, where f_k is the number of k-gonal faces of the polyhedron? A famous equality which is a direct consequence of Euler's formula states that if the polyhedron is trivalent (that is, if three edges meet at each vertex), then

$$3f_3 + 2f_4 + f_5 = 12 + \sum_{k>6}(k - 6)f_k. \quad (11.1)$$

This is a condition that the face sequence of a polyhedron must satisfy, but it does not guarantee that a polyhedron with such a face sequence exists.

Euler's formula has been generalized to polyhedra of higher dimension, and of other genera. For a polyhedron of genus g, the formula becomes

$$V - E + F = 2 - 2g. \quad (11.2)$$

Thus for the torus the right-hand side is zero. A great deal of effort has been, and is being, devoted to finding analogues, for polyhedra of genus greater than one, of *Eberhard's theorem*, which is closely related to Eq. (11.1): For every finite sequence of nonnegative integers $[f_k, k \geq 3, k \neq 6]$ satisfying Eq. (11.1) there are values of f_6 such that a polyhedron with face sequence $[f_k]$ exists. Peter Gritzmann found a toroidal analogue in 1983 (see Chapter 18).

Next we might ask: "How many distinct combinatorial types of polyhedra belong to each sequence $[f_k]$?" This is an unsolved problem. The numbers of combinatorial types belonging to each f-vector are known only for polyhedra with eleven or fewer faces.[8] There is no apparent pattern to these numbers, but Eli Goodman and Richard Pollack have recently found surprisingly low bounds for the number of combinatorially distinct polytopes (of a certain type) with n vertices in d-dimensional space.[9]

Geometric Realization. Even when a polyhedron network exists, it may happen that polyhedra with straight edges and planar faces cannot be constructed according to these plans. If the edges of the digonal networks, for example, are straightened out, they all collapse to a single line. This raises the question of determining the conditions under which a polyhedron with certain properties can be *realized geometrically*.

In studying the problem of realization, a fundamental theorem is that of Steinitz, which characterizes the types of planar graphs[10] which correspond to convex polyhedra in three-dimensional space. Among the unsolved problems in polyhedron theory are several concerning realizations of combinatorial polyhedra of higher genera. The long-range goal of such research is of course to find appropriate analogues of Steinitz's theorem.

Realization questions lead to the study of other properties of polyhedra. Polyhedra were originally studied from the metric point of view. Later, Euclid explained how to construct the regular solids. Descartes' work on polyhedra was concerned with the relations among the polygonal angles of polyhedral faces and the dihedral angles between faces. Most polyhedron models have definite angles and edge lengths; most buildings are built to precise architectural plans. It would be easy to cite other cases in which the metric properties of polyhedra play an indispensable role. But because of the power and comparative simplicity of the combinatorial approach, the metrical theory has been relatively neglected for many years.

The theory of the rigidity of polyhedral frameworks is enjoying a renaissance in this decade, due in large measure to the work of the Structural Topology Research Group. Older theories are being reexamined from a modern point of view, and new techniques are being applied to them. The work is well represented in this book.

A Word of Warning

Specialized discussions of problems in polyhedron theory use technical mathematical terminology. If you are a nonspecialist, it may be helpful if we explain, in intuitive language, some of the less familiar mathematical terms that you will encounter. (You should skip the next two paragraphs if you are already understand the italicized words.)

We already defined the word *polygon* (Definition 11.1). Since this word has metric connotations (e.g., angles and edge-lengths can be defined) which are not needed in the purely combinatorial theory of polyhedra, some authors prefer to speak of a *2-cell*, a floppy polygon that can be deformed into a circle. You can think of a *2-manifold* as a *two-dimensional surface in three-dimensional space,* such as a plane, an infinitely long cylinder, a sphere, or a torus, possibly deformed. Similarly, the *2-sphere* is the ordinary sphere in three dimensions (surface only, of course). The prefix "2-" emphasizes the fact that the surface is two-dimensional, as opposed to three-, four-, or *n*-dimensional. A *2-cell complex* is a 2-manifold every point of which belongs to the interior or boundary of a 2-cell.

The terms "closed," "bounded," and "boundary" can lead to confusion unless you remember that each of them has a precise mathematical meaning. A subset of space is *bounded* if it is of finite extent, that is, if it can be entirely enclosed in some sphere of finite radius (even if the radius has to be very large). On the other hand, to say that a set is bounded does not mean that it contains or even has a boundary! The word *boundary* refers to an intrinsic property of the set, whereas "bounded" refers to the space in which the set lies. A set of points is *closed* if it contains all its boundary points. For example, the set of points in the plane *interior* to a circle is not closed, although it is bounded, because it contains points arbitrarily close to the circle itself, but the points of the circle do not belong to it. A set is *compact* if it is closed and bounded; for example a circle together with its interior points is a compact set. But the (infinite) plane is not compact because it is not bounded. Finally, you can think of an *orientable* manifold as a two-sided surface; the plane, the sphere, and the torus are orientable, but the Möbius band is not. As mentioned earlier, the *genus* of a manifold is its number of holes: "hole" is used here in the sense that a torus has a hole; it is not the kind of hole which punctures the manifold.

You can test your understanding of this terminology by trying to decipher and interpret the definitions of the word "polyhedron" that are given below (all taken from papers in this section). We leave it to you to determine which shapes are walled in by each of them, and which are walled out. Study the definitions carefully before deciding! It is easy to be misled. For example, Delta did not seem to realize that her definition did not exclude polyhedra like our toroidal hat, which she despised (she called them "non-Eulerian pests")!

Definition 11.3. A (convex) polyhedron is a bounded subset of E^3 which can be expressed as the intersection of a finite number of closed half-spaces.

Definition 11.4. A polyhedron is a cell complex whose point set is a closed orientable 2-manifold, each of whose 2-cells is an affine polygon that is not coplanar with any adjacent 2-cell.

Definition 11.5. A polyhedron (in three-dimensional space) is a compact 2-manifold that has no boundary and can be expressed as a finite union of plane polygonal regions.

Notes

[1] Imre Lakatos, *Proofs and Refutations* (New York: Cambridge University Press, 1976)

[2] B. Grünbaum, "Regular Polyhedra, Old and New," *Aequationes Mathematicae 16* (1977):1–20.

[3] The term *compact* is defined later in the chapter.

[4] A prismatic, or antiprismatic, polygon is a finite nonplanar zigzag polygon whose vertices are the vertices of a prism or antiprism, respectively.

[5] The genus of a polyhedron is the number of its holes. Convex polyhedra have genus zero, like the sphere; toroidal polyhedra have genus one, and so forth. Note that the plural of genus is genera.

[6] I am delighted that the Shaping Space Conference provided an opportunity for the authors to complete this research!

[7] For more details, see B. Grünbaum and G. C. Shephard, *Tilings and Patterns* (San Francisco: W. H. Freeman, 1986).

[8] See P. Engel, "On the Enumeration of Polyhedra," *Discrete Mathematics 41* (1982):215–18.

[9] J. E. Goodman and R. Pollack, "There Are Asymptotically Far Fewer Polytopes Than We Thought," *Bulletin (New Series) American Mathematical Society 14,* no. 1 (January 1986):127–29.

[10] A planar graph is a network, or 1-*skeleton* of edges and vertices. The 1 indicates the one-dimensionality of the edges, which can be drawn in the plane without any unintended crossing of edges. Schlegel diagrams are planar graphs; star polygons are not.

12

Combinatorial Prototiles

EGON SCHULTE

Tiling problems have been investigated throughout the history of mathematics, leading to a vast literature on the subject. Our present knowledge of tilings of the plane is quite good, though there are of course many open problems even in the comparatively elementary and easily accessible levels. Much of the work on plane tilings has been done by Grünbaum and Shephard in the last ten years; it will be summarized in their forthcoming book about tilings and patterns.[1]

As soon as we raise the dimension of the space from two to three or higher, our knowledge about tilings becomes comparatively poor. This is probably due to the fact that it is much harder to visualize the situation. In particular, considerations about the local structures of tilings are needed.

Before turning to the subject of this chapter, let us recall some definitions and notations. Although most of the results I will discuss can be extended to higher dimensions, I will restrict my considerations to ordinary three-dimensional space. Thus the underlying space for our tilings will be Euclidean 3-space, E^3, and we will tile this space by convex 3-polytopes, that is, with bounded convex polyhedra. A tiling of Euclidean 3-space is a family of convex 3-polytopes, called the tiles of the tiling, which cover the space without gaps or overlaps. This means that every point in space is contained in a tile, and no two tiles have common interior points.

To avoid pathological situations, we will always assume that our tiling is locally finite. By a *locally finite* tiling, we mean a tiling that has the following property: every point in space has a neighborhood that meets only finitely many tiles. Of natural interest are those tilings which respect the facial structure of the tiles or, more precisely, the facial structure of the boundaries of the tiles. These are exactly the face-to-face tilings. A tiling is called *face-to-face* if the intersection of any two tiles is either empty or a face of each; this means that two tiles may share a vertex, an edge, or a facet. Note that this definition of a face is slightly different from the usual one. Here a face can be 0-, 1-, or 2-dimensional. I will use the word "facet" to mean a two-dimensional face of a 3-polytope. (More generally, a facet of a $(d + 1)$-polytope will be a d-dimensional face.)

A tiling of E^d is called *normal* if its tiles are uniformly bounded, that is, if there are two positive real numbers r_1 and r_2 such that each tile contains a ball of radius r_1 and is contained in a ball of radius r_2. Obviously, normal tilings are necessarily locally finite. If each tile in a tiling happens to be congruent to one of the tiles in a finite family of k polytopes, then we say that the tiling has k isometric prototiles. Clearly such a tiling must be normal.

Finally, we recall that two polytopes P and Q are isomorphic, or *combinatorially equivalent,* or of the same combinatorial type, if there is an inclusion preserving bijection between the set of faces of P and the set of faces of Q (for 3-polytopes, that means, between the set of vertices, edges, and facets of P, and the set of vertices and edges and facets of Q). For example, the polytope in Fig. 12-1 is combinatorially equivalent to the octahedron. Its faces are 3-gons and they fit together exactly the same way as the faces of the octahedron. Thus combinatorially these polytopes are the same, although they have totally different shapes. Notice, for instance, that the base of the polyhedron does not lie in a plane. If every tile of a tiling is combinatorially equivalent to a given 3-polytope P, we say that the tiling is *monotypic*. The type of the polytope is the *combinatorial prototile* of the tilings.

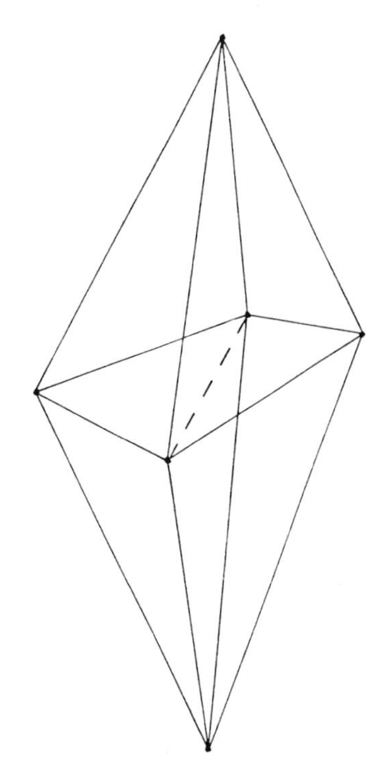

Fig. 12-1. A combinatorial octahedron. The dotted line indicates that the four median vertices are not coplanar.

Nontiles

One of the main problems in the theory of tilings of Euclidean 3-space is to characterize those convex polytopes, congruent copies of which tile space in a face-to-face manner. In other words, we are interested in finding those polytopes which play the role of triangles, quadrangles, pentagons, and hexagons in the plane. The answer to this extremely difficult problem is completely out of reach at the moment; we don't even know whether there are only finitely many combinatorial types of 3-polytopes that give such a tiling. Now whenever people cannot solve a problem, they study related problems and hope that by doing so they will get additional information about the original one.

Thus we will discuss the three-dimensional analogue of the well-known fact, first observed by Schlegel in 1883, that for each $n \geq 3$, the Euclidean plane can be tiled by convex n-gons. Now the tiles in a tiling of the plane by n-gons, although they all have the same number of edges, cannot be congruent if $n > 6$ (and indeed the tiling cannot be normal). Thus the tiles in such a tiling are combinatorially equivalent, but not congruent.

The combinatorial analogue of the tiling problem for higher dimensions was posed by Ludwig Danzer at a symposium[2] on convexity in 1975. He suggested replacing the requirement of congruence for the tiles of a tiling of E^d by the much weaker requirement of combinatorial equivalence of the tiles. This raises the following question:

Given a convex 3-polytope P, is there a locally finite tiling of space by convex polytopes isomorphic to P?

In other words, we would like to know whether every polytope is combinatorial prototile of a monotypic tiling of three space. In particular, we would be interested to find face-to-face tilings—tilings that respect the facial structure.

Just as in the plane, this problem would be nonsense with combinatorial equivalence replaced by congruence of the tiles. But as long as we are only interested in combinatorial isomorphism, we have much freedom in the choice of the particular metrical shape of the tiles, so this is actually a reasonable problem. The general belief was that the answer to this problem should be positive, even in the strongest sense. That is, every convex three polytope was assumed to be a combinatorial prototile of a monotypic face-to-face tiling of three space. However, this is not true; in fact, the cuboctahedron, which is a very well-known polyhedron (Fig. 12-2), is not the combinatorial prototile of a face-to-face tiling: *There is no (locally finite, face-to-face) tiling of space by convex polytopes combinatorially equivalent to the cuboctahedron.*

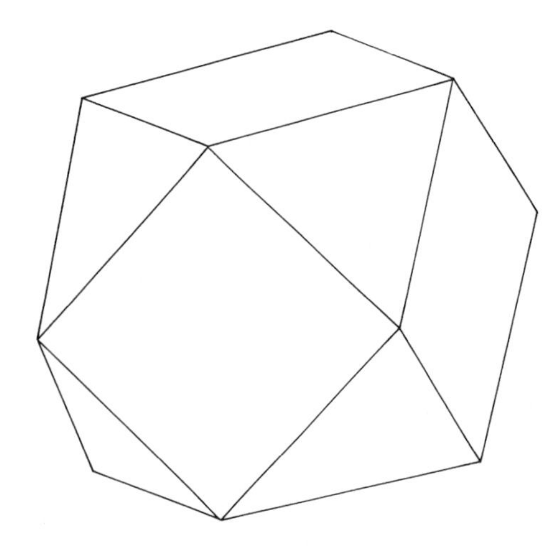

Fig. 12-2. The cuboctahedron.

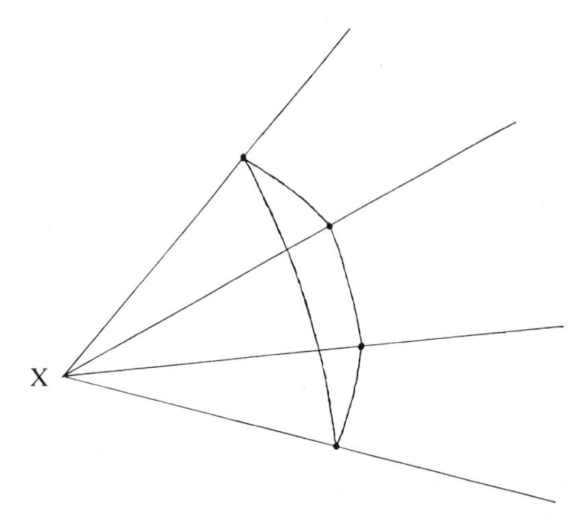

Fig. 12-3. The facet of a spherical complex determined by a cuboctahedron with vertex X.

The proof is as follows. Let us look at the vertices of the cuboctahedron. They are all 4-valent and they are all surrounded by triangles and quadrangles in an alternating way: if we go around a vertex then we meet a triangle, a quadrangle, a triangle, and a quadrangle. Thus it is natural to associate to the vertices of the cuboctahedron the type [3, 4, 3, 4]. It is important that it is of alternating type.

Now let us assume that there is a tiling of Euclidean space by polytopes that are combinatorial cuboctahedra and that fit together in a locally finite face-to-face manner. We fix one particular vertex of the tiling, that is, a vertex of one of the tiles, and call it X. This vertex is contained in many tiles. We choose a sufficiently small sphere centered at that vertex and consider the intersection of the sphere with the tiling. Now the tiling cuts out a spherical complex on our sphere. The facets of the spherical complex are just the intersections of the sphere with the tiles of the tiling which contain the fixed vertex X.

We will show that this spherical complex has quadrangular spherical facets and that each vertex is even-valent. This will immediately give us a contradiction, since Euler's theorem implies that a spherical complex without triangular facets must have a 3-valent vertex. (To see this, write Euler's formula in the form $2F - 2E + 2V = 4$, or $(2F - E) + (2V - E) = 4$. If the complex has no triangular facets, then counting the edges of the complex by going around each facet, and taking into account that each edge is shared by two facets, we have $2E \geq 4F$, so $2F - E \leq 0$. Similarly, if the spherical complex has no 3-valent vertices, then $2V - E \leq 0$. But these inequalities are not simultaneously compatible with Euler's formula.) This means that our spherical complex cannot exist and so the tiling cannot exist. So all we have to do is to show that our spherical complex has spherical quadrangles and even-valent vertices.

The first is easy. Why does the spherical complex have quadrangular facets? Recall that the facets are just the intersections of the sphere with the tiles that contain the fixed vertex X. Since the polytopes in our tiling are combinatorially cuboctahedra, the vertex X has valence 4 in each polytope that contains it. Thus the sphere cuts out a spherical quadrangle in each polytope, and so the spherical complex has spherical quadrangles as facets (Fig. 12-3).

Next, why are the vertices even-valent? The vertices of the spherical complex are just the intersections of the sphere with the edges of the polytopes that contain the fixed vertex X. What are the edges of the spherical complex? They are just the intersections of the sphere with the facets of the tiles that contain X. The facets of our tiles are triangles and quadrangles because the tiles are all combinatorial cuboctahedra and so, since the tiling is face to face, we can assign to each edge of the spherical complex one of the labels 3 or 4, according to the number of vertices of the facet which defines the edge. And now comes the main point: since the vertices of the cuboctahedron are surrounded by triangles and quadrangles in an alternating way, the vertices of the spherical complex must also be surrounded in an alternating way. That means the edges which come together in the vertex X can be labeled 3, 4, 3, 4, . . . in an alternating way. But this means the complex has even-valent vertices, and, as we have seen, quadrangular facets. This spherical complex cannot exist, and so the tiling cannot exist.

In fact, this is a very strange result, because one would expect that such a nice symmetric polytope should give such a tiling. Indeed, the tetrahedron, octahedron, icosahedron, and dodecahedron, which are not isometric space-fillers, are combinatorial prototiles of monotypic tilings. The cube is even an isometric space-filler. Our counterexample reveals a very strange and very interesting aspect of the theory of tilings by convex polytopes: there seems to be no intrinsic relation between the regularity or symmetry properties of the polytope and its tiling properties. In fact, in higher dimensions (seven or higher) it turns out that the regular crosspolytope (the higher dimensional analogue of the octahedron) does not give a face-to-face tiling by combinatorially equivalent polytopes.

Once we have a counterexample to our problem, we can ask to what extent we *can* expect positive results. The ideal situation would clearly be to give a characterization of all those polytopes which give a tiling, but this seems to be rather hopeless. The next best thing would be to try to determine certain classes of polytopes which do or do not give such tilings. For example, with the techniques used in the proof above we can prove the following generalization[3]:

Theorem 12.1. Let P be a convex 3-polytope and $x_1, x_2, \ldots, x_k$ the vertices of P, all of even valence. Assume that it is possible to assign to each x_i ($i = 1, 2, \ldots, k$) its type $[p_{i,1}, p_{i,2}, \ldots, p_{i,2m}]$ in such a way that

$$\left\{ \bigcup_{i=1}^{k} \{p_{i,1}, p_{i,3}, \ldots, p_{i,2m-1}\} \right\} \cap \left\{ \bigcup_{i=1}^{k} \{p_{i,2}, p_{1,4}, \ldots, p_{i,2m}\} \right\} = \emptyset.$$

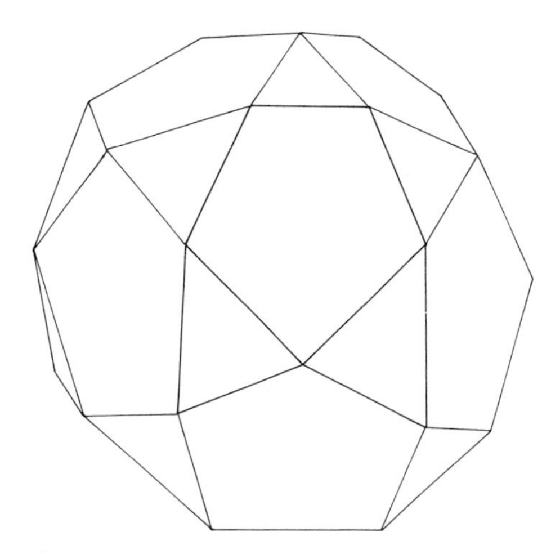

Fig. 12-4. The icosidodecahedron.

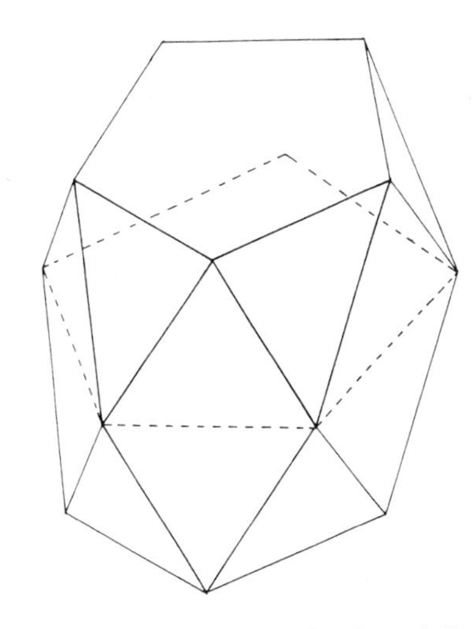

Fig. 12-5. The polytope P_5. The dotted lines indicate a median cross section (they are not hidden edges).

Then P is a nontile. That is, P does not give a locally finite face-to-face tiling.

With the help of this theorem, it is easy to construct many nontiles. If we start with a simple 3-polytope without triangular facets and cut off its vertices up to the midpoints of the edges incident with them, we obtain a nontile. (This operation, due to Steinitz, is denoted $I(G)$ in Grünbaum.[4]) For example, starting in this way from the octahedron and icosahedron we obtain the cuboctahedron and icosidodecahedron (see Figs. 12-2 and 12-4). Also, an infinite sequence of nontiles P_n can be obtained by applying the operation $I(G)$ to prisms over n-gons, when $n \geq 4$; see Fig. 12-5.

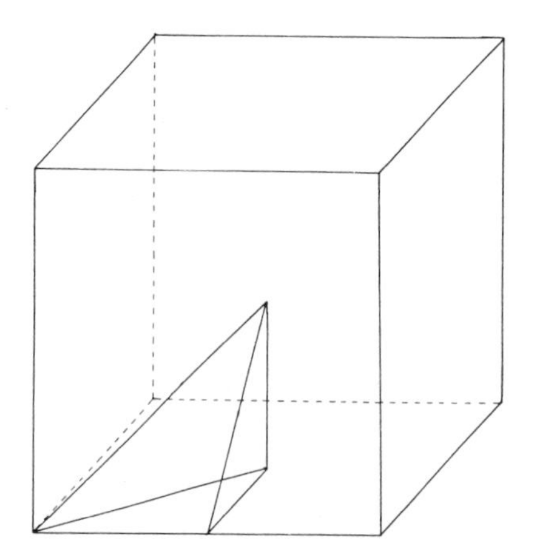

Fig. 12-6. A fundamental region for the symmetry group of the regular tessellation of E^3 by cubes.

The resulting polytopes have $3n$ vertices and $3n + 2$ facets. All vertices are 4-valent and of type $[3, 4, 3, n]$. For $n = 4$ we get a polytope combinatorially equivalent to the cuboctahedron. One might conjecture that the cuboctahedron (with 12 vertices and 14 facets) is the "smallest" nontile in three dimensions. It is worth noting that no simple or cubical nontiles are known yet. (A 3-polytope is called simple or cubical, if all its vertices are 3-valent or all its facets are quadrangles, respectively.) I conjecture here that simple nontiles exist, but I doubt the existence of cubical nontiles.

As an example of a class of 3-polytopes which *do* tile E^3, Branko Grünbaum, Peter Mani, and Geoffrey Shephard have shown that all simplicial polytopes give locally finite face-to-face tilings.[5]

Constructions of Monotypic Tilings

One can also find very nice tilings of E^3 by projections of convex 4-polytopes. Intuitively we expect certain connections between the properties of 3-polytopes which are the combinatorial prototiles of monotypic tilings of E^3 and 3-polytopes which are the 3-facet types of equifaceted 4-polytopes in E^4. (A $(d + 1)$-polytope is said to be equifaceted of type P if all its facets are isomorphic to a single d-polytope P. P is a nonfacet if it is not the facet-type of an equifaceted $(d + 1)$-polytope.)

Certainly a combinatorial prototile will not be a facet type in general, since this does not even hold in dimension 2; in fact, by Euler's theorem, a 3-polytope cannot have all its facets n-gons with $n > 5$, while on the other hand there are no restrictions on n for tilings of the plane by convex n-gons. However, it has been proved[3] that the reverse is true: every 3-polytope that is the facet-type of an equifaceted 4-polytope is also the combinatorial prototile of a locally finite face-to-face tiling of E^3. (It follows that all of the nontiles described above are also nonfacets!) The monotypic tiling of E^3 is obtained from the 4-polytope by an infinite sequence of projections. The construction works equally well in higher dimensions. Unfortunately this projection method produces non-normal monotypic tilings. However, in some instances another projection method provides normal face-to-face tilings with only finitely many isometric prototiles:

Theorem 12.2. Let the convex 3-polytope P be realized as the facet-type of an equifaceted convex 4-polytope Q with at least one 4-valent vertex. Let m denote the number of facets of Q. Then P is the combinatorial prototile of a monotypic face-to-face tiling of E^3 with only $m - 4$ isometric prototiles.

To prove this, let X be a 4-valent vertex of Q and T the 3-simplex whose vertices are the four neighboring vertices of X in the boundary complex of Q. By projecting Q centrally from X onto the affine hull of T, we get a face-to-face dissection of T into convex polytopes isomorphic to P. The 3-polytopes in the dissection are the images of facets of Q under the projection.

Next, we make use of the well-known fact (see, e.g., Coxeter[6]) that the fundamental region for the symmetry group of the regular tessellation of E^3 by cubes is a 3-simplex T' (see Fig. 12.6). Mapping T affinely onto T' we turn the dissection of T into a dissection of T'. Then, if we apply all the symmetries of the tessellation, we obtain a tiling of the whole space, in which each tile is congruent to one of the 3-polytopes in the dissection of T'. Since the number of 3-polytopes in this dissection equals the number of facets of Q not containing X, that is, $m - 4$, the tiling has at most $m - 4$ isometric prototiles. In particular, the tiling is monotypic of type P, since P is the facet-type of the equifaceted 4-polytope Q. The tiling is face-to-face because T' is a fundamental region for the symmetry group of the tessellation of E^3 by cubes, and that group is generated by the reflections in the planes bounding T'.

Another generalization of the tiling problem would be to relax the condition that the tiling be face-to-face. That is, we can ask: Is every polytope the combinatorial prototile of a tiling that is not necessarily face-to-face? And here we get the surprising result that the answer is definitely positive. But the construction of these tilings is too complicated to discuss it in detail here; for further discussion, see the first reference in note 3.

Finally, if we are willing to relax the conditions that our tiles must be convex, then we get some nice things. Figure 12-7 is a tiling by hexagonal pyramids, derived from the hexagonal tiling of the plane. It has only three types of pyramids; one of them is not convex.

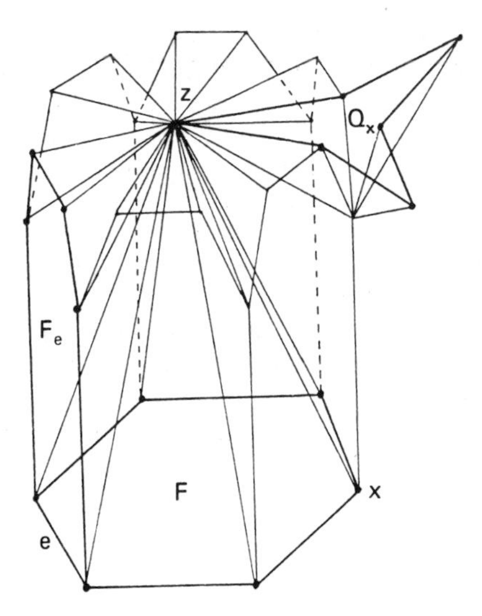

Fig. 12-7. A face-to-face tiling of E^3 by hexagonal pyramids with three prototiles, derived from the usual hexagonal tessellation of the plane. To each hexagon F of the tessellation belong seven pyramids with a common apex Z. One has base F and is surrounded by six congruent pyramids, whose bases F_e share an edge e with F and lie in planes orthogonal to F. To each vertex X of the tessellation corresponds a nonconvex hexagonal pyramid with base Q_X. By taking suitable layers of this arrangement a tiling of E^3 by hexagonal pyramids arises.

Related Problems

So far our investigations of monotypic tilings were restricted to the case where the tiles of the tilings were topological balls, generally convex polytopes. But of course, we can as well consider combinatorial prototiles of other homeomorphism types, and we can ask if they admit locally finite tilings of the Euclidean space or not. Of particular interest is the case where the polytope P is a "polyhedron" in E^3 bounded by a closed polyhedral 2-manifold, for example a toroid. There are examples[7] of space-filling toroids! On the other hand, the existence of toroidal nontiles is almost trivial. In fact, if a toroid has only vertices of valence > 5, then isomorphic copies of it will not fit together to form a vertex-figure of a locally finite face-to-face tiling: this is an easy consequence of Euler's theorem applied to the spherical complex determined by the vertex-figure. Polyhedra with this property have recently been studied.[8]

Another direction for research is offered by replacing the Euclidean space as the underlying space of the tilings by a topological 3-manifold M, and investigating monotypic tilings of M by topological polytopes or tiles of another homeomorphism type.[9]

Notes

Acknowledgments: I would like to thank Professors Ludwig Danzer and Branko Grünbaum for many helpful suggestions.

[1] B. Grünbaum and G. C. Shephard, *Tilings and Patterns* (San Francisco: W. H. Freeman, 1986).

[2] L. Danzer, B. Grünbaum, and G. C. Shephard, "Does Every Type of Polyhedron Tile Three-space?" *Structural Topology* 8 (1983):3–14. D. G. Larman and C. A. Rogers, "Durham Symposium on the Relations between Infinite-dimensional and Finitely-dimensional Convexity," *Bulletin of the London Mathematical Society 8* (1976):1–33.

[3] E. Schulte, "Tiling Three-space by Combinatorially Equivalent Convex Polytopes," *Proceedings of the London Mathematical Society 49,* no. 3,

(1984):128–40. E. Schulte, "The Existence of Nontiles and Nonfacets in Three Dimensions," *Journal of Combinatorial Theory,* ser. A, *38* (1985):75–81. E. Schulte, "Nontiles and Nonfacets for the Euclidean Space, Spherical Complexes and Convex Polytopes," *Journal für die Reine und Angewandte Mathematik 352* (1984):161–83.

[4] Branko Grünbaum, *Convex Polytopes* (New York: John Wiley and Sons, 1967).

[5] B. Grünbaum, P. Mani-Levitska, and G. C. Shephard, "Tiling Three-dimensional Space with Polyhedral Tiles of a Given Isomorphism Type," *Journal of the London Mathematical Society 29,* no. 2 (1984):181–91.

[6] H. S. M. Coxeter, *Regular Polytopes,* 3rd ed. (New York: Dover Publications, 1973).

[7] D. Wheeler and D. Sklar, "A Space-filling Torus," *The Two-Year College Mathematics Journal 12,* no. 4 (1981):246–48.

[8] P. McMullen, C. Schulz, and J. M. Wills, "Polyhedral Manifolds in E^3 with Unusually Large Genus," *Israel Journal of Mathematics 46* (1983):127–44.

[9] E. Schulte, "Regular Incidence-polytopes with Euclidean or Toroidal Faces and Vertex-figures," *Journal of Combinatorial Theory,* ser. A, *40* (1985):305–30.

13

Duality of Polyhedra

BRANKO GRÜNBAUM AND G. C. SHEPHARD

An author who wishes to use material from mathematical folklore[1] faces two unpleasant alternatives: either to quote the result (qualified by a phrase such as "it is well known that") or to prove it. The latter course may lead a referee or reviewer to ridicule the effort, and possibly identify it with an ancient result from the *Upper Slobbovian Journal of Recreational Mathematics,* or some other equally obscure source. Usually the situation is even worse because much folklore is imprecisely formulated (if, indeed, one can say that it is formulated at all) and quite frequently it is definitely wrong. The purpose of this note is to show that many of the "well-known facts" about *duality of polyhedra* are of latter kind, and it is well worth some effort to clarify the situation and arrive at the truth.

So far as we can ascertain, some, if not all, of the following "facts" are generally accepted among working mathematicians:

1. For every polyhedron P there exists a dual polyhedron P^*, and the dual of P^* is equal to P, or at least similar to it.
2. If a polyhedron P has any convexity, symmetry, transitivity, or regularity properties, then the same is true, possibly in an appropriately modified form, for P^*.
3. The dual of a polyhedron can always be obtained by reciprocation with respect to a suitable sphere or more general quadric.
4. Duality between polyhedra is consistent with the combinatorial duality of their boundary complexes, that is, the cell complexes whose cells are the proper faces of the polyhedron. Moreover, in the particular case of polyhedra in three-dimensional Euclidean space, duality is consistent with the duality of planar graphs.
5. With appropriate interpretations, projective duality, duality in algebra and duality (conjugacy) in functional analysis are consistent with duality for polyhedra.

Before we examine these statements in detail, it is necessary to define some of the terms that we shall use. For simplicity we restrict attention to polyhedra in Euclidean space of three dimensions, that is to say, to compact 2-manifolds in E^3 which have no boundary and can be expressed as a finite union of plane polygonal regions. If these regions are such that no two adjacent ones are coplanar, then they are called the *faces* of the polyhedron; the *edges* and *vertices* of the polyhedron are the edges and vertices of its faces. We shall sometimes use words like "convex" and "star-shaped" to describe a polyhedron P though, strictly speaking, these terms apply to the polyhedral solid bounded by P. A vertex, or an edge, of P is said to be *convex* if the intersection of the polyhedral solid bounded by P with a sufficiently small spherical ball, centered at the vertex or at an interior point of the edge, is a convex set.

By the *elements* of a polyhedron P we mean the family consisting of all the faces, edges, and vertices of P. Two polyhedra P_1 and P_2 are *isomorphic* (or *combinatorially equivalent,* or *of the same type*) if there exists a one-to-one correspondence (bijection) between the elements of P_1 and the elements of P_2 which preserves the relation of inclusion between the elements. In a similar manner, P_1 and P_2 are called *combinatorial duals* of each other if there exists a bijection that is inclusion-reversing; this is the concept referred to in statement 4. It extends in a natural way to topological complexes more general than polyhedra (in particular, to planar graphs or maps, and to maps on other 2-manifolds). The notions have their roots in the eighteenth-century works of Euler (see Federico) and Meister (see Brückner).[2] Dual polyhedra are called *reciprocals* with respect to a sphere S (see statement 3) if each face of one is the polar[3] with respect to S of the dually corresponding vertex of the other.

Now let us examine statements 1 to 5. The first of these is formulated in a misleading way since it seems that, in general, it is impossible to define in any useful or canonical way a *unique* polyhedron P^* as *the dual* to a given polyhedron P. In other words, though polyhedra that are combinatorial duals of P may exist, there appears to be no reasonable way in which one of these can be singled out and called *the dual* polyhedron to P. At best, we must therefore think of duality as a relation between isomorphism classes of polyhedra rather than between individual polyhedra. In this generalized sense, the second part of statement 1 is true, but most parts of statement 2 become vacuous.

If we consider only convex polyhedra—which may be thought of as a very simple special case—reciprocation can always be applied; as the center of the reciprocating sphere S we may take any point in the interior of the polyhedral solid enclosed by the polyhedron. In this way we can construct a convex polyhedron P^* dual to P, but even this does not lead to a unique dual since there is arbitrariness in the choice of the center and the radius of the sphere S. In fact all the duals obtained in this way are projectively equivalent to each other, so the same difficulties as before still arise except that projective equivalence classes, rather than isomorphism classes, need to be considered.

In this connection we remark that reciprocation is the *only* known method of actually *constructing* a polyhedron P^* dual to a given polyhedron P. In some special cases which we shall now examine, reciprocation can lead to an essentially unique dual polyhedron, and then statements 1 to 4 become true for this restricted meaning of duality. It seems likely that the existence of these special cases, and the emphasis on them by many authors, has led to the misconception that these statements are true more generally.

The first special case is when P is one of the five regular polyhedra. Such polyhedra have a natural center O which is the circumcenter, incenter, and centroid of P. If we choose S as any sphere centered at O, then the reciprocal is a dual polyhedron P^* (defined within a similarity, its size depending upon the radius of S) which is also regular. The reciprocal of P^* with respect to the same sphere S is P, and assertion 1 is true. In particular, if S is chosen so that the edges of P are tangent to it, then the edges of P^* have the same property, and intersect those of P at right angles. This leads many authors to describe a certain regular octahedron as *the* dual of a given cube, a certain regular icosahedron as *the* dual of a given regular dodecahedron, and a regular tetrahedron as self-dual. Similar is the situation concerning the Archimedean (uniform) polyhedra and the polyhedra reciprocal to them.

There are more general cases in which a natural center of P exists. Suppose P is *isogonal;* that is, the symmetry group of P is transitive on its vertices. Then all the vertices of P lie on a sphere S which may be used in the process of reciprocation. To each vertex of P corresponds a face of the reciprocal P^*, and the planes of these faces are the tangent planes to S at the vertices of P; moreover, P^* is *isohedral;* that is, its symmetry group is transitive on its faces. Similar considerations apply if P is isohedral and then P^* is isogonal, or if P is *isotoxal* (the symmetry group of P is transitive on the edges of P, which are therefore tangent to a sphere) and then P^* is isotoxal as well. In a similar way, if P has an axis of rotational symmetry R, or a plane of reflective symmetry E, then reciprocation with respect to a sphere centered at a point of R, or of E, will lead to a dual polyhedron P^* which has the same symmetry as P. Therefore in all these cases statements 1 to 4 are true.

It should be carefully noted that the discussion in the previous two paragraphs depended essentially on the fact that only convex polyhedra are under consideration. All the situations in which statement 5 can be successfully applied also deal with such polyhedra only. If we drop the convexity restriction, then despite some encouraging signs, things go sadly awry.

These encouraging signs appear when we consider examples such as the following, which are culled from the rather meager literature on nonconvex polyhedra. The icosahedron in Fig. 13-1a (see Jessen[4]) and the dodecahedron in Fig. 13-1b (see Ounsted, Stewart, and Grünbaum and Shephard[5]) are duals of each other and have the same group of symmetries. The first is isogonal and the second is isohedral; nonconvex edges of both correspond to each other. Another dual pair consists of the well-known Császár polyhedron (see, for example, Császár, Grünbaum, Gardner, and Stewart[6]) and the less well-known but remarkable Szilassi polyhedron (see Szilassi, Gardner, Stewart and Gritzmann[7]). We recall that the Császár polyhedron is a triangulation of the torus with 7 vertices, 21 edges, and 14 triangular faces and the Szilassi polyhedron is toroidal with 7 hexagonal faces, 21 edges, and 14 vertices. These polyhedra are not only duals of each other but also have analogous symmetry and convexity properties. Such examples may seem to vindicate the folklore and imply that it may be possible to prove the statements listed at the beginning of this paper once we have learned how to deal with nonconvex polyhedra and, in particular, how to construct their duals.

This last is, in a sense, the nub of the problem: there is no difficulty finding topological complexes that are duals of any given polyhedron, but finding a *dual polyhedron* is a much more elusive goal. An indication that this goal may be unattainable is implied by recent results concerning isohedral and isogonal polyhedra (Grünbaum and Shephard[8]): isohedral polyhedra are always star-shaped and have star-shaped faces, whereas isogonal polyhedra have convex faces but need not even be simply connected. Hence these kinds cannot be related by duality.

To illustrate some of the difficulties we shall consider a very simple example. In Fig. 13-2a we show a polyhedron P which may be described as a cube with a four-sided pyramid adjoined to one of its faces. Reciprocation with respect to a suitable sphere (for example, the circumsphere of the cube) leads to the truncated octahedron of Fig. 13-2b, which is

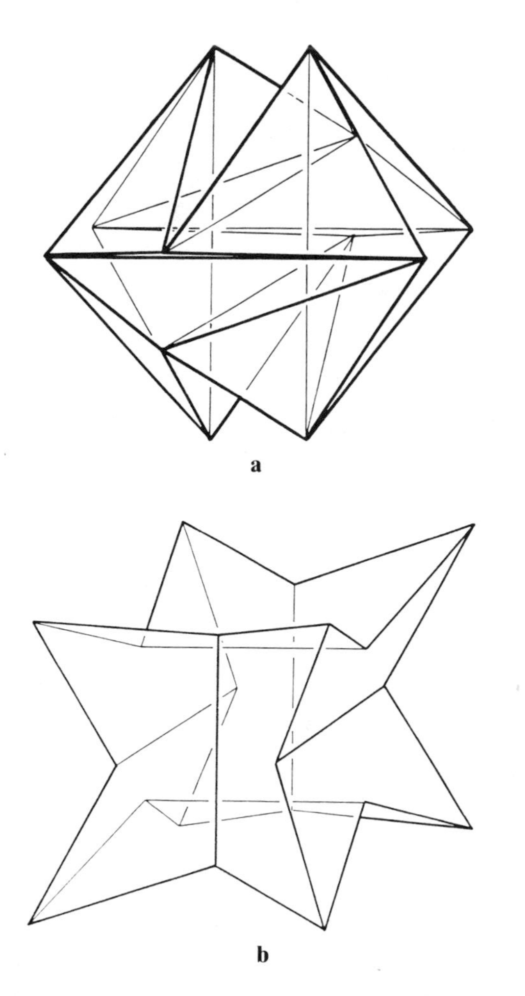

Fig. 13-1. A dual pair of nonconvex polyhedra: (a) an isogonal icosahedron; (b) an isohedral dodecahedron.

therefore a dual P^* of P. In this particular case it happens that P and P^* are isomorphic, so P is self-dual, though this fact is only incidental to the following discussion. Now consider the nonconvex polyhedron P_1 of Fig. 13-2c. Since this is isomorphic to P, every dual of P_1 will be isomorphic to P^*. However it is not hard to see that no such dual has corresponding convexity properties: P_1 has four nonconvex edges meeting at a vertex, so a dual *ought* to have four nonconvex edges bounding a quadrangular face. Further, the four congruent nonconvex vertices of P_1 ought to correspond to four congruent nonconvex quadrangular faces of its dual. Examination of the various possibilities, such as those shown in Fig. 13-2d and 13-2e, shows that no such dual exists.

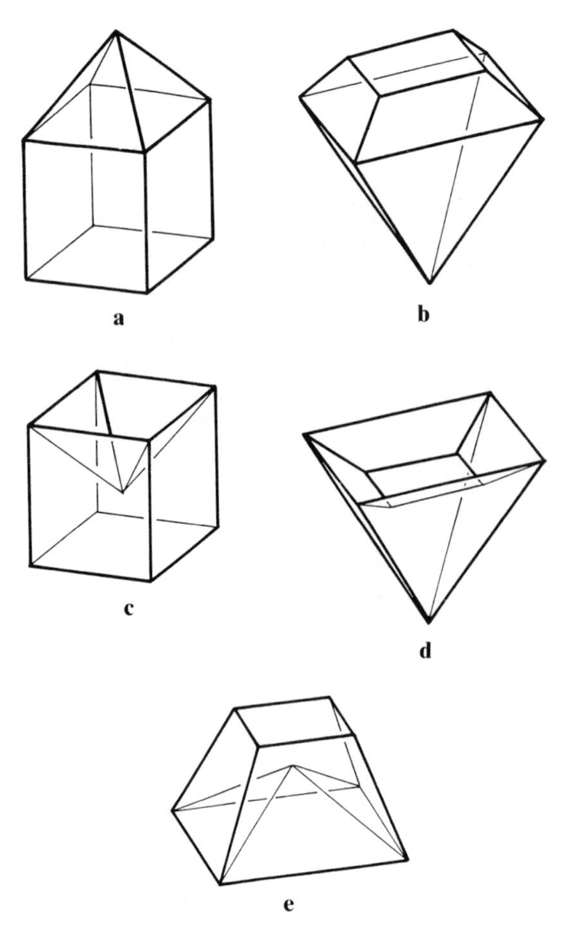

Fig. 13-2. An illustration of the problem of duality for nonconvex polyhedra. The convex polyhedra illustrated in (a) and (b) are duals of each other, but the isomorphic nonconvex polyhedra in (c), (d), and (e) have no duals which preserve convexity properties.

Hence, even in this very simple case, convexity properties cannot be preserved in duality.

It may seem that there is a very simple way out of this impasse. Let us start with any polyhedron P and then construct Q in the following way. After choosing a sphere S centered at an interior point of P, we define the vertices of V_j^* of Q as the polars with respect to S of the faces F_j of P. If two faces F_1 and F_2 of P meet in an edge, then the join of corresponding vertices V_1^* and V_2^* is defined to be an edge of Q. If F_1, F_2, . . . , F_r is a circuit of adjacent faces round a vertex of P, then the corresponding vertices of V_1^*, V_2^*, . . . , V_r^* are coplanar and so may be used to define a face of Q. Proceeding in this way, all the elements of Q may be defined, and so Q is completely determined. (Alternatively, and equivalently, the construction may be reversed: consider the set of planes which are the polars of the vertices of P, and then the edges and vertices of Q are defined as intersections of suitable subsets of these planes.) It is easy to verify that this procedure can be carried out for all polyhedra P, and that Q has the same symmetry and convexity properties as P. So it appears that Q is an obvious candidate for the dual of P. Unfortunately this is not the case since in general (and, in particular, if P is not convex) *Q will not even be a polyhedron as we have defined the word*. The union of the faces of Q *will not form a manifold either because they are not polygons or because they are mutually intersecting.*

We illustrate these assertions by an example that is chosen so as to be computationally and graphically easy to follow. It is not, in any sense, unique as the reader will discover by using the same procedure to find "duals" of the polyhedra in Figs. 13-2c, 13-2d, and 13-2e or of the toroidal isogonal polyhedra described in our paper cited in note 5. Consider the polyhedron P in Fig. 13-3a. It is obtained from the octagonal prism in Fig. 13-3b by replacing its mantle of 8 rectangular faces by one of 16 triangles. The polyhedron P has convex faces and is isogonal, but since some of its edges and all of its vertices are nonconvex, so a dual P^* (if it exists) should have convex vertices, nonconvex faces, and be isohedral. Let us attempt to find such a dual by applying the construction for Q described in the previous paragraph. Take S (the reciprocating sphere) as the sphere that passes through the vertices of P. The set of planes tangent to S at these vertices (their polars) is easily visualized; it is the set of planes determined by the faces of a regular octagonal bipyramid. These will be the face planes of Q. To determine the edges and vertices of Q we proceed as described above. There is an immediate simplification: the fact that P is isogonal implies that Q will be isohedral, so it is only necessary to determine *one* face of Q. The other faces will then arise by applying the symmetries of P to this face. Let the plane T be tangent to S at the vertex A′ of P. Apart from the tangent plane to S at the point E, which is parallel to T, each of the

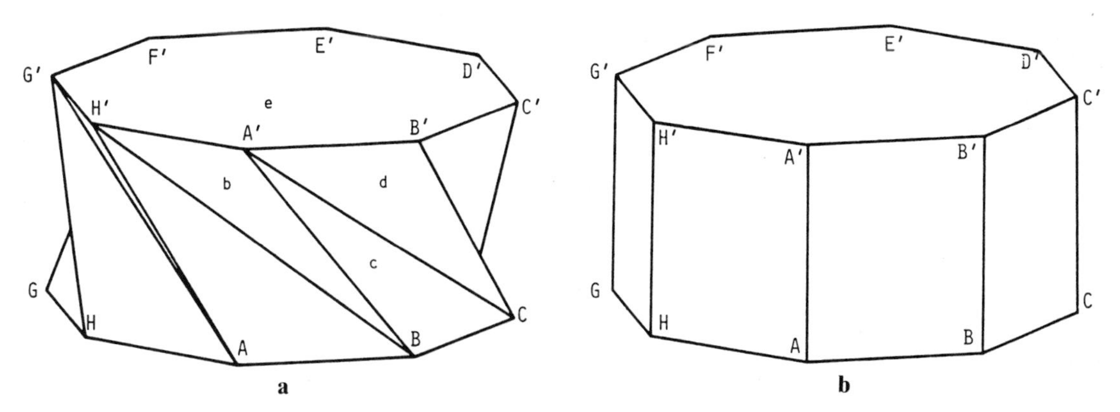

Fig. 13-3. A nonconvex isogonal antiprism (a), with the same vertices as a prism (b). There exists no polyhedron dual to the antiprism, if "polyhedron" is understood in the usual sense.

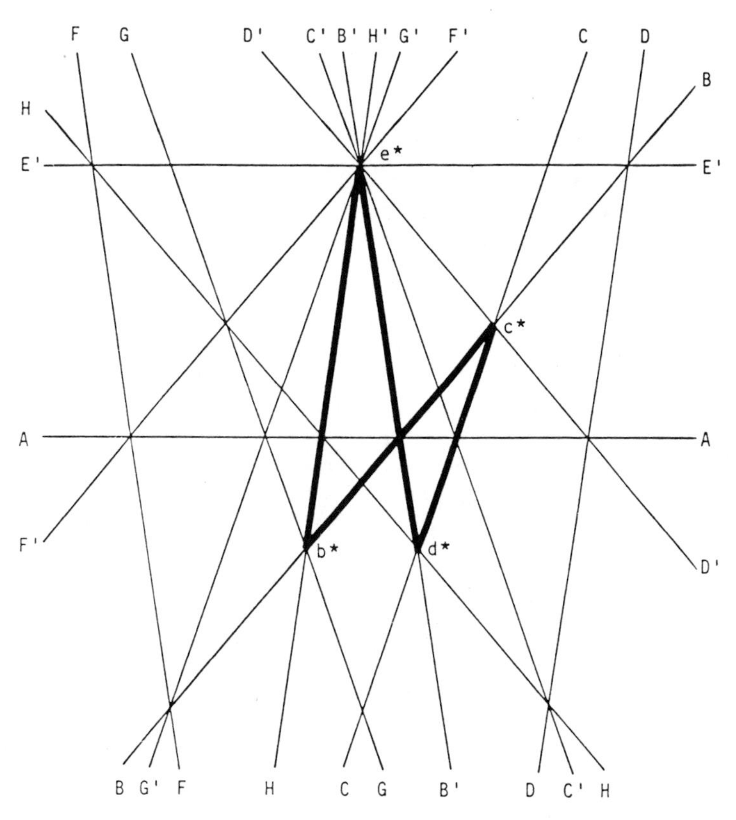

Fig. 13-4. The arrangement of 14 lines in a plane, which can be used to prove the nonexistence of a polyhedron dual to the antiprism in Fig. 3(a).

other 14 face planes will meet T in a line. In Fig. 13-4 we show the arrangement of 14 lines in T obtained in this way. Each line is marked with a letter indicating the vertex of P at which the corresponding plane touches S. The four faces b, c, d, and e of P contain a vertex A′, so the corresponding points b*, c*, d*, and e* (defined as intersections of appropriate sets of lines in T) are the vertices of Q that lie in T. Since b, c, d, e form a circuit of faces at the vertex A′ of P, the points b*, c*, d*, e* (in this order) should form a circuit of vertices around the corresponding face of Q. This is the nonconvex quadrangular face that we have been seeking. Unfortunately, it is not only nonconvex, but it is also self-intersecting. Hence it is not a polygon and so it is not acceptable as a face of *any* polyhedron. This Q is not a polyhedron, and our attempt to find a dual of P has failed.

We can summarize the situation by saying that in trying to establish a dual of each three-dimensional polyhedron we can *either* prescribe the character of the polyhedron as a 2-manifold, *or* we can adopt a wider definition of "polyhedron" admitting mutual intersections and self-intersections of the polygonal faces. In the later case, as we have seen, duals can always be constructed by polarity. Then, if interpreted in a suitable manner, statement 2 at the beginning of this chapter will be true. It is interesting to note that this generalization of the concept of a polyhedron was adopted in part some 120 years ago by Möbius, and has often been applied (without any explicit definitions) in the study of regular, uniform, or other very special polyhedra. (See, for example, Hess, Coxeter, Coxeter, Longuet-Higgins, and Miller, Skilling, Wenninger, Brückner, and Norgate.[9]) However it seems that no systematic study of "polyhedra" in this sense has ever been carried out, although there seems to be no reason why a completely satisfactory theory could not be developed.

So it appears that the various statements concerning duality made at the beginning of this chapter are mutually incompatible, and that the folklore is only a superstition! It may well be that this is the case; the brunt of our thesis is that, with the approaches to duality followed so far, this incompatibility is inescapable. It is not impossible, of course, that with some suitable generalizations, a theory might be formulated in which all the different aspects will fall into place so that folklore will be vindicated. We would like to suggest that the chief need at present is a theory of topological 2-manifolds in which faces may mutually intersect and even self-intersecting polygons are admissible as faces. We know of no attempts at such a theory, but it seems that it would be worth the effort to develop it, especially as it could have far-reaching applications and implications for other branches of mathematics.

Notes

Acknowledgment: This research was supported by the National Science Foundation Grant MCS8301971.

[1] The expression "mathematical folklore" refers to results that most mathematicians take for granted, but which may never have been proved to be true. (Indeed, some of them are not true.) It is widely believed that the first person to call attention to this phenomenon was the French mathematician Jean Dieudonné.

[2] L. Euler, "Elementa doctrinae solidorum," *Novi Comm. Acad. Sci. Imp. Petropol. 4* (1752–53):109–40; see also in *Opera Mathematica,* vol. 26, pp. 71–93. P. J. Federico, *Descartes on Polyhedra* (New York: Springer-Verlag, 1982), pp. 65–69. A. L. F. Meister, "Commentatio de solidis geometricis," *Commentationes soc. reg. scient. Gottingensis,* cl. math. 7 (1785). M. Brückner, *Vielecke und Vielfläche* (Leipzig: Teubner, 1900), p. 74.

[3] For a definition of *polar* see H. S. M. Coxeter, *Regular Polytopes,* 3rd ed. (New York: Dover Publications, 1973), p. 126.

[4] B. Jessen, "Orthogonal Icosahedron," *Nordisk Matematisk Tidskrift 15* (1967):90–96.

[5] J. Ounsted, "An Unfamiliar Dodecahedron," *Mathematics Teaching, 83* (1978):45–47. B. M. Stewart, *Adventures among the Toroids,* 2nd ed. (Okemos, Mich.: B. M. Stewart, 1980), p. 254. B. Grünbaum and G. C. Shephard, "Polyhedra with Transitivity Properties," *Mathematical Reports of the Academy of Science* [Canada] *6,* (1984):61–66.

[6] A. Császár, "A Polyhedron without Diagonals," *Acta Scientiarum Mathematicarum 13* (1949):140–42. B. Grünbaum, *Convex Polytopes* (London: Interscience, 1967), p. 253. M. Gardner, "On the Remarkable Császár Polyhedron and its Applications in Problem Solving," *Scientific American 232,* no. 5 (May 1975):102–107. Stewart, *Adventures among the Toroids,* p. 244.

[7] L. Szilassi, "A Polyhedron in Which Any Two Faces Are Contiguous" [In Hungarian, with Russian summary], *A Juhasz Gyula Tanarkepzo Foiskola Tudomanyos Kozlemenyei,* 1977, Szeged. M. Gardner, "Mathematical Games, in Which a Mathematical Aesthetic Is Applied to Modern Minimal Art," *Scientific American 239,* no. 5 (November 1973):22–30. Stewart, *Adventures among the Toroids,* p. 244. P. Gritzmann, "Polyedrische Realisierungen geschlossener 2-dimensionaler Mannigfaltigkeiten im R^3," Ph.D. thesis, Universität Siegen, 1980.

[8] B. Grünbaum and G. C. Shephard, "Polyhedra with Transitivity Properties."

[9] A. F. Möbius, "Über die Bestimmung des Inhaltes eines Polyeders" (1865), in *Gesammelte Werke,* vol. 2 (1886), pp. 473–512. E. Hess, *Einleitung in die Lehre von der Kugelteilung* (Leipzig: Teubner, 1883). Brückner, *Vielecke und Vielfläche,* p. 48. H. S. M. Coxeter, *Regular Polytopes.* H. S. M. Coxeter, M. S. Longuet-Higgins, and J. C. P. Miller, "Uniform Polyhedra," *Philosophical Transactions of the Royal Society* [London], sec. A, *246* (1953–54):401–450. J. Skilling, "The Complete Set of Uniform Polyhedra," *Philosophical Transactions of the Royal Society* [London], sec. A, *278* (1975):111–135. M. J. Wenninger, *Polyhedron Models* (London: Cambridge University Press, 1971). M. Norgate, "Non-Convex Pentahedra," *Mathematical Gazette 54* (1970):115–24.

14

Polyhedral Analogues of the Platonic Solids

J. M. WILLS

In this chapter we investigate polyhedra in Euclidean 3-space, E^3, without self-intersections and with some local and global properties related to those of the Platonic solids. A *polyhedron* is the geometric realization of a compact 2-manifold in E^3 such that its 2-faces are (not necessarily convex) plane polygons bounded by finitely many line segments. Adjacent faces and edges are not coplanar. A *flag* of a polyhedron P is any triple consisting of a vertex, an edge, and a face of P, all mutually incident.

Perhaps the most important property of the Platonic solids is that the set of their flags is transitive under the corresponding full Platonic symmetry group, consisting of all the rotations and reflections. There are no other compact (i.e., finite) and self-intersection-free polyhedra in E^3 with this property, so in order to carry the theory further one has to weaken this strict condition. A first step in this direction is to replace the global algebraic property of flag transitivity by the local property that all flags are combinatorially equivalent:

Definition 14.1. An equivelar manifold (i.e., with equal flags) is a polyhedron with the property: All faces are p-gons; all vertices are q-valent ($p \geq 3$, $q \geq 3$). Notation $\{p, q; g\}$, where g denotes the genus of the manifold.

For $g = 0$ (the sphere) one obtains the five Platonic solids, and for $g = 1$ (the torus) one obtains infinite series of tori, which were investigated long before. So in the following, all new polyhedra have genus $g > 1$.

In the definition it is not required that the faces are regular or congruent to each other, and indeed all known equivelar manifolds contain at least one nonregular face. It has been shown[1] that there exist infinitely many equivelar manifolds (see also Problems at end of this chapter). So equivelarity alone is too weak to yield close analogues to the Platonic solids and one has to find appropriate further conditions. It turns out that global algebraic conditions seem to be the most successful conditions, namely transitivity properties under certain symmetry and automorphism groups. (*Symmetries* are isometries of E^3 which map the polyhedron onto itself, and *automorphisms* are combinatorial isomorphisms.) Among the various possibilities we choose one in the following section which leads to nice polyhedra.

Platonohedra

Definition 14.2. A Platonohedron is an equivelar manifold such that a group isomorphic to its symmetry group acts transitively on its vertices or faces.

Simple combinatorial arguments show that there are only finitely many Platonohedra. Seven have been found so far: $\{3, 8; 3\}$, $\{3, 8; 5\}$, $\{4, 5; 7\}$, $\{5, 4; 7\}$, $\{3, 9; 7\}$, $\{9, 3; 7\}$, and $\{3, 8; 11\}$. Figures 14-1–14-4 show some of them. We let $f = (f_0, f_1, f_2)$ denote the number of vertices, edges, and faces. Because of their equivelarity the Platonohedra can be represented in a flag diagram (Fig. 14-13), explained later. The Platonohedra have the same rota-

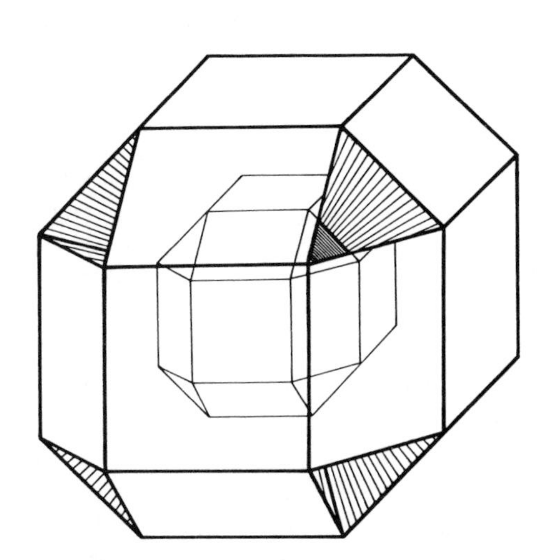

Fig. 14-1. The Platonohedron {4, 5; 7}; $f = 12(4, 10, 5)$.

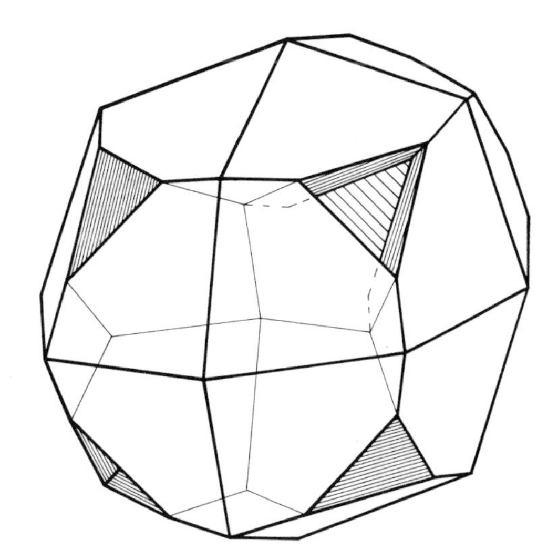

Fig. 14-2. The Platonohedron {5, 4; 7}; $f = 12(5, 10, 4)$.

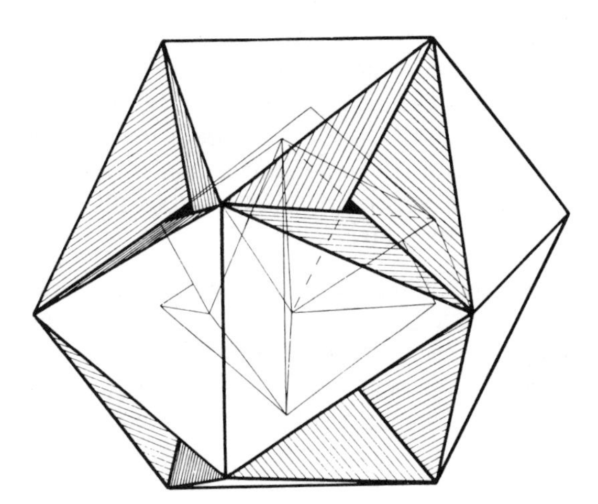

Fig. 14-3. The Platonohedron {3, 9; 7}; $f = 12(2, 9, 6)$.

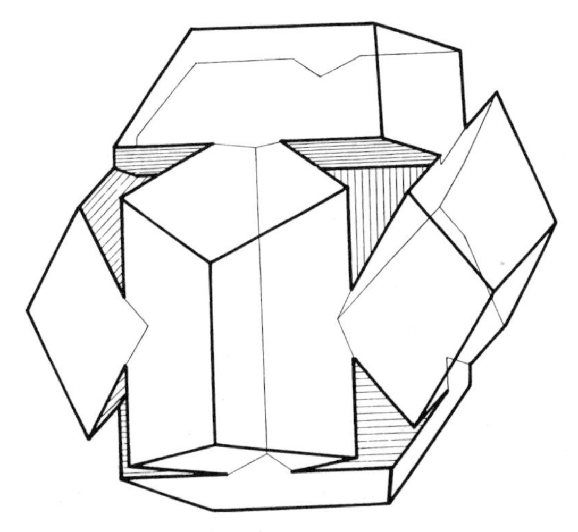

Fig. 14-4. The Platonohedron {9, 3; 7}; $f = 12(6, 9, 2)$.

tion group as the corresponding Platonic solids, whereas the usual reflection in a plane is replaced, in the case of vertex-transitivity, by a reflection in a plane and a simultaneous inside-outside inversion. For the face-transitive Platonohedra the analogous reflection-inversion can be described for the normal-vector.

Let us consider the simplest cases, {4, 5; 7} and {5, 4; 7} (see Figs. 14-1 and 14-2). The 48 vertices of the {4, 5; 7} lie pairwise on 24 rays which have their common endpoint at the rotation center. The inside-outside inversion interchanges the "inner" and "outer" vertices. The situation for the {5, 4; 7} is analogous: its 48 faces fall into two classes of 24 "outer" and 24 "inner" faces and each "outer" face corresponds to one "inner" face. The inside-outside inversion interchanges the corresponding "inner" and "outer" faces. Clearly the inside-outside inversion is no isometry. Nevertheless it has a geometric meaning and is not a purely combinatorial automorphism. This corresponds to the fact that the usual reflection is an improper movement.

If one restricts the conditions in Definition 14.2 and requires transitivities only under symmetries, then there exist no face-transitive

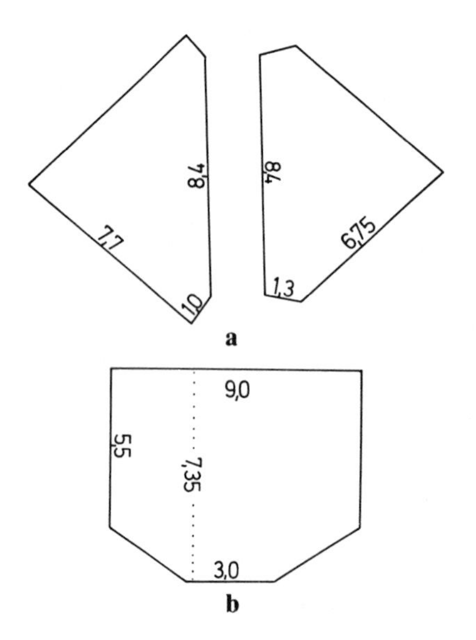

Fig. 14-5. (a) The two kinds of pentagons used in construction of the Platonohedron {5, 4; 7}. (b) The nonregular hexagonal face used in the construction of the regular polyhedron {6, 4; 6}.

polyhedra with $g > 0$, as Grünbaum and Shephard have shown.[2] But they found three remarkable vertex-transitive Platonohedra {3, 8; g}, $g = 3, 5, 11$.

The first vertex-transitive polyhedron with $g > 0$ to be discovered was the flat torus ($g = 1$), which Brehm found in 1978; it is vertex-transitive under the dihedral group.[3] It is three-dimensional; its name comes from the fact that the Gauss-curvature in its vertices is zero. It was rediscovered by Grünbaum and Shephard, who found two more polyhedra with vertex-transitivity under symmetries[4]; those polyhedra have two combinatorially distinct types of faces, and so they are not considered here.

In contrast with the three Platonohedra of genus 3, 5, and 11, the four others (of genus 7) have two metrically different types of vertices and faces. This "disadvantage" corresponds to the "advantage" that they occur in dual pairs and that both their vertex-figures and their faces have one additional symmetry.

Construction of the Platonohedra

Figures 14-1–14-4 give an impression of four Platonohedra,[5] but clearly models of cardboard make it easier to understand them. Here is a brief description of their construction.

- {4, 5; 7} is the simplest Platonohedron. It consists of 18 exterior squares of edge-length a, 18 interior squares of edge-length b, and 24 trapezoids (in the tunnels) of edge-length a, b, and c. Suitable choice: $a = 9$ cm, $b = 6$ cm, $c = 4.2$ cm.
- {5, 4; 7} consists of 24 outer and 24 inner pentagons. Their shape is shown in Fig. 14-5a. One should start with blocks of four outer and four inner pentagons and then fit the six blocks together.
- {3, 9; 7} is easy to construct if one regards the following: the outer 12 vertices are those of a regular icosahedron; the 12 inner vertices are of a distorted icosahedron. Both types of vertices lie on two concentric cubes, where the exterior has, say, twice the edge-length of the interior one. From this it is easy to determine the coordinates of the vertices and so the five different edge-lengths of {3, 9; 7}.
- {9, 3; 7} This "disdodecahedron" consists of 12 outer and 12 inner nonagons. It is much harder to construct than the three others, so we omit the construction.

The remaining Platonohedra belong to the family {3, 8; g}, $g = 3, 5, 11$. Grünbaum and Shephard give a figure of {3, 8; 5} from which a three-dimensional construction is possible.[6] A precise construction of {3, 8; 3} has been described.[7]

Regular Polyhedra

In this section we consider equivelar manifolds with flag transitivities under certain automorphism groups; this has been done by many authors in spaces other than E^3 and in abstract configurations. We use the same notation as above. The case $g = 0$ corresponds to the five Platonic solids and $g = 1$ to the regular toroidal polyhedra, which were found by Coxeter and Moser. Thus we consider $g > 1$; seven of them are shown in Figs. 14-6–14-12. If one considers only polyhedra in E^3 and requires further that the polyhedron has, besides its automorphism group, a nontrivial symmetry group (for example, a Platonic group or a nor-

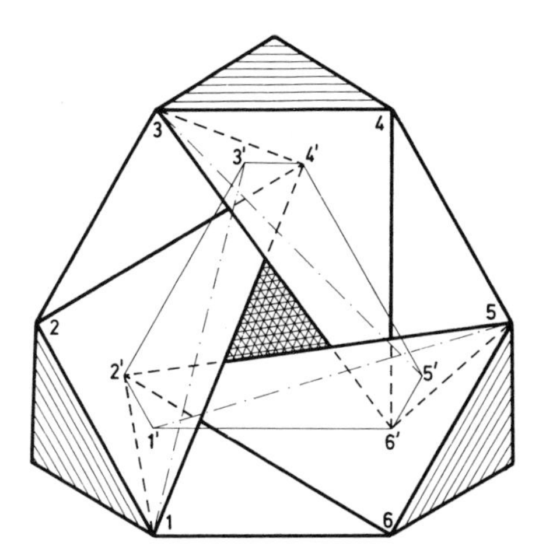

Fig. 14-6. The regular polyhedron $\{3, 7; 3\}$; $f = 4(6, 21, 14)$.

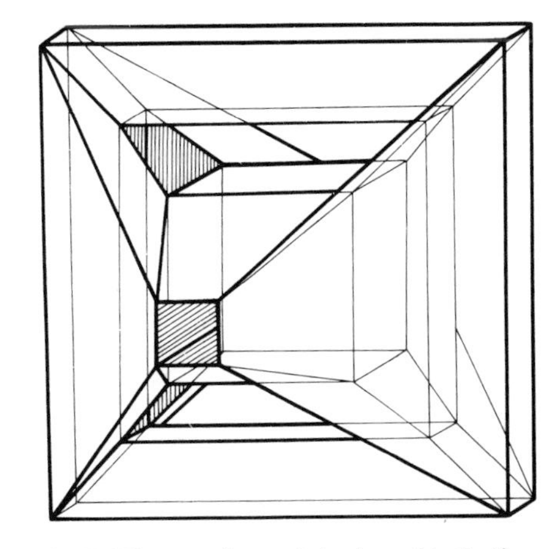

Fig. 14-7. The regular polyhedron $\{4, 5; 5\}$; $f = 8(4, 10, 5)$.

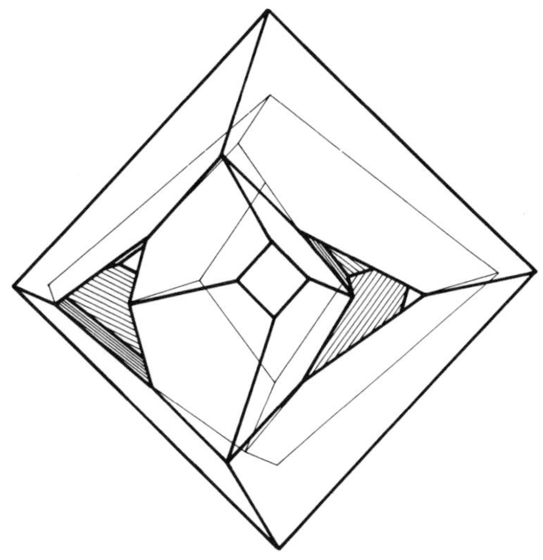

Fig. 14-8. The regular polyhedron $\{5, 4; 5\}$; $f = 8(5, 10, 4)$.

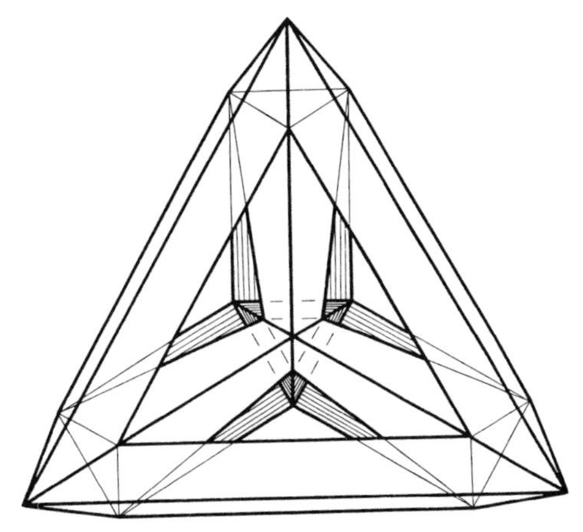

Fig. 14-9. The regular polyhedron $\{4, 6; 6\}$; $f = 10(2, 6, 3)$.

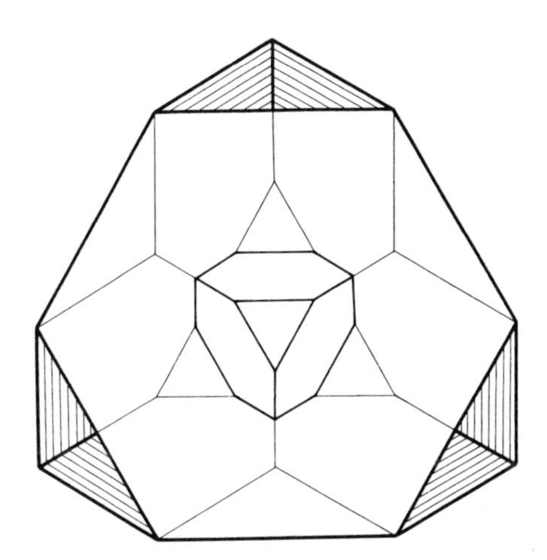

Fig. 14-10. The regular polyhedron $\{6, 4; 6\}$; $f = 10(3, 6, 2)$.

Fig. 14-11. The regular polyhedron $\{4, 8; 73\}$; $f = 144(1, 4, 2)$.

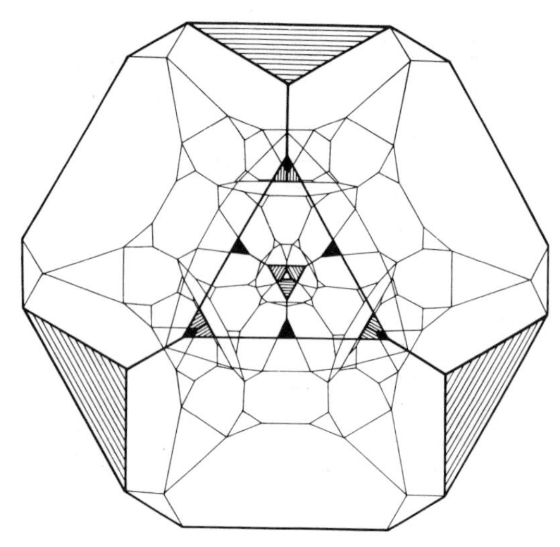

Fig. 14-12. The regular polyhedron {8, 4; 73}; $f = 144(2, 4, 1)$.

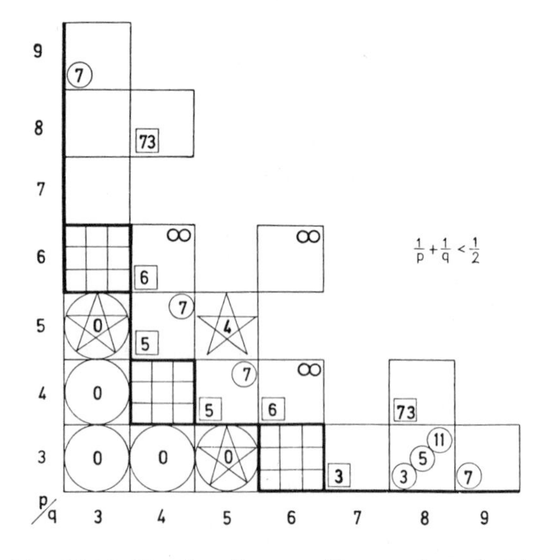

Fig. 14-13. The flag diagram. The numbers denote the genera; large circles denote the five Platonic solids ($g = 0$), the pentagrams denote the four Kepler–Poinsot polyhedra, the frames denote the three regular tilings of the plane and tori ($g = 1$), small circles denote the seven known Platonohedra with $g > 1$, and squares denote seven of the known regular polyhedra with $g > 1$.

mal subgroup of it), then at least six of Coxeter's finite skew polyhedra can be realized in E^3 as equivelar manifolds, namely by suitable projections. If A and S denote the orders of their automorphism group and of their symmetry group, respectively, we find:

{4, 5; 5} and {5, 4; 5}, $A = 320$, $S = 8$
{4, 6; 6} and {6, 4; 6}, $A = 240$, $S = 24$
{4, 8; 73} and {8, 4; 73}, $A = 2{,}304$, $S = 48$.

Coxeter[8] showed that the last four can be realized in E^4 and the first two in E^5 with the appropriate symmetry group. The geometric construction traces back to Alicia Boole-Stott.[9] The projections into E^3 were first given by McMullen, Schulz, and Wills[10] in 1982; it has since been shown that they are the projections of Coxeter's regular skew polyhedra.[11]

The seventh and most spectacular example is the polyhedral realization of Felix Klein's famous quartic (a complex algebraic curve) as a polyhedron of genus 3. Because it is not easy to explain the construction we refer for more details to a paper by Schulte and Wills.[12] It is remarkable that this polyhedron has the same number of vertices, edges, and faces (and so the same genus) as our Platonohedron {3, 7; 3}, although the polyhedra differ combinatorially. This close coincidence was one motivation for finding the "Klein polyhedron." So far we know one infinite series and five single combinatorially regular polyhedra of genus $g > 1$. (See also Problems at end of chapter).

We now describe the construction of the {6, 4; 6} (Fig. 14-10). This polyhedron has four regular hexagonal faces of edge-length a and four regular hexagonal faces of edge-length $3a$. Further it has 12 nonregular hexagonal faces which are all congruent to each other. In Fig. 14-5b we show one of these faces (with $a = 3$). Three of these hexagons are fitted together along their edges of length 5.5 to make four tunnels. The four tunnels are first joined with the small regular hexagons (along the edges of length 3), and then the tunnels must be joined with the four large regular hexagons (see Fig. 14-10).

The Flag Diagram

A survey of equivelar and regular polyhedra is given by the (p, q)-diagram or flag diagram (Fig. 14-13). On the p-axis of the diagram the values p of the p-gons ($p \geq 3$) are plotted; on the q-axis are the valences q of the vertices ($q \geq 3$), the number of edges incident with the vertex. Thus, for example the tetrahedron can be found in $p = q = 3$; the usual 3-cube can be found in $p = 4$, $q = 3$, and so forth. (Clearly, polyhedra with different types of faces or vertices—as, for example, the Archimedean solids—cannot be shown in the flag diagram.) The labels in the flag diagram denote the genera: thus $g = 0$ for the Platonic solids. In particular, the flag diagram shows in a very suggestive way the three types of polyhedral geometry, due to the values of p and q.

The hyperbola $1/p + 1/q = 1/2$ dissects the lattice points (p, q), $p \geq 3$, $q \geq 3$ of the diagram into three subsets:

- The elliptic case: $1/p = 1/q > 1/2$. Here we find the five Platonic solids and the two Kepler–Poinsot star polyhedra of genus 0 (which are isomorphic to the dodecahedron and the icosahedron, respectively).
- The parabolic or Euclidean case: $1/p + 1/q = 1/2$. Here we find the three regular tilings of the Euclidean plane, and the regular tori.
- The hyperbolic case: $1/p + 1/q < 1/2$. Here we find the two Kepler–Poinsot star polyhedra of genus 4 and the three infinite regular Petrie–Coxeter polyhedra. Further we find here the Platonohedra and the regular polyhedra of genus $g > 1$ mentioned previously.

Problems

We end this chapter with five open problems on equivelar manifolds and regular polyhedra. Although all of these problems are easy to understand, their solution seems to be not too straightforward. For more details we refer to the original papers.

1. Do equivelar manifolds exist with $p \geq 5$ and $q \geq 5$?[13]
2. Do equivelar manifolds exist with all faces being regular? (It has been shown[14] that for $q = 4$ no such manifold exists.)
3. In McMullen, Schulz, and Wills' "Polyhedral 2-Manifolds in E^3 with Unusually Large Genus" (cited in note 1), an equivelar manifold with $g = 577$ and number of vertices $f_0 = 576 < g$ is constructed. Are these the smallest possible numbers? (In E^5 there is one with $g = 20$ and $f_0 = 19$.)
4. Are there more than seven Platonohedra of genus $g > 1$?
5. Are there other combinatorially regular polyhedra (in E^3 without self-intersection) for $g > 1$? In particular: Does the dual of Klein's quartic exist as an intersection-free polyhedron?

Notes

[1] B. Grünbaum and G. Shephard, "Polyhedra with Transitivity Properties," *Mathematical Reports of the Academy of Science* [Canada] *6* (1984):61–66. P. McMullen, C. Schulz, and J. M. Wills, "Polyhedral 2-Manifolds in E^3 with Unusually Large Genus," *Israel Journal of Mathematics 46* (1983):127–44.

[2] Grünbaum and Shephard, "Polyhedra with Transitivity Properties."

[3] U. Brehm and W. Kühnel, "Smooth Approximation of Polyhedral Surfaces Regarding Curvatures," *Geometriae Dedicata 12* (1982):438.

[4] Grünbaum and Shephard, "Polyhedra with Transitivity Properties."

[5] J. M. Wills, "Semi-Platonic Manifolds" in P. Gruber and J. M. Wills, eds., *Convexity and Its Applications* (Basel: Birkhäuser, 1983). J. M. Wills, "On Polyhedra with Transitivity Properties," *Discrete and Computational Geometry 1* (1986):195–99.

[6] B. Grünbaum and G. C. Shephard, "Polyhedra with Transitivity Properties."

[7] E. Schulte and J. M. Wills, "Geometric Realizations for Dyck's Regular Map on a Surface of Genus 3," *Discrete and Computational Geometry 1* (1986):141–53.

[8] H. S. M. Coxeter, "Regular Skew Polyhedra in Three and Four Dimensions, and Their Topological Analogues," *Proceedings of the London Mathematical Society,* ser. 2, *43* (1937):33–62.

[9] A. Boole-Stott, *Geometrical Reduction of Semiregular from Regular Polytopes and Space Fillings* (Amsterdam: Ver. d. K. Akademie van Wetenschappen, 1910).

[10] P. McMullen, C. Schulz, and J. M. Wills, "Equivelar Polyhedral Manifolds in E^3," *Israel Journal of Mathematics 41* (1982):331–46.

[11] E. Schulte and J. M. Wills, "On Coxeter's Regular Skew Polyhedra," *Discrete Mathematics 60* (1986):253–62.

[12] E. Schulte and J. M. Wills, "A Polyhedral Realization of Felix Klein's Map $\{3, 7\}_8$ on a Riemann

Surface on Genus 3," *Journal of the London Mathematical Society 32* (1985):539–47.

[13] McMullen, Schulz, and Wills, "Polyhedral 2-Manifolds in E^3 with Unusually Large Genus." McMullen, Schulz, and Wills, "Equivelar Polyhedral Manifolds in E^3." P. McMullen, C. Schulz, and J. M. Wills, "Two Remarks on Equivelar Manifolds," *Israel Journal of Mathematics 52* (1985):28–32.

[14] McMullen, Schulz, and Wills, "Two Remarks on Equivelar Manifolds."

15

Uniform Polyhedra from Diophantine Equations

BARRY MONSON

A simple set of coordinates eases the study of metrical properties of uniform polyhedra. For instance, the six vertices of the regular octahedron $\{3,4\}$ have Cartesian coordinates $(\pm 1, 0, 0)$, etc. where the "etc." means "permute the coordinates in all possible ways." I find it pleasing in such examples that the coordinates are given by systematic choices.[1] Observe further that the coordinates provide all *integral* solutions to the Diophantine equation

$$x^2 + y^2 + z^2 = N, \qquad (15.1)$$

when $N = 1$. If instead $N = 3$, we obtain the eight vertices $(\pm 1, \pm 1, \pm 1)$ of the cube $\{4,3\}$. Less obviously, we get vertices for the cuboctahedron $\left\{\begin{smallmatrix}3\\4\end{smallmatrix}\right\}$* when $N = 2$ and the truncated octahedron $t\{3,4\}$ when $N = 5$. Clearly, Eq. (15.1) is unchanged by the eight possible sign changes or six possible permutations of x, y, and z. Thus the $48 = 6 \cdot 8$ geometric symmetries in the octahedral group are represented as algebraic symmetries of Eq. (15.1). In fact, for any N, we may thus construct a polyhedron with octahedral symmetry although it may be uninteresting; there usually is no uniform way of defining its edges and faces.

Interlude

The remaining uniform polyhedra with octahedral symmetry pose another problem. For example, the truncated cube $t\{4,3\}$ has typical vertex $(1, 1, \sqrt{2} - 1)$, but the irrational $\sqrt{2} - 1$ is not the sort of integer required in Eq. (15.1). We have tackled these cases with some (untidy) success using the ideas exploited in the next section. Also, Eq. (15.1) is invariant under the central inversion $(x, y, z) \to (-x, -y, -z)$; thus a homogeneous quadratic must be replaced by some other equation when describing polyhedra without central symmetry, such as the tetrahedron or the snub cube.

Uniform Polyhedra with Icosahedral Symmetry

The icosahedron $\{3,5\}$ and its relatives have fivefold rotational symmetry (Fig. 15-1). Hence, our coordinates must somehow involve the number

$$\cos(2\pi/5) = 1/2\tau = (\sqrt{5} - 1)/4,$$

where the *golden ratio* $\tau = (1 + \sqrt{5})/2$ satisfies

$$\tau^2 = \tau + 1. \qquad (15.2)$$

To reconcile this irrational with the integral nature of our equations we replace the rational field Q and its ring of *ordinary integers* Z by the quadratic number field $Q(\sqrt{5})$ and its ring of algebraic integers $Z[\tau]$. The ring $Z[\tau]$ consists of all polynomials in τ with integral coefficients; using Eq. (15.2), any integer $x \in Z[\tau]$ is uniquely expressed as $x = x_1 + \tau x_2$, $(x_1, x_2 \in Z)$. See Hardy and Wright[2] for the number-theoretic properties of the Euclidean domain $Z[\tau]$; note that the arbitrary magnitude of the *units* $\pm\tau^n$, $(n \in Z)$, complicates the solution of Diophantine equations over $Z[\tau]$.

* Coxeter's notation for a quasiregular polyhedron in which p-gons and q-gons alternate at each vertex is $\left\{\begin{smallmatrix}p\\q\end{smallmatrix}\right\}$.

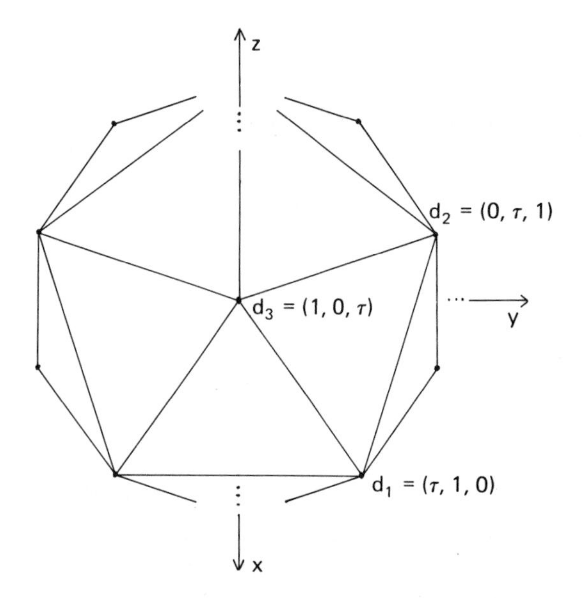

Fig. 15-1. The icosahedron {3,5}.

The icosahedron has 12 vertices with Cartesian coordinates $(\pm\tau, \pm 1, 0)$ and its *cyclic* permutations only. Now let us solve Eq. (15.1) for

$$N = \tau^2 + 1^2 + 0^2 = \tau + 2. \quad (15.3)$$

Letting $x = x_1 + \tau x_2$, $y = y_1 + \tau y_2$, $z = z_1 + \tau z_2$, we split Eq. (15.1) into its rational and irrational parts, then solve two simultaneous ordinary Diophantine equations in six variables:

$$x_1^2 + x_2^2 + y_1^2 + y_2^2 + z_1^2 + z_2^2 = 2; \quad (15.4)$$

$$2x_1x_2 + x_2^2 + 2y_1y_2 + y_2^2 + 2z_1z_2 + z_2^2 = 1. \quad (15.5)$$

Disappointingly, we find 24 solutions, namely *all* permutations of $(\pm\tau, \pm 1, 0)$. But then we recall that our solution must have 48 octahedral symmetries whereas the icosahedron has 120 symmetries. Since $120/48 = 2.5$ we should have expected this incompatibility; the convex hull of the 24 points is a nonuniform truncation of {3,4} with two naturally inscribed {3,5}'s.

Thus we must abandon Cartesian coordinates in favor of some system of oblique coordinates referred to a basis $d_1d_2d_3$. Much effort leads to the obvious choice of three vertices of a triangular face of {3,5}, say $d_1 = (\tau, 1, 0)$, $d_2 = (0, \tau, 1)$, $d_3 = (1, 0, \tau)$ in Cartesian coordinates (Fig. 15-1).

A typical point $u = xd_1 + yd_2 + zd_3$ thus has squared length $u \cdot u = N$, that is

$$(x^2 + y^2 + z^2)(2 + \tau) + 2\tau(xy + xz + yz) = N. \quad (15.6)$$

This equation has built-in icosahedral symmetry, since each of the 120 symmetries preserves points with coordinates in the ring $Z[\tau]$. We solve Eq. (15.6) by splitting it into rational and irrational parts; some results are shown in Table 15-1. In the last case we find 60 superfluous solutions.

It is unclear what is merely fortuitous in the last example. A more insightful account may appear elsewhere. Perhaps, however, the reader has enjoyed yet another duet played by geometry and number theory.

Notes

[1] H. S. M. Coxeter, *Regular Polytopes* (New York: Dover Publications, 1973), pp. 50–53, 156–162.

[2] G. H. Hardy and E. M. Wright, *An Introduction to the Theory of Numbers*, 4th ed. (London: Oxford University Press, 1960), pp. 221–222.

Table 15-1. Solutions of Eq. (15.6).

Polyhedron	Symbol	Vertices	N	Number of Solutions to 15.6
Icosahedron	{3,5}	12	$2 + \tau$	12
Icosidodecahedron	$\left\{ {3 \atop 5} \right\}$	30	4	30
Dodecahedron	{5,3}	20	3	20
Truncated icosahedron	$\tau\{3,5\}$	60	$10 + 9\tau$	60
Rhombicosidodecahedron	$r\left\{ {3 \atop 5} \right\}$	60	$6 + \tau$	60
Truncated dodecahedron	$t\{5,3\}$	60	$7 + 4\tau$	60
Truncated icosidodecahedron	$t\left\{ {3 \atop 5} \right\}$	120	$14 + 5\tau$	180

16

Torus Decompositions of Regular Polytopes in 4-Space

THOMAS F. BANCHOFF

When a regular polyhedron in ordinary 3-space is inscribed in a sphere, then a decomposition of the sphere into bands perpendicular to an axis of symmetry of the polyhedron determines a corresponding decomposition of the polyhedron. For example, a cube with two horizontal faces can be described as a union of two horizontal squares and a band of four vertical squares, and an octahedron with a horizontal face is a union of two horizontal triangles and a band formed by the six remaining triangles.

We may approach the study of regular figures in 4-space in a similar way. The corresponding statement one dimension higher says that if a regular polytope in 4-space has its vertices on a hypersphere such that a symmetry axis coincides with the axis perpendicular to ordinary 3-space, then the polytope can be described as a union of polyhedra arranged in "spherical shells." For example, a 4-cube with one cubical face parallel to 3-space can be described as a union of two cubes and a shell made from the remaining six cubes. Similar shell decompositions have become a standard means of describing the way various three-dimensional faces fit together in 4-space to form regular polytopes. (See for example D. M. Y. Sommerville's description of the regular polytopes in 4-space, or H. S. M. Coxeter's treatment.[1]) In this chapter we examine an alternative way of describing regular figures in 4-space, presenting them as unions not of spherical shells but of rings of polyhedra known as *solid tori*. Such *torus decompositions* are especially convenient for studying symmetries of these figures and for investigating their topological properties. A valuable tool in this project is a remarkable mapping discovered by Heinz Hopf which relates the geometry of circles on the hypersphere in 4-space to the geometry of points on the ordinary sphere in 3-space. One of the aims of this chapter is to give additional geometric insight into the *Hopf mapping* by describing its relationship to torus decompositions of regular polytopes.

Decompositions

Decompositions of objects are often easier to visualize when we project them into lower-dimensional spaces. For a regular polyhedron inscribed in a 2-sphere centered at the origin, if we use central projection from the North Pole to the horizontal plane which passes through the origin, the images of the vertices and edges of the polyhedron form a *Schlegel diagram* of the polyhedron. In such a diagram we may identify the convex cells in a decomposition of the polyhedron corresponding to the decomposition of the 2-sphere into horizontal bands (Fig. 16-1).

If we follow the same procedure one dimension higher, we project centrally from a point on the hypersphere to our three-dimensional space and the images of the vertices and edges of a regular figure determine its Schlegel diagram. Just as the central projection of a cube to the plane leads to a "square-within-a-square," the central projection of a hypercube may appear as a "cube-within-a-cube" with

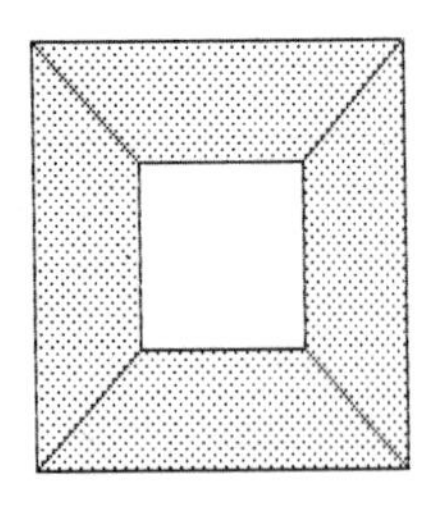

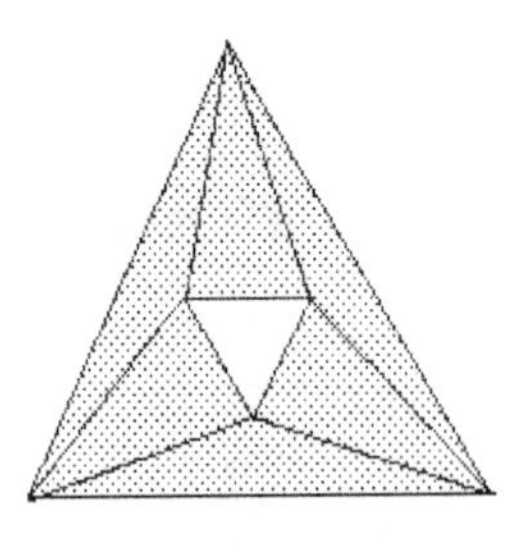

Fig. 16-1. Band decompositions of the cube and octahedron.

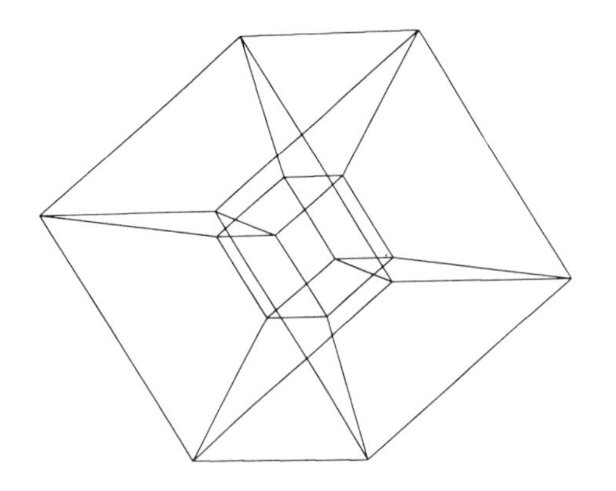

Fig. 16-2. Cube-within-a-cube projection of the hypercube.

corresponding vertices connected in each case. The annular band in the plane formed by four trapezoids separating two squares corresponds to the region separating two cubes which is decomposed into six congruent truncated square pyramids (Fig. 16-2).

The four vertical truncated pyramids in the central projection of the hypercube fit together to form a solid torus which is the image of a ring of four cubes on the hypercube in 4-space. The remaining four cubes also form a ring and the common boundary of these two rings is the surface of a torus formed by 16 squares in the hypercube. This is the prototype of a torus decomposition, and it is this sort of analysis we wish to carry out with respect to other regular polytopes in 4-space. In this chapter we pay special attention to the 24-cell, a polytope formed from 24 regular octahedra. This polytope is complicated enough to exhibit most of the interesting phenomena of torus decompositions and it is still relatively easy to visualize, especially when we use the techniques of computer graphics.[2]

The Cube and Its Associated Polyhedra

Associated with the cube are other polyhedra with vertices at the centers of faces or edges of the cube. If we take the six centers of square faces, we obtain the regular *octahedron* inscribed in the cube. If on the other hand we take the midpoints of the 12 edges of the cube, we obtain a *cuboctahedron,* a semiregular polyhedron with faces of two types: squares determined by the midpoints of the edges of the cube's square faces and triangles determined by the midpoints of the three edges coming from each vertex of the cube.

If we project the cube into the plane by central projection, we can identify the projections of the octahedron and the cuboctahedron by joining images of centers of faces and edges of the cube (Fig. 16-1). The cuboctahedron in particular can be expressed as a union of a large square surrounding a small square, with the region between them subdivided into four squares and eight triangles (Fig. 16-3). This gives a "band decomposition" corresponding to the decomposition of the cube itself into two horizontal squares and a band of four vertical squares.

The Hypercube and Its Associated Polytopes

In a similar manner we may identify regular and semiregular polytopes associated with a hypercube by taking the midpoints of faces of certain dimensions. If we take the midpoints of the eight cubical faces, we obtain the vertices of a *cube-dual* determining the 16 tetrahedral faces of the *16-cell.* If we take the midpoints of all 32 edges, we obtain a semiregular polytope with 24 cells: 16 tetrahedra connecting the midpoints of quadruples of edges emanating from each of the vertices of the hypercube, and eight cuboctahedra determined by the midpoints of edges of each of the cubical faces of the hypercube.

On the other hand, if we take the centers of all square faces of the hypercube we obtain a polytope which is regular. Each vertex of the hypercube is a vertex of six squares, each with a pair of sides chosen from among the four

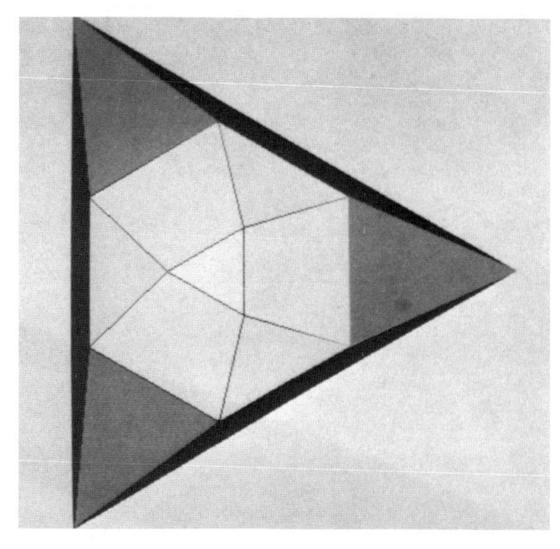

Fig. 16-3. Band decomposition of the cuboctahedron.

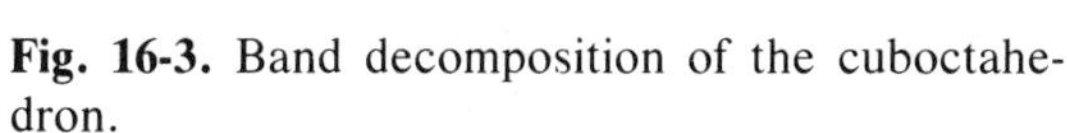

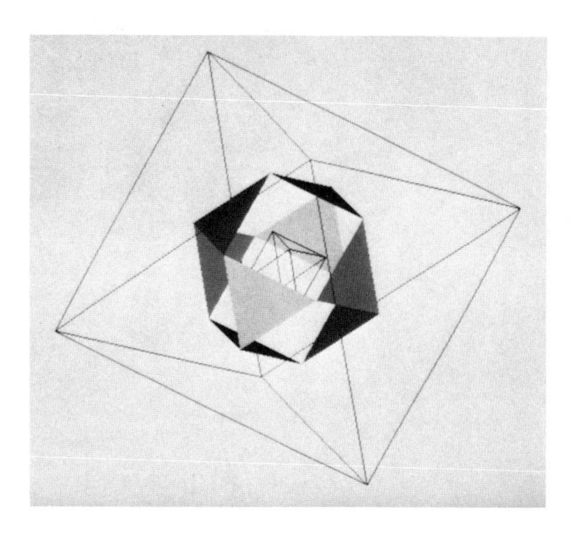

Fig. 16-4. Projection of the 24-cell.

a

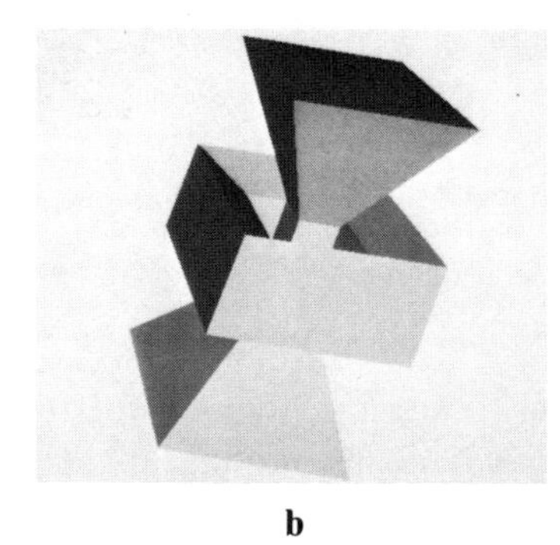

b

c

Fig. 16-5. Torus decompositions of the hypercube.

edges emanating from the vertex. The midpoints of these six squares will be vertices of an octahedron. Moreover, the midpoints of the six square faces of a cube in the hypercube also determine an octahedron. We thus obtain a polytope with 24 octahedral faces, 16 corresponding to vertices of the hypercube and eight corresponding its dual polytope. This polytope is called the *24-cell,* and it is the main object of our study in this chapter.

As in the lower-dimensional situation, we may identify the projections of these semiregular and regular polytopes by referring to the central "cube-within-a-cube" projection of the hypercube (Fig. 16-2). In particular, the 24-cell may be presented as a large octahedron surrounding a small octahedron, with the region between them decomposed into six octahedra meeting vertex-to-vertex and 16 octahedra each meeting either the large octahedron or the small one along a triangular face. We may think of this polytope as consisting of one octahedron on each face of the small octahedron and one on each face of the larger one. Each octahedron of the first set shares a triangle with one octahedron of the other set. The gaps left between these 16 octahedra determine the places for the remaining six octahedra (Fig. 16-4).

The hypercube may be decomposed into two solid torus rings, each a cycle of four cubes meeting along square faces. In 4-space

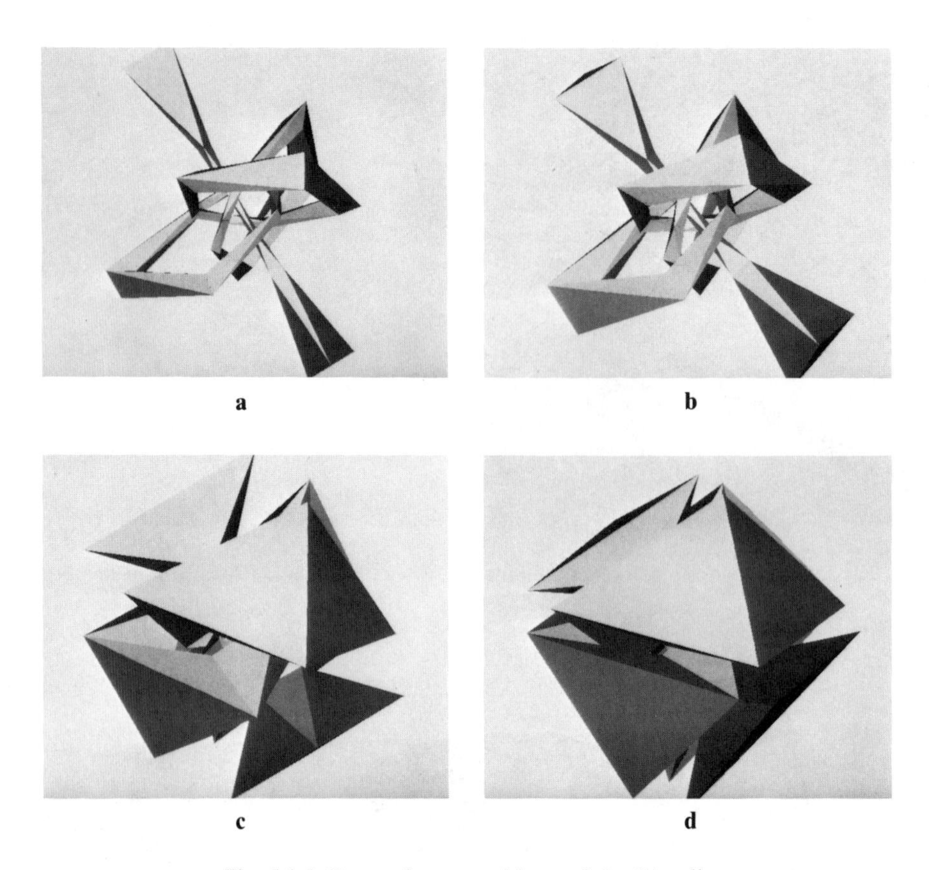

Fig. 16-6. Torus decompositions of the 24-cell.

the centers of the cubes in one of these rings will form the vertices of a square. In the cube-within-a-cube projection, the centers of four of these cubes lie on a vertical straight line and the other four centers are vertices of a horizontal square. The vertical line meets the horizontal square disc in exactly one point, and this fact implies that the line and the square are *linked* (Fig. 16-5a–c).

Analogously we may express the 24-cell as a union of four solid tori, each a cycle of six octahedra meeting along triangular faces. The centers of the six octahedra in a ring will form a planar hexagon, with four vertices lying in two parallel edges of the hypercube and two opposite vertices from the dual 16-cell. Any two of these hexagons are linked, so that any hexagon meets the disc bounded by any other hexagon in exactly one point. We can identify such hexagonal cycles in the central projection of a hypercube and its dual polytope. One of the hexagons includes a vertex at infinity, so it is a straight line. The other three hexagons are arranged symmetrically about this line (Fig. 16-6a–d).

Fold-Out Decomposition of the Hypercube and 24-Cell

The decomposition of a hypercube into two solid tori with a common polyhedral boundary can be described in a different way by "folding the figure out into 3-space." We may express a cube folded out into the plane by giving two squares together with a strip of four squares. The ends of the strip are to be identified to form a cylinder with two boundary square polygons, which will match up with the boundaries of the remaining two squares. The analogous decomposition of the hypercube starts with two solid stacks of four cubes. The ends of the stacks are to be identified by folding up in 4-space to obtain the two solid tori. The common boundary of these solid tori is a polyhedral torus which can be expressed as a square subdivided into 16 squares, with its left and right edges identified and its top and bottom edges identified (Fig. 16-7).

The corresponding fold-out description of the 24-cell starts with a solid stack formed by six octahedra. The top and bottom triangles of the stack can be identified by folding up into 4-space to obtain a solid torus. The boundary of this solid torus can be expressed as a union of 36 triangles arranged in a polygonal region in the plane, to be folded up in 4-space so that its left and right edges are identified and its top and bottom edges are identified (Fig. 16-8a–b).

The remaining 18 octahedra in the 24-cell can be arranged into three other stacks, each with six octahedra. We can place the four stacks together in 3-space to indicate the way the four solid tori will be linked when the stacks are folded together in 4-space (Fig. 16-9).

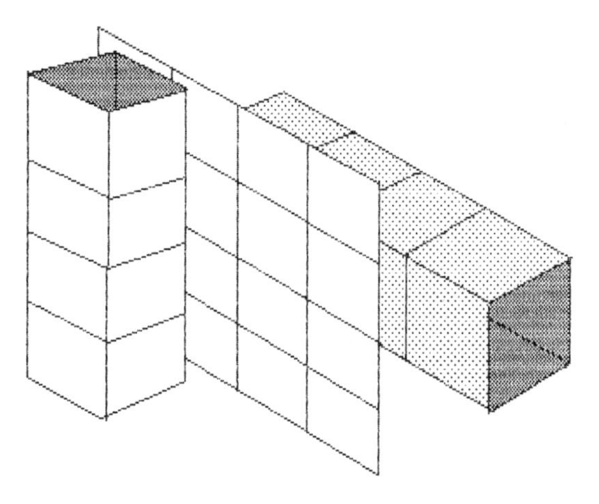

Fig. 16-7. Fold-out decomposition of the hypercube.

a

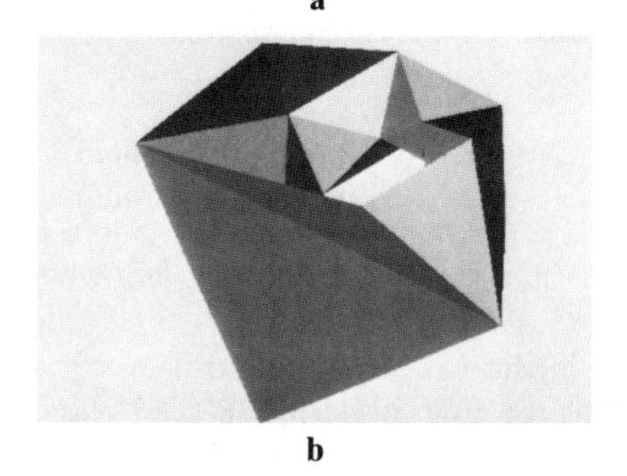

b

Fig. 16-8. A stack of six octahedra, (a) unfolded, (b) folded.

Cartesian and Torus Coordinates

To describe polyhedra in 3-space in coordinates, it is most convenient to parametrize the unit sphere by longitude and *co*-latitude (measured down from the North Pole instead of up from the Equator). A point on the unit sphere then has Cartesian coordinates

$$(\cos(\theta)\sin(\varphi),\ \sin(\theta)\sin(\varphi),\ \cos(\varphi)).$$

Stereographic projection from the North Pole to the horizontal plane which passes through the origin sends a point to the intersection of the line through the North Pole and the point with the horizontal plane. The point whose coordinates are given above is then sent to

$$\begin{aligned}&(\cos(\theta)\sin(\varphi)/(1 - \cos(\varphi)),\\ &\qquad \sin(\theta)\sin(\varphi)/(1 - \cos(\varphi)),\ 0)\\ &\qquad = (\cos(\theta)\cot(\varphi/2),\ \sin(\theta)\cot(\varphi/2),\ 0)\end{aligned}$$

(Fig. 16-10). Circles of latitude are sent to circles centered at the origin and semicircles of longitude are sent to straight lines passing through the origin. A rotation or reflection of the sphere about the axis in 3-space corresponds to a rotation or reflection in the plane.

Fig. 16-9. Four stacks of octahedra.

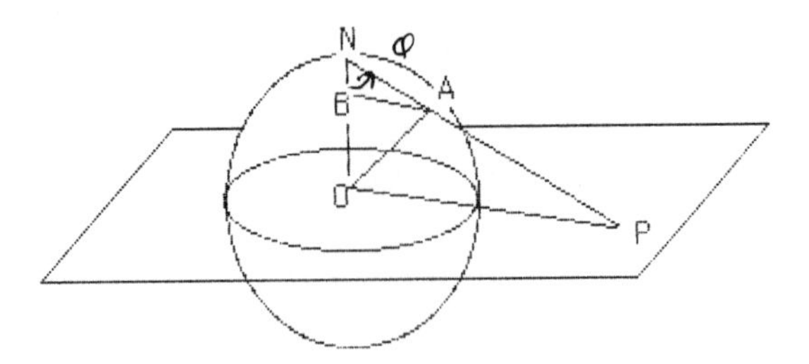

Fig. 16-10. Stereographic projection of the 2-sphere.

A regular polyhedron with a vertical axis of symmetry will have its vertices on certain parallels of latitude and the symmetries of the polyhedron preserving the axis will lead to symmetries of its Schlegel diagram in the plane.

In 4-space the sphere of points at unit distance from the origin can be parametrized in different ways.[4] In this exposition, we concentrate on a coordinate system on the 3-sphere which has especially interesting geometric properties. In this *torus coordinate system* the coordinates of a point are

$$(x,y,u,v) = \{(\cos(\theta)\sin(\varphi),\ \sin(\theta)\sin(\varphi),\ \cos(\psi)\cos(\varphi),\ \sin(\psi)\cos(\varphi)\}$$

where θ and ψ run from 0 to 2π, and where $0 \leq \varphi \leq \pi/2$. The points with $\varphi = 0$ give the unit circle in the (u, v) plane and if $\varphi = \pi/2$, the locus is the unit circle in the (x,y) plane.

Points on the 3-sphere corresponding to other values of φ give tori on the 3-sphere. For example, if $\varphi = \pi/4$ we get a symmetric torus,

$$(1/\sqrt{2})(\cos(\theta),\ \sin(\theta),\ \cos(\psi),\ \sin(\psi)).$$

This torus is the Cartesian product of two circles, one lying in the xy-plane and the other in the uv-plane.

As φ moves from 0 to $\pi/2$, these tori sweep out the region between the two linked circles. In particular, the entire 3-sphere is then displayed as a union of two tori, one corresponding to negative values of φ and the other corresponding to positive values.

If we position the vertices of a regular polytope symmetrically with respect to this coordinate system, we obtain a *torus decomposition* of the regular polytope. The torus coordinate system on the 3-sphere is particularly well-suited to the study of the Hopf mapping, the final topic of this chapter.

Coordinates for Polyhedra and Polytopes

In three-dimensional space we may describe the cube by the eight vertices $(\pm t, \pm t, \pm t)$ for a positive constant t. These points lie on the sphere of radius $\sqrt{3}t$. The centers of cubical faces will then be the points $(\pm t, 0, 0)$, $(0, \pm t, 0)$ and $(0, 0, \pm t)$, the vertices of a regular octahedron inscribed in a sphere of radius t. The midpoints of edges of the cube will have coordinates $(\pm t, \pm t, 0)$, $(0, \pm t, \pm t)$ and $(\pm t, 0, \pm t)$, forming the vertices of a cuboctahedron inscribed in a sphere of radius $\sqrt{2}t$.

For the hypercube we may choose 16 vertices $(\pm t, \pm t, \pm t, \pm t)$ situated on the hypersurface of a 3-sphere of radius $2t$. In torus coordinates, these 16 points

$$2t(\cos(\theta)\sin(\varphi),\ \sin(\theta)\sin(\varphi),\ \cos(\psi)\cos(\varphi),\ \sin(\psi)\cos(\varphi))$$

all lie on the torus with $\varphi = \pi/4$. The coordinates are given by letting θ and ψ take on the values $\pi/4 + k\pi/2$ for $k = 0, 1, 2, 3$. Just as the 3-sphere is expressed as a union of two solid tori with a common boundary torus, the boundary of the hypercube is expressed as a union of two solid tori with a common boundary. The boundary polyhedral torus can be expressed as the Cartesian product of two square polygons. It includes all 16 vertices and all 32 edges of the hypercube as well as 16 of its squares.

The centers of the eight three-dimensional cubical faces of the hypercube have coordinates $(\pm t, 0, 0, 0)$, $(0, \pm t, 0, 0)$, $(0, 0, \pm t, 0)$ and $(0, 0, 0, \pm t)$, lying on a hypersphere of radius t. This gives coordinates for the regular 16-cell in 4-space. The midpoints of edges of the hypercube will be the 32 points $(0, \pm t, \pm t, \pm t)$, $(\pm t, 0, \pm t, \pm t)$, $(\pm t, \pm t, 0, \pm t)$ and $(\pm t, \pm t, \pm t, 0)$, lying on a hypersphere of radius $\sqrt{3}t$, giving the coordinates of the vertices of a semiregular polytope.

The vertices of the regular 24-cell in 4-space can be given by the midpoints of square faces of the hypercube, with coordinates $(\pm t, \pm t, 0,$

0), $(0, \pm t, 0, \pm t)$, $(\pm t, 0, \pm t, 0)$, $(\pm t, 0, 0, \pm t)$, $(0, 0, \pm t, \pm t)$ and $(0, \pm t, \pm t, 0)$, lying on a hypersphere of radius $\sqrt{2}t$. Another set of coordinates for the 24-cell is given by taking the vertices of a hypercube $(\pm s, \pm s, \pm s, \pm s)$ together with the vertices of a dual 16-cell $(\pm 2s, 0, 0, 0)$, $(0, \pm 2s, 0, 0)$, $(0, 0, \pm 2s, 0)$ and $(0, 0, 0, \pm 2s)$. These 24 coordinates lie on a hypersphere of radius $2s$.

The Hopf Mapping

Torus coordinates are especially well suited for describing the Hopf mapping, a mapping from the 3-sphere to the 2-sphere for which every point of the 3-sphere lies on a circle that is the preimage of a point on the 2-sphere. One the easiest ways to describe the Hopf mapping is to think of the 3-sphere as a collection of pairs of complex numbers with the squares of their lengths adding up to 1. We then have

$$S^3 = \{(x + iy, u + iv), x^2 + y^2 + u^2 + v^2 = 1\}$$
$$\equiv \{[z,w], z^2 + w^2 = 1\}.$$

Hereafter we shall adopt the convention of using square brackets to indicate the description of a point of S^3 by pairs of complex numbers and parentheses to indicate the usual description in terms of quadruples of real numbers.

To describe the Hopf mapping we send a pair of complex numbers to their quotient; that is, $h[z,w] = w/z$ if $z \neq 0$ and $h[0, w] = \infty$, the infinite point in the extended complex plane. If we write a point in S^3 in polar coordinates, then we have

$$[z,w] = [\sin(\varphi)e^{i\theta}, \cos(\varphi)e^{i\psi}] \text{ and}$$

$$h[z,w] = \cot(\varphi)e^{i(\psi-\theta)}$$

if $\varphi \neq \pi/2$ and $h[0,w] = \infty$ as before.

To complete the description of the Hopf mapping we use inverse stereographic projection to map the extended complex plane to the 2-sphere. The effect of this is to map the point $[z,w] = (x,y,u,v)$ to

$$h[z,w] = [2\bar{z}w, w\bar{w} - z\bar{z}] = (2xu + 2yv, 2xv - 2yu, -x^2 - y^2 + u^2 + v^2, 0).$$

Here $\bar{w} = u - iv$ indicates the complex conjugate. The coordinate $w\bar{w} - z\bar{z}$ is a real number and the image of each point of S^3 lies in three-dimensional space.

In torus coordinates, the Hopf mapping is given by

$$h[\sin(\varphi)e^{i\theta}, \cos(\varphi)e^{i\psi}] = [\sin(2\varphi)e^{i(\varphi-\theta)}, \cos(2\varphi)].$$

It is clear from this form that the image of each point lies on the 2-sphere of radius 1 in 3-space, so we write

$$h: S^3 \to S^2.$$

Under this mapping the unit circle in the xy-plane corresponding to $\psi = \pi/2$ is sent to the point $(0, 0, -1)$ and the unit circle in the uv-plane corresponding to $\varphi = 0$ is sent to point $(0, 0, 1)$. The middle torus $(1/\sqrt{2})[e^{i\theta}, e^{i\varphi}]$ corresponding to $\varphi = \pi/4$ is sent to the Equator $[e^{i(\psi-\theta)}, 0]$. The preimage of the point $(1, 0, 0)$ on the Equator is the set of points with $\varphi = \pi/4$ and $\theta = \psi$. This curve lies on the sphere and in the plane given by $x = u$ and $y = v$, so it must be a circle. Similarly any point $[e^{i\beta}, 0]$ on the Equator will be a circle determined by the conditions $\alpha = \pi/4$ and $\varphi = \theta + \beta$. More generally for any point $[\sin(\gamma)e^{i\beta}, \cos(\gamma)]$ with $\sin(\gamma)$ positive on the 2-sphere, the preimage under the Hopf mapping will be a circle determined by the conditions $\varphi = \gamma/2$ and $\psi = \theta + \beta$. All of these circles will be great circles on the 3-sphere and no two of them will have a point in common. Since the discs bounded by these circles will meet only at the origin in 4-space, the circles will be linked.

The Hopf Decomposition of the Hypercube

We now consider the hypercube from the point of view of the Hopf mapping. In complex coordinates, the 16 vertices of the hypercube on the unit hypersphere may be given by $1/2[\pm 1 \pm i, \pm 1 \pm i]$.[3] Under the Hopf mapping the images of these vertices will be four points on the Equator of S^2, namely $[\pm 1, 0]$ and $[\pm i, 0]$ (Fig. 16-11).

We indicate on the unfolded torus diagram two of the four quadrilaterals containing the vertices of the hypercube. These four quadrilaterals are squares in 4-space which we may call "Hopf polygons." The edges of these

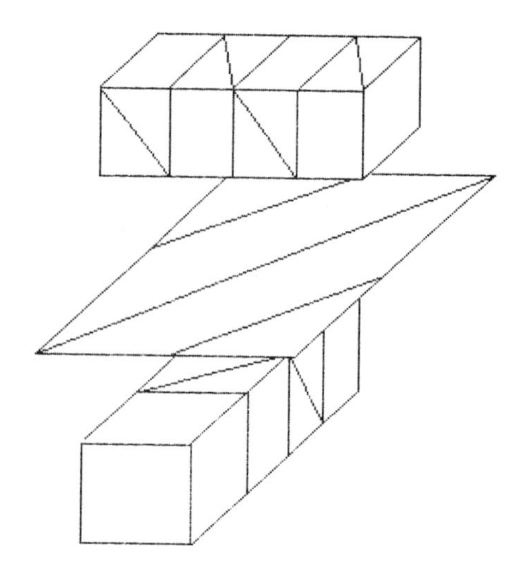

Fig. 16-11. Hopf polygons on the hypercube.

Hopf polygons are diagonals in the square faces of the flat torus which is the preimage of the Equator under the Hopf mapping. One of the solid tori is the preimage of the upper hemisphere and the other is the preimage of the lower hemisphere.

Torus Decomposition of the 24-Cell

We may now attempt a similar decomposition of the 24-cell. In the fold-out version we may identify three hexagonal helices on each stack of six octahedra corresponding to the four quadrilateral helices on the stacks of four cubes in the hypercube. Unfortunately these hexagons are not preimages of points of S^2 under the Hopf mapping, in either of the two coordinate systems we have described for the 24-cell. The coordinates which we have given for the hypercube are sent by the Hopf mapping to four points on the Equator and the eight vertices of the 16-cell are sent to the North and South Poles of S^2. Thus the 24 coordinates of one coordinate system on the 24-cell are sent to six points of S^2 situated at the vertices of a regular octahedron. Similarly if we use the coordinates of the 24-cell obtained by taking midpoints of square faces, again the images under the Hopf mapping are the same six vertices of an octahedron. In order to obtain the decomposition of the 24-cell into four solid tori each with six octahedra, we need to reposition the 24-cell so that its vertices are sent to the four vertices of tetrahedron under the Hopf mapping.

To determine a rotation that will align the 24-cell so that it is situated well with respect of the Hopf mapping, we carry out a closer examination of the previously given coordinates of the 24-cell where we now consider the case $t = 1$ so the polytope lies on a sphere of radius equal to 2. If we project stereographically from the point $(0, 0, 0, \sqrt{2})$ then the image of a point (x,y,u,v) is $\{\sqrt{2}/(\sqrt{2} - v)\}(x,y,u)$. If $v = 0$, then the image of $(x,y,u,0)$ is (x,y,u). Thus the images of the 12 vertices of the 24-cell of the forms $(\pm 1, \pm 1, 0, 0)$, $(\pm 1, 0, \pm 1, 0)$ and $(0, \pm 1, \pm 1, 0)$ will be the vertices of a cuboctahedron in 3-space. The six vertices of the form $(\pm 1, 0, 0, 1)$, $(0, \pm 1, 0, 1)$, and $(0, 0, \pm 1, 1)$ will be sent to the vertices of large octahedron containing the cuboctahedron, and the six vertices with fourth coordinate -1 will be sent to a small octahedron contained within the cuboctahedron.

The cuboctahedron has eight triangular faces, each only lying in one distorted octahedron with its opposite triangle on the small octahedron and another with opposite triangle on the large octahedron. This accounts for 18 of the octahedra in the 24-cell. The remaining six each have four of the vertices on the square faces of the cuboctahedron, and one vertex on the large and one on the small octahedron.

We may then identify one of the four tori in the toroidal decomposition by taking the small octahedron and two adjacent octahedra with their opposite triangles on the semiregular polyhedron. The octahedra opposite these three complete a cycle of six octahedra on the 24-cell. We could for example take the large octahedron, with center $(0, 0, 0, 1)$, the small one with center $(0, 0, 0, -1)$, two others adjacent to the small one with centers $1/2(-1, -1, -1, -1)$ and $1/2(1, 1, 1, -1)$, and their opposite octahedra with center $1/2(1, 1, 1, 1)$ and $1/2(-1, -1, -1, 1)$. These six vertices may be arranged in a hexagon so that the angle between any two adjacent vertices is 120°, as shown in Fig. 16-12.

We may label the vertices of the projected 24-cell so that this cycle of six octahedra ap-

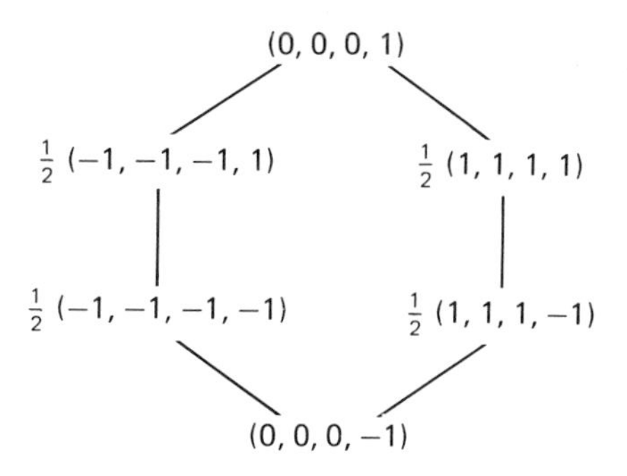

Fig. 16-12. A hexagon of centers of octahedra in a Hopf cycle.

pears as five octahedra in a vertical stack, together with the large octahedron. The remaining solid tori are obtained by taking a pair of octahedra meeting the central polyhedron in opposite square faces and then connecting them by two other pairs each with one octahedron inside and one outside the central polyhedron. Once we have one of these we obtain the other two by rotating about the vertical line by angles of 120° and 240°.

Fortunately it is possible to find a rotation of 4-space which realigns the vertices so that their images are situated at the vertices of a regular tetrahedron in 3-space! To find one such rotation, we look for a linear transformation T which sends the hexagon to a regular hexagon in the uv-plane in 4-space. We can let T fix the vector (0, 0, 0, 1) and let T send 1/2(1, 1, 1, 1) to (0, 0, 3/2, 1/2) and 1/2(−1, −1, −1, 1) to $(0, 0, -\sqrt{3}/2, 1/2)$. It follows that $T(1/\sqrt{3}, 1/\sqrt{3}, 1/\sqrt{3}, 0) = (0, 0, 1, 0)$. The vectors sent to (1, 0, 0, 0) and (0, 1, 0, 0) by T must be a pair of mutually orthogonal unit vectors which are perpendicular to (0, 0, 0, 1) and $(1/\sqrt{3}, 1/\sqrt{3}, 1/\sqrt{3}, 0)$ and we may choose these preimages to be $(1/\sqrt{2}, -1/\sqrt{2}, 0, 0)$ and $(1/\sqrt{6}, 1/\sqrt{6}, 2/\sqrt{6}, 0)$. This completely determines the matrix of T, and we may then check that after this rotation, the vertices of the 24-cell do indeed lie in four (planar) regular hexagons in 4-space which are mapped to the vertices of a regular tetrahedron inscribed in the unit 2-sphere in 3-space under the Hopf mapping.[5]

The preimages of the four triangular faces of this spherical tetrahedron correspond to the four cycles of six octahedra described in the previous paragraph. To see how these four rings of octahedra fit together to fill out the 24-cell, we may shrink each ring toward the hexagon, with vertices at the centers of the six triangles between adjacent octahedra. We may interpolate linearly between the 24-cell and this union of four hexagons, constantly projecting the vertices centrally to the hypersurface of the 3-sphere. Illustrations of several stages of its deformation are shown in Figs. 16-8–16-11.

Note that the 24 centers of four hexagons may be obtained from the centers of the 24 octahedra by a rotation in 4-space which moves each Hopf circle along itself by 60°. The comparable treatment of the hypercube shrinks the two rings of four cubes to the quadrilaterals determined by the centers of the eight squares where adjacent cubes meet in the two rings as shown in Figs. 16-5–16-7. These eight points may also be obtained from the centers of the cubes in the two rings by a rotation in 4-space moving each Hopf circle along itself by 45°.

Conclusion

The familiar central projection of the hypercube suggests a decomposition of the hypersphere into solid tori, and this decomposition carries over to other polytopes as well, in particular the 24-cell. This investigation gives additional geometric insight into the properties of these polytopes and at the same time it elucidates some of the geometry of the Hopf mapping.

Notes

Acknowledgment: I acknowledge with gratitude the help I received from correspondence and conversations with Professor Coxeter during the course of preparation of this paper.

Computer images in this chapter have been generated in collaboration with David Laidlaw and David Margolis, and I acknowledge as well the cooperation of the entire graphics group at Brown University. Several of the illustrations are taken from the film *The Hypersphere: Foliation and Projection* by Huseyin Koçak, David Laidlaw, David Margolis, and myself.

[1] D. M. Y. Sommerville, *Geometry of n Dimensions* (London: Methuen, 1929). H. S. M. Coxeter, *Regular Polytopes,* 3rd ed. (New York: Dover Publications, 1973).

[2] Similar treatments of the 120-cell and the 600-cell are implicit in the work of several mathematicians, notably Coxeter whose *Regular Complex Polytopes* (Cambridge, England: Cambridge University Press, 1974) is the primary source for all material of this sort.

[3] Compare p. 37 of Coxeter, *Regular Complex Polytopes*.

[4] One direct analogy with the usual coordinate system on the 2-sphere in 3-space would be to use

$$(x,y,u,v) = (\cos(\theta)\sin(\varphi)\sin(\psi),\ \sin(\theta)\sin(\varphi)\sin(\psi),\ \cos(\varphi)\sin(\psi),\ \cos(\psi)),$$

which would suggest the same sort of decomposition of the 3-sphere into "parallel 2-spheres of latitude." Such a decomposition has been carried out by several authors (including D. M. Y. Sommerville), and an early computer generated film by George Olshevsky uses such an approach to display the slices of regular polyhedra in 4-space by sequences of hyperplanes perpendicular to various coordinate axes.

[5] The coordinates for the 24-cell obtained in this way are very similar to those which appear in Coxeter's discussion of the 24-cell in *Twisted Honeycombs* (Regional Conference Series in Mathematics, no. 4, American Mathematical Society, Providence, 1970), although he does not explicitly use the Hopf mapping in any of his constructions. Professor Coxeter pointed out that these coordinates also appear in a slightly different form in the 1951 dissertation of G. S. Shephard.

17

Convex Polyhedra, Dirichlet Tessellations, and Spider Webs

PETER ASH, ETHAN BOLKER, HENRY CRAPO, AND WALTER WHITELEY

Plane pictures of three-dimensional convex polyhedra, plane sections of three-dimensional Dirichlet tessellations, and flat spider webs with tension in all the threads are essentially the same geometric objects. At the root of this remarkable coincidence is a single geometric diagram that permits us to offer a unified image of the connections among these and other objects. Some hints of these connections are more than a century old, but others are very recent. We begin with an historical sketch.

In the nineteenth century, mathematicians and engineers investigated frameworks built from iron bars and pins to determine when they were rigid.[1] Their studies led them to consider static stresses: tensions and compressions in the bars in internal static equilibrium. In 1864, James Clerk Maxwell discovered a geometric tool for studying the static equilibrium of forces on a plane framework with a planar graph: the *reciprocal figure,* a drawing of the dual planar graph with the dual edges perpendicular to the original edges and forces (see Fig. 17-1).[2] Maxwell built this reciprocal by patching together the polygons of forces expressing the vector equilibrium at each joint. He then observed that this construction yields a polyhedron in space which projects onto the framework. These results belong to the field of *graphical statics,*[3] a branch of graphical and mechanical science which withered around the turn of the century, along with much of projective geometry.

Recent work on the statics of frameworks grows from these geometric roots.[4] In particular, we now know that a convex polyhedron, projected from a point on one face onto a plane parallel to this face, corresponds to a *spider web:* a framework with no crossing edges and some edges going to infinity, which has an internal static equilibrium formed entirely with tension in the members. In the plane, the spider webs are frameworks with *convex reciprocals:* reciprocals in which the convex polygons have disjoint (that is, nonoverlapping) interiors (Fig. 17-2). Other recent work has extended some hints in the work of Maxwell and a conjecture by Janos Baracs, a modern structural engineer and geometer, to show that three-dimensional projections of convex 4-polytopes correspond to some, but not all, spider webs in 3-space.

In the 1970s, computer scientists sought algorithms to recognize and draw correct pictures of objects in space. Several workers independently observed[5] that the existence of a reciprocal figure was the natural geometric condition for a correct *picture of a polyhedron,* noting that the reciprocal figure records the normals to the faces (Fig. 17-3). At first some critical topological details were not properly addressed, but this construction of plane reciprocals has now been refined to give a necessary and sufficient condition for correct pictures of any oriented polyhedron.[6]

At about the same time, computer scientists were studying *Dirichlet tessellations* (also known as Voronoi diagrams): subdivisions of the plane (and of n-space) into the polygonal (or polyhedral) regions of points closest to given centers (Fig. 17-4). In 1979, Brown observed that a Dirichlet tessellation in the plane corresponds to a convex polyhedron with all

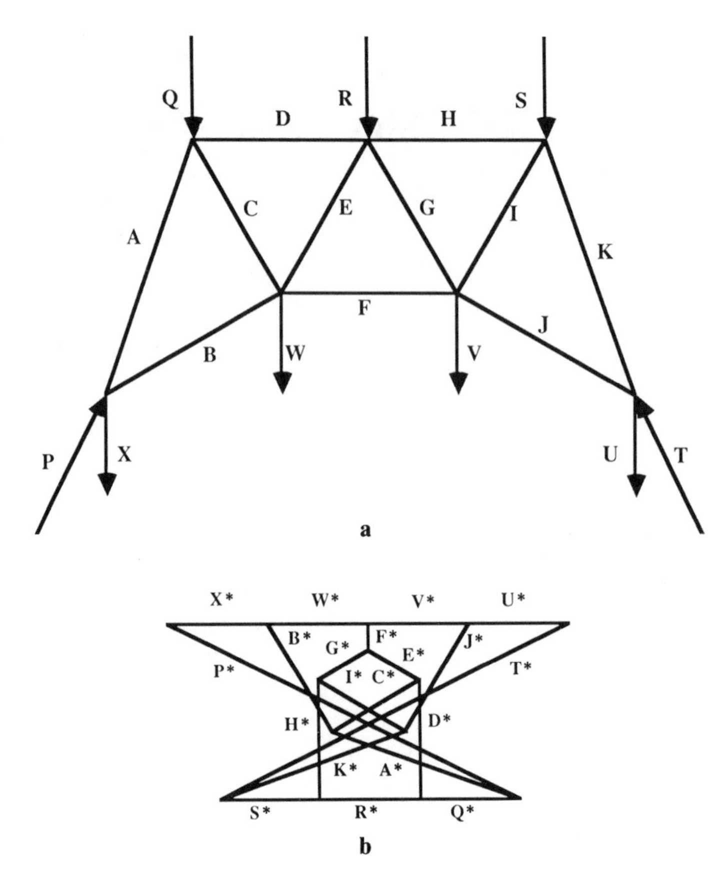

Fig. 17-1. A framework (a) in static equilibrium with a set of external forces (*the arrows*), has a Maxwell reciprocal figure (b) with dual edge Z* perpendicular to the original edge Z, and a polygon dual to the edges at each vertex of the original.

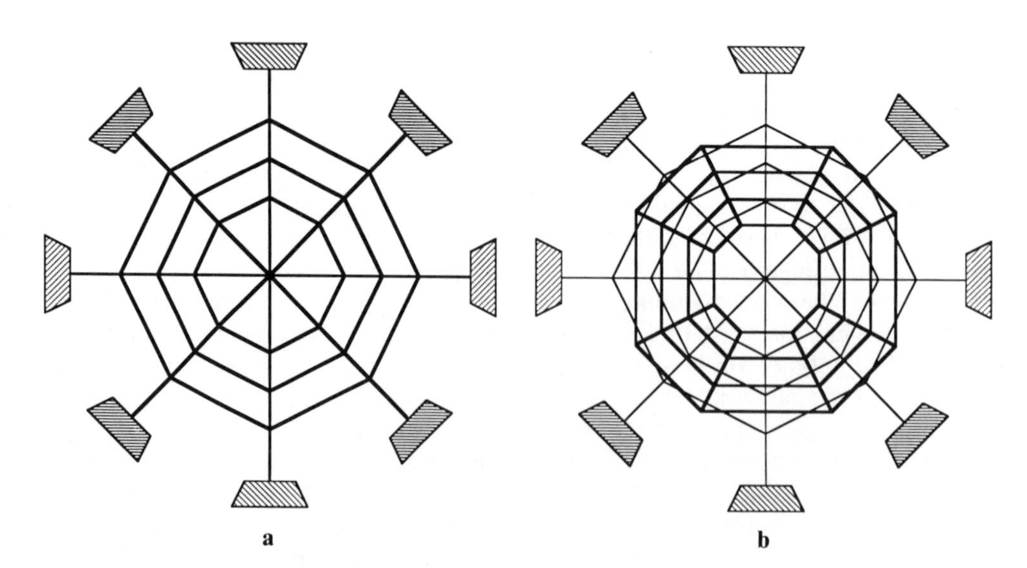

Fig. 17-2. A plane spider web (a) has an internal static equilibrium with tension in all members, and a convex reciprocal figure derived from this equilibrium (b).

faces tangent to a sphere, projected from the point of tangency of one face onto a plane parallel to this face.[7] He used this observation to develop efficient algorithms to compute the Dirichlet tessellation for a set of centers. Ash and Bolker observed that the diagram of centers forms a classical reciprocal figure for its Dirichlet tessellation.[8] More generally, a *sectional Dirichlet tessellation,* a plane section of a Dirichlet tessellation in 3-space, has a plane reciprocal formed by the orthogonal projection of the spatial centers. At the 1984 Shaping Space Conference, Whiteley and Bolker forged the last link in the proof that sectional Dirichlet tessellations and plane spider webs coincide.[9] This says, implicitly, that convex polyhedra projected from one face are just sectional Dirichlet tessellations, and conversely. Independently, Edelsbrunner and Seidel gave an explicit construction of a polyhedron which will correspond to a given sectional Dirichlet tessellation.[10]

Thus many of the results we present are not new. However, the unified picture is, and some new results follow from this unification. We highlight the reciprocal figure as the central geometric construction and some direct geometric arguments replace previous, seemingly accidental coincidences. We will sketch proofs when they are simple or illuminating; otherwise we will refer the reader to the literature.

In the next section, we carefully describe the equivalence of the following finite geometric objects in the plane:

- A plane section of a Dirichlet tessellation of 3-space.
- A plane section of a furthest-point Dirichlet tessellation of 3-space.
- A projection of a convex polyhedron in 3-space from a point inside one of its faces onto a plane parallel to this face.
- A plane framework, without self-intersection, with a static equilibrium using tension in all members.
- A plane drawing of a planar graph, with a planar reciprocal figure of disjoint convex polygons.

We also describe the special correspondence between Dirichlet tessellations in the plane and convex polyhedra with all faces tangent to a sphere (or a paraboloid). Later we survey some infinite analogues and point out how most but not all equivalences remain true. Such infinite, but locally finite, structures occur in the study of circle packings of the entire plane and in both periodic and aperiodic tilings of the plane.

All of the questions we raise, and many of the answers, generalize to n-space, but in this

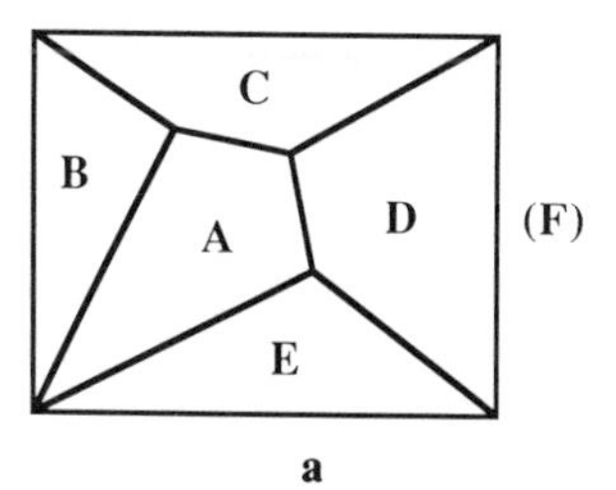

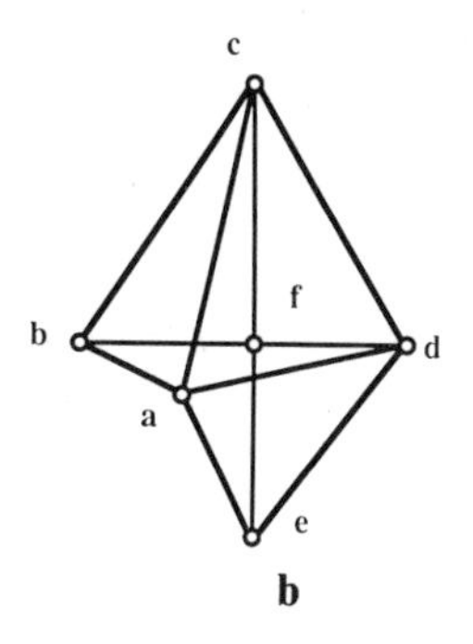

Fig. 17-3. The projection of a polyhedron (a) has a reciprocal figure (b) which places the vertex dual to the face C at the point given by the gradient c of this plane.

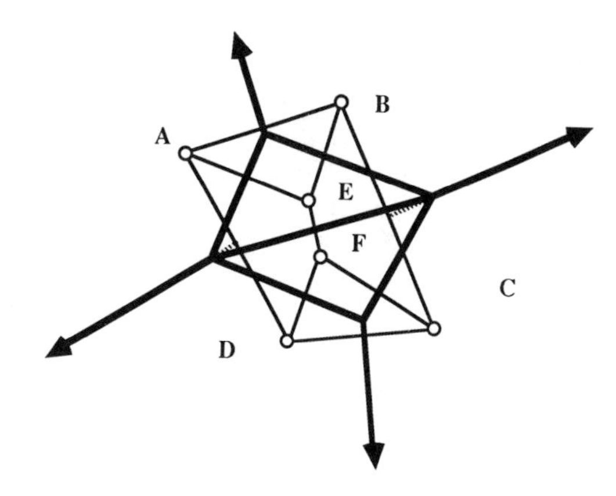

Fig. 17-4. A set of centers (*the circles*) defines a Dirichlet tessellation (*heavy lines*) and a reciprocal diagram of centers (*lighter lines*).

chapter we limit ourselves to dimensions 1, 2, and 3.

Each of the fields we touch on has its own favorite questions, results, and unsolved problems. The connections we establish among these fields have important implications for those preoccupations. For example the question: "Which graphs can be realized (that is, constructed) as a Dirichlet tessellation in the plane?" now coincides with the classical problem: "Which spherical polyhedra can be realized with all faces tangent to an insphere?" The question: "Which graphs can be realized as a sectional Dirichlet tessellation in the plane?" is answered by Steinitz's theorem: adding the polygon at infinity must create a triply connected planar graph! We will draw out some of these implications as we proceed through the correspondences.

This study represents geometry in what is for us its best sense: recognizing the kinship among classes of tangible, visible objects that one can draw, build, and manipulate. We feel pleasure in seeing a spider weave a many-faceted diamond, excitement in discovering the geometric basis of common algorithms to recognize or create these patterns, horror in seeing a building shake because it contains a 4-polytope, and satisfaction in knowing that a circle-packing is locally maximally dense because the graph of the centers is a rigid spider web.

Cell Decompositions and Reciprocal Figures

Imagine cutting the plane into a finite number of convex polygons and unbounded convex polygonal regions (see Fig. 17-5a). A *proper cell decomposition* of the plane is a finite set of convex polygons and unbounded convex polygonal regions, called the cells, such that

(i) every point in the plane belongs to at least one cell;
(ii) the cells have disjoint interiors;
(iii) the decomposition is edge-to-edge; that is, every edge of a cell is a complete edge of a second cell.

For example, the *Dirichlet tessellations* described above are proper cell decompositions. Each proper cell decomposition D of the plane (henceforth in this section we shall omit "of the plane") has an abstract *dual graph* D*. The vertices of D* are the cells of D; two vertices c and c′ are joined by an edge just when the corresponding cells C and C′ share an edge. It is clear that D* is always a planar graph, because it can be drawn in the plane simply by choosing a point inside each of the cells and joining the points in cells which share an edge (see Fig. 17-5b). For a Dirichlet tessellation, the centers are the vertices of a planar embedding of D* and the edge separating two cells C and C′ is the perpendicular bisector of the segment cc′ in this drawing of the dual (see Figs. 17-4 and 17-8b).

This example suggests a way to try to draw the dual graph of a cell decomposition. A *reciprocal figure* for D is a plane drawing of D* in which the edges are straight-line segments which are (when extended) perpendicular to the (extended) edges of D. Figure 17-6 shows additional examples of cell decompositions with reciprocal figures. Note that we do *not* demand that the vertices of a reciprocal figure lie in the cells to which they correspond.

So far our discussion of the reciprocal figure has concentrated on the graph: the edges and the vertices. The cell decomposition has vertices, edges, and cells, and we now restore this symmetry to the reciprocal. Around each vertex of the original decomposition we have a cycle of exiting edges and a corresponding polygon of orthogonal edges in D*. If all these dual edges have nonzero length, and the resulting polygons are convex and have disjoint interiors, we say we have a *convex reciprocal figure*. Figure 17-6a shows a convex reciprocal figure, while the cell decomposition in Fig. 17-6b has no convex reciprocal, because we have turned the edge between cells C and D. Figures 17-6c and 17-6d show a single cell decomposition with a convex reciprocal (C) and a nonconvex reciprocal, in which the convex polygons are not disjoint (D).

A set of parallel lines and the strips between them is a *trivial cell decomposition* which has no vertices (Fig. 17-7). Such decompositions are just perpendicular translations of cell decompositions of the line. They have reciprocal figures in which all the dual vertices lie on some line perpendicular to the edges. For the

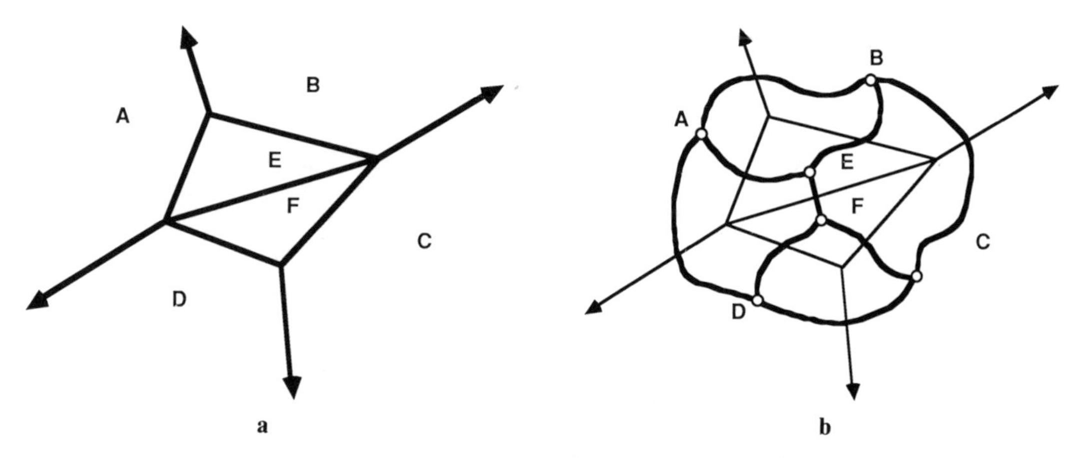

Fig. 17-5. Any proper cell decomposition of the plane (a) has a dual graph (b).

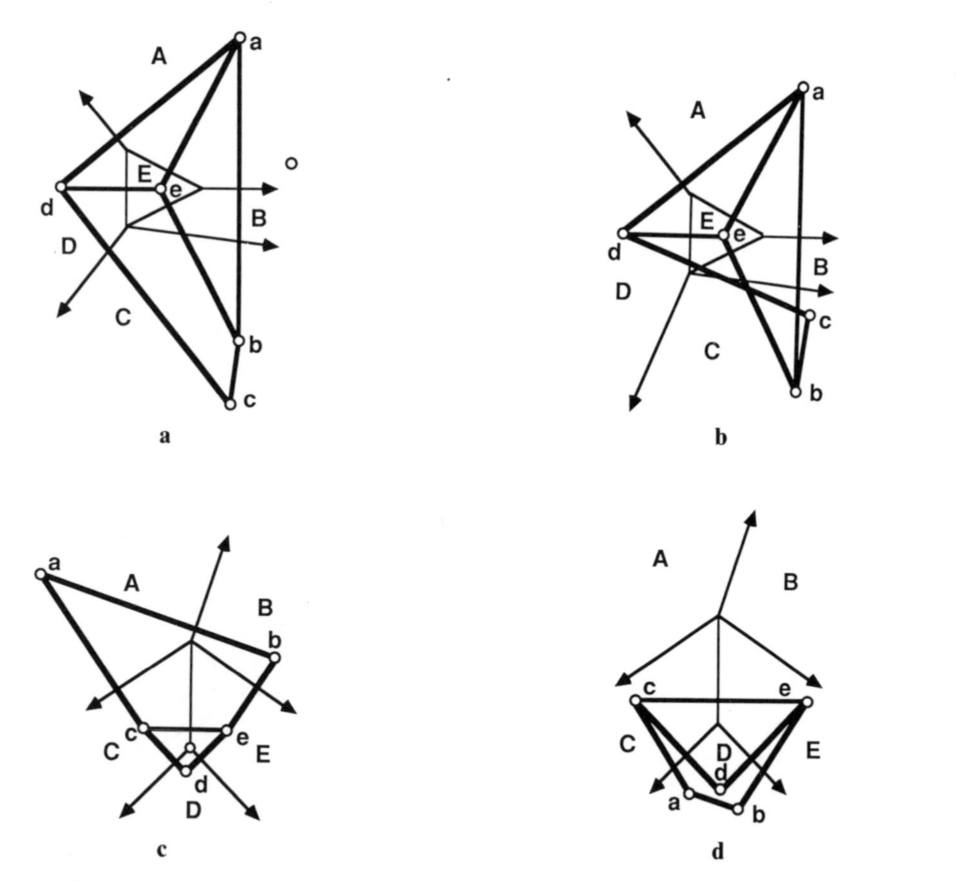

Fig. 17-6. Some cell decompositions (*light lines*) with reciprocal figures (*heavy lines*). The cell decomposition in (a) has only convex reciprocals, but that in (b) has only nonconvex reciprocals. A single cell decomposition may have both convex (c) and nonconvex (d) reciprocals.

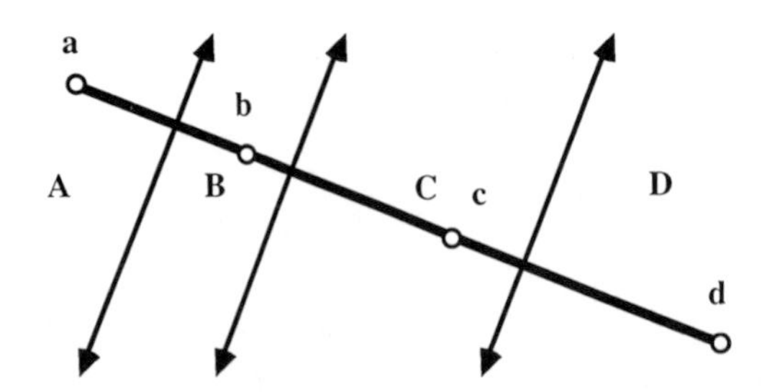

Fig. 17-7. A trivial cell decomposition of the plane (*light lines*) has a convex reciprocal which lies on a perpendicular line (*heavy line*).

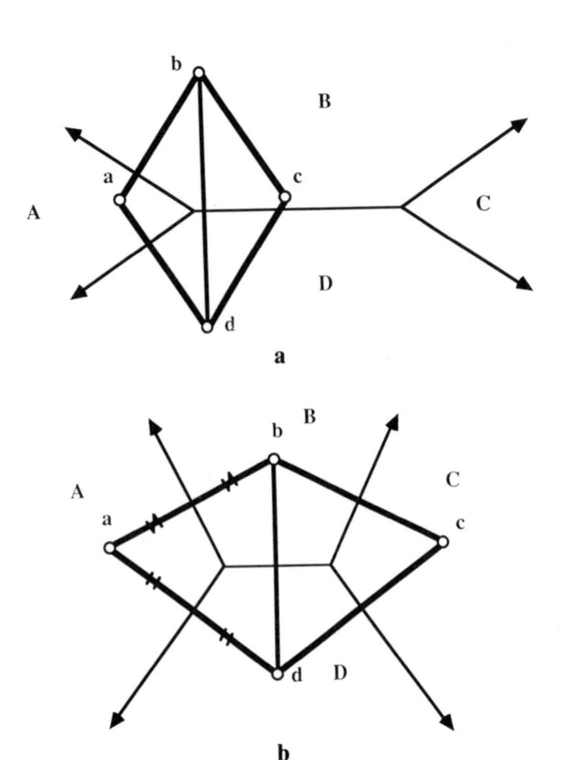

Fig. 17-8. Some cell decompositions with a convex reciprocal cannot be a Dirichlet tessellation (a), while others with a similar structure are Dirichlet tessellations (b).

purposes of our theorems and constructions we call such a trivial reciprocal figure convex if the order of the dual vertices along the line matches the order of the strips along the line. We note that any nontrivial proper cell decomposition in the plane has at least one vertex on each edge.

These trivial cell decompositions hint at the fact that our entire theory can be restated simply for figures on the line. A cell decomposition of the line is a set of line segments with disjoint interiors. A convex reciprocal is just a set of reciprocal points for these cells, which respect the order of the cells (see Figs. 17-9 and 17-13). From time to time we shall use figures on the line to illustrate the concepts we are exploring for figures in the plane.

Returning to the example of a Dirichlet tessellation, we see that the diagram of centers is a convex reciprocal figure. If all the vertices are 3-valent, the reciprocal will be a triangulation known as the Delauney triangulation.[11] This observation is our first theorem.

Theorem 17.1. A Dirichlet tessellation has a convex reciprocal figure. The converse of this statement is false.

Figure 17-8a shows a proper cell decomposition which has a convex reciprocal figure but is not a Dirichlet tessellation: there is no way to position the centers so that the edges of the original decomposition bisect the edges of the dual (as in Fig. 17-8b). Experimental evidence is quite convincing; a proof will be found in Ash and Bolker's "Recognizing Dirichlet Tessellations."[12]

To find a converse, we must broaden our search.

A sectional Dirichlet tessellation is a plane section of a Dirichlet tessellation of 3-space. (We define a Dirichlet tessellation of 3-space by replacing "the plane" by "space" in our previous definition.) If we throw away any centers in space whose cells do not meet the slicing plane in a nonempty open set, the remaining centers are in one-to-one correspondence with the cells of the sectional Dirichlet tessellation. Their orthogonal projections onto the slicing plane form a convex reciprocal figure for the sectional Dirichlet tessellation (see Fig. 17-9 for the analogue on the line). This gives the result of Ash and Bolker's "Generalized Dirichlet Tessellations"[13]:

Theorem 17.2. A sectional Dirichlet tessellation has a convex reciprocal figure.

These sectional Dirichlet tessellations, also called power Voronoi diagrams or Voronoi diagrams in Laguerre geometry, model a simple biological phenomenon.[14] Suppose bacteria start to grow at center c at time t_c, with growth rate at the boundary inversely proportional to the distance from the center. If the colonies cannot overlap, the cells occupied by the colonies form the sectional Dirichlet tessellation on the plane $z = 0$ of the spatial tessellation with centers (c, $\sqrt{t_c}$). If all the bacteria start at the same time, we have a Dirichlet tessellation. In this model each cell contains its center; this need not always be true. Other examples and a more complete bibliography are given by Ash and Bolker.[15]

We shall prove the converse of Theorem 17.2 and discuss the furthest-point Dirichlet tessellations after we examine spider webs and projections of convex polyhedra.

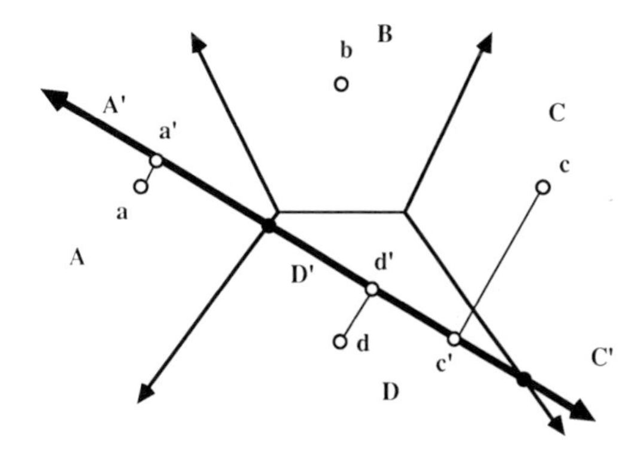

Fig. 17-9. Any section of a plane Dirichlet tessellation creates a sectional Dirichlet tessellation on the line (*the black dots on the heavy line*), with a convex reciprocal (*the circles*) given by the orthogonal projection of the plane centers.

Spider Webs and Projections

A proper cell decomposition is a *spider web* if it supports a spider web stress: a set of nonzero tensions in the edges which leads to mechanical equilibrium at each vertex (Fig. 17-10a). More specifically, a spider web stress is a nonzero force F_{VE} in each edge E at a vertex V, directed from V out along the edge, such that

(i) for a finite edge E joining V and V′, the forces at the two ends are equal in size: $F_{VE} = -F_{V'E}$;
(ii) for each vertex V, the vector sum of the forces on the edges leaving V is zero.

Spider webs are interesting and important. If they are built with cables, and pinned to the ground on the infinite edges, they are rigid in the plane.[16] In fact they are the basic building blocks of all rigid cable structures in the plane. At the other extreme, if a plane bar-and-joint framework has the minimum number of bars needed to restrain $|V|$ joints ($|E| = 2|V| - 3$), then the appearance of a spider web signals that it is shaky.[17] Finally, recent work on packing circles of a fixed radius, without overlapping, into a convex polygon has shown that such a packing cannot be made denser by a small jiggle if, and only if, the associated graph of centers and contact points is an infinitesimally rigid spider web.[18]

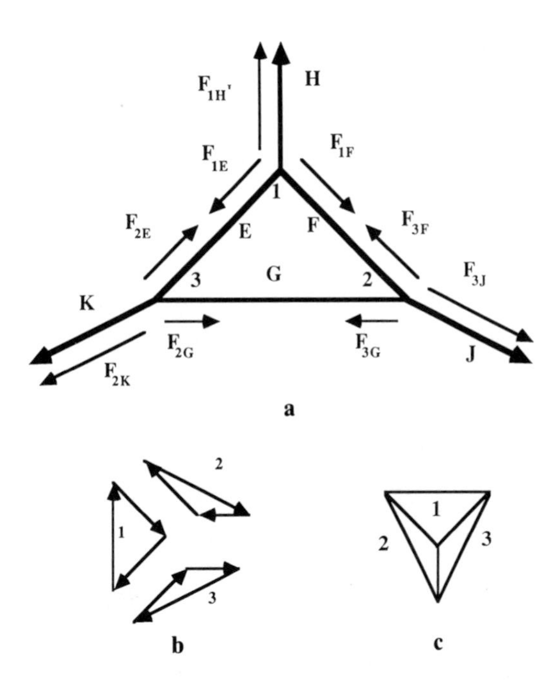

Fig. 17-10. The arrows in (a) show the tensions of a spider web stress on a cell decomposition. The polygons of forces for the equilibria at the vertices (b) are pieced together and rotated 90° to form a convex reciprocal figure (c).

If a cell decomposition has a spider web stress, then the vanishing of the vector sum of forces at each vertex says that these forces can be drawn as a closed convex polygon (Fig. 17-10b). If we rotate each such polygon clockwise by 90° then each edge is perpendicular to the edge of the original figure to which it corre-

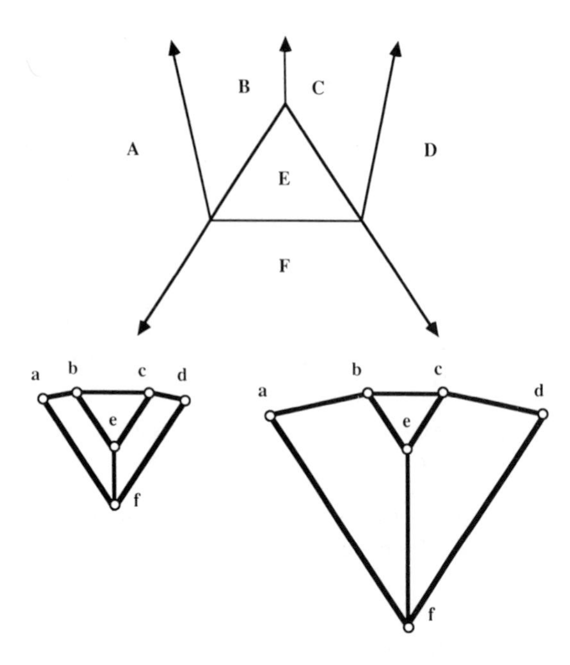

Fig. 17-11. A cell decomposition (*light lines*) can have two dissimilar convex reciprocal figures (*heavy lines*) given by free choices for the lengths of some reciprocal edges.

sponds. Since the forces at the two vertices of a finite edge are equal in size and opposite in direction, these polygons can be glued together to make a planar drawing of the dual graph D* (Fig. 17-10c); we have constructed a convex reciprocal.[19] Conversely, assume that a cell decomposition has a convex reciprocal. We turn this diagram 90° counterclockwise and use the length of each dual edge to define the size of the tension in the corresponding edge of the decomposition. The convex polygons of the reciprocal imply the vector equilibrium of these tensions at each vertex. Thus we have proved[20]:

Theorem 17.3. A proper cell decomposition is a spider web if and only if it has a convex reciprocal figure. The convex reciprocal determines the spider web stress, and the spider web stress determines the convex reciprocal, up to translation and rotation by 180°.

The trivial cell decompositions satisfy this theorem in an appropriately trivial way. They all have convex reciprocals, and require no tensions for edges with no vertices!

Theorem 17-3 shows that if a proper cell decomposition has one convex reciprocal figure, it has many. Any translation of a reciprocal produces another reciprocal; clearly this translation has no effect on the tensions. If we turn the reciprocal by 180°, we also get a reciprocal. In the classical literature, this turn corresponds to a switch from tensions to compressions, but we have chosen to concentrate on the tensions. Any dilation converts a convex reciprocal to a new reciprocal; the stress corresponding to the new reciprocal is a scalar multiple of the one corresponding to the original.

For some figures, we can freely choose the lengths of several different reciprocal edges and still complete the reciprocal. For the example in Fig. 17-11 the lengths of the reciprocal edges bc and ab are independent choices. The existence of such dissimilar convex reciprocals reflects the fact that the set of stresses, which is a vector space, has dimension greater than 1.

We now turn to study the projections of polyhedra. Consider the intersection of the upper half spaces of a finite set of nonvertical planes. The faces, edges, and vertices of this intersection form a *convex polyhedral bowl.* (Our choice of upturned bowls is simply a convenient convention, as you will see below.) The vertical projection of a convex polyhedral bowl is a proper cell decomposition of the plane.

To construct a convex reciprocal figure for such a projection, suppose that the boundary planes P and P′ which meet at an edge E have equations

$$Ax + By - z - C = 0,$$
$$A'x + B'y - z - C' = 0.$$

Then the line joining the points (A,B) and (A',B') in the plane is perpendicular to the vertical projection of the edge E, because those points are the intersections of the plane and the normals to P and P′ drawn from the point (0, 0, 1) (Fig. 17-12). The set of points (A,B), one for each face of the bowl (and thus one for each cell of the projected cell decomposition), form a reciprocal figure for the projection (see Fig. 17-13 for an example on the line).

The convexity of the polygons in the reciprocal follows from the convexity of the vertices in the polyhedral bowl. Finally, observe that this reciprocal is also the vertical projection of a dual object in 3-space. This *dual polyhedral bowl* has vertices (A,B,C) corresponding to the planes $Ax + By - z - C = 0$ of the original and boundary planes $Px + Qy - z - R = 0$, one for each vertex (P,Q,R) of the original bowl. This dual bowl is created by a projective polarity about the *Maxwell paraboloid* [21] $x^2 + y^2 - 2z = 0$. (Notice that a dual polyhedral bowl is also convex, but it has a cylinder of vertical planes dual to the points at infinity on the unbounded edges of the original bowl.)

The converse is also true. If a proper cell decomposition has a convex reciprocal, then there is a convex polyhedral bowl projecting to this decomposition, with the normals to the faces given by the reciprocal vertices (or by the reciprocal turned 180°). Any single boundary plane of the bowl can be chosen freely (perpendicular to its known normal). Then the positions of the remaining planes can be deduced (see Crapo and Whiteley[22] for a proof). Therefore:

Theorem 17.4. A proper cell decomposition has a convex reciprocal if and only if it is the vertical projection of a convex polyhedral bowl. The convex reciprocal can be reconstructed from the bowl by taking the normals to the faces, and the convex reciprocal determines the bowl, up to vertical translation.

The trivial cell decompositions are projections of trivial convex polyhedral bowls formed by planes parallel to a line. The normals to the faces of such a bowl yield the trivial convex reciprocals as we defined them.

A vertical scaling of a convex polyhedral bowl projecting to our cell decomposition (that is, changing all the z-coordinates by a positive constant factor) corresponds to a similarity transformation of the reciprocal by the same factor. The translations of a reciprocal come from a more subtle "rolling" of the angles of the planes (see Fig. 17-14a). If we reflect the bowl in the xy-plane, the reciprocal turns 180°. We turn all our bowls up so that the reciprocals created match those for sectional

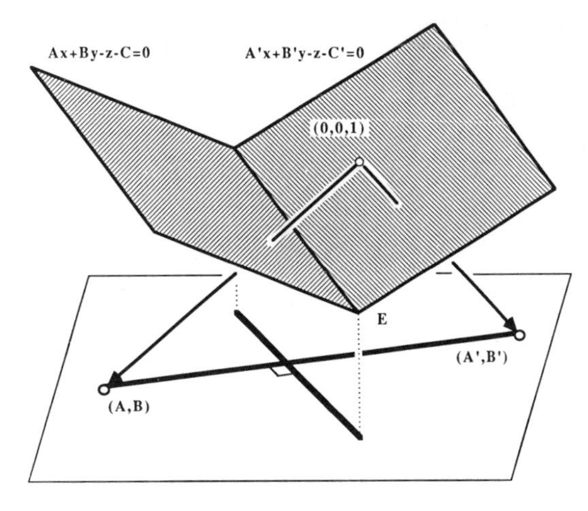

Fig. 17-12. These normals to the face planes at a polyhedral edge section to the gradients, and to a reciprocal edge perpendicular to the projection of the polyhedral edge.

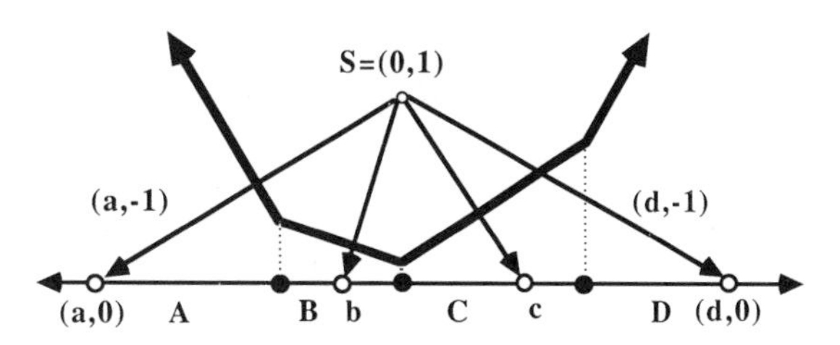

Fig. 17-13. A polygonal bowl (*heavy lines*) projects to a cell decomposition of the line (given by the *black dots*), and the normals to the edges produce a convex reciprocal figure (*the circles*).

Dirichlet tessellations, for which the dual edge cc′ is oriented so that it crosses from cell C to cell C′.

While the reciprocal figure is given by an Euclidean construction, the existence of a reciprocal is an essentially affine geometric property: any affine transformation of a cell decomposition with a convex reciprocal figure extends to an affine transformation of the corresponding polyhedral bowl, which then gives the new reciprocal figure. (Alternatively, if the affine transformation of the cell decomposition is AX, then the reciprocal vertices are transformed by $(A\dagger)^{-1}X^*$.) In fact, the invariance has projective overtones as well. If we add the plane at infinity to a nontrivial convex polyhedral bowl, we get a closed polyhedron in projective space, with the point of projection on this added face. We can make this a

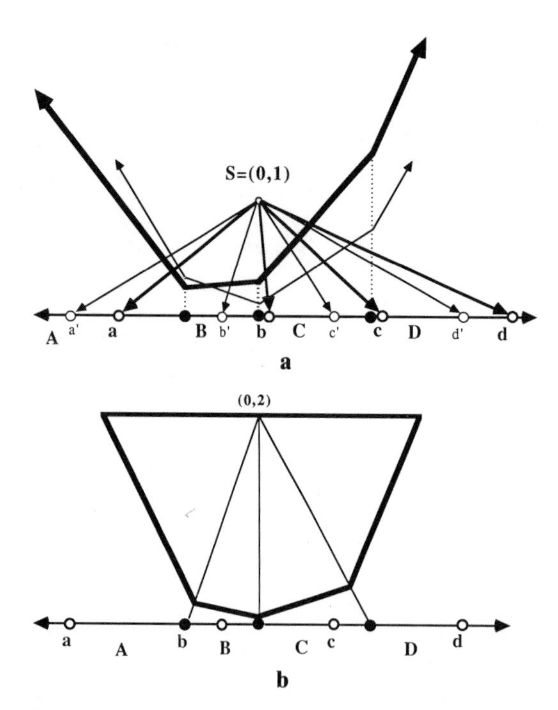

Fig. 17-14. If a reciprocal figure to a projected polygonal bowl (*heavy lines* in (a)) is translated (b to b′) then the bowl is rolled (*light lines* in (a)). The same cell decomposition is also the projection of a closed polygon from a point on one edge (b).

finite polyhedron by a projective transformation that brings the plane at infinity to the plane $z = 2$, and leaves the xy-plane unchanged (Fig. 17-14b). The cell decomposition is now the projection of this polyhedron from a point on the face which is parallel to the xy-plane. This polyhedron is convex if the vertical direction is enclosed by the cone of normals to the faces of the bowl, or equivalently, if the origin is in the convex hull of the vertices in the reciprocal. This can be arranged by a simple translation of the plane, so we have the following corollary to Theorem 17.5:

Corollary 17.5. A nontrivial proper cell decomposition has a convex reciprocal figure if and only if it is the projection of a convex polyhedron from a point inside a face that is parallel to the projection plane.

The Main Result

We are now ready to work toward the converse of Theorem 17.2. Consider a proper cell decomposition with a convex reciprocal. This decomposition is the projection of a convex polyhedral bowl with faces of the form $Ax + By - z - C = 0$. Choose the centers in space to be the points[23] $(A,b, \sqrt{2C - A^2 - B^2})$; if necessary, the bowl can be lowered by adding a constant d to all the C to make all these z-coordinates well-defined. It is a simple exercise to check that a point (x,y) is in the projection of a face if and only if it is closest to the corresponding center (see Fig. 17-15), and therefore is in the cell of the sectional Dirichlet tessellation.[24]

Theorem 17.6. Each sectional Dirichlet tessellation corresponds to a convex polyhedral bowl, and each convex polyhedral bowl can be vertically translated to correspond to a sectional Dirichlet tessellation.

If we translate the bowl by a constant d, a center at height h moves to one at height $h' = \sqrt{2d + h^2}$ (provided that $2d + h^2$ is positive for all vertices). We are also free to choose the sign $\pm h$ for each center independently. This completes the converse[25] to Theorem 17.2:

Corollary 17.7. A cell decomposition has a convex reciprocal if and only if it is a sectional Dirichlet tessellation. The centers in space can be chosen over the vertices of the convex reciprocal, and these centers are unique up to a vertical scaling of the form $h' = \sqrt{2d + h^2}$.

If we reflect the bowl through the xy-plane, and therefore turn the reciprocal by 180°, then the point (x,y) is in the region corresponding to $(A',B') = (-A,-B)$ if the associated plane gives the minimum value of z. After rescaling all C' by a vertical translation, our construction shows that $(A', B', \sqrt{2C' - A'^2 - B'^2})$ gives the maximum distance from $(x,y,0)$. Our cell decomposition is thus the section of the *furthest-point Dirichlet tessellation* of centers;[26] that is, each point belongs to the cell of the center which is furthest from the point (Fig. 17-16). Our entire theory of reciprocal diagrams applies to these *sectional furthest-point Dirichlet tessellations*. In particular, our inversion of the bowl gives the following result:

Theorem 17.8. Given a proper cell decomposition D, the following are equivalent: (i) D is a sectional Dirichlet tessellation with projected centers P. (ii) D is the sectional furthest-point Dirichlet tessellation with projected centers $-P$.

This completes our chain of equivalences. We summarize:

Theorem 17.9. Given a proper cell decomposition D in the plane, the following are equivalent: (i) D has a convex reciprocal figure. (ii) D is a spider web. (iii) D is the vertical projection of a convex polyhedral bowl in 3-space. (iv) D is a sectional Dirichlet tessellation. (v) D is a sectional furthest-point Dirichlet tessellation. If D is nontrivial, these are also equivalent to (vi) D is the vertical projection of a convex polyhedron from a point inside one face into a plane parallel to this face.

These equivalences have some interesting consequences. For example, if we build a proper cell decomposition from rubber bands, pin down the edges that go to infinity, and let the tensions position the vertices at a mechanical equilibrium, we will have drawn a picture of a convex polyhedral bowl; spiders really draw such pictures. Moreover, any plane picture of a spider web in space is also a plane spider web, even if the spatial web is warped. Conversely, if we wish to design a rigid cable structure in the plane with a planar graph, we must build a spider web, so we can just use the picture of some convex polyhedron.[27]

The plane Dirichlet tessellations we first studied have very special reciprocals: ones whose dual edges are bisected by the edges of the tessellation. We will see that the polyhedral bowl which projects to a Dirichlet tessellation is correspondingly special; its faces are tangent to the Maxwell paraboloid. Consider such a *Maxwell bowl.* The point of tangency on each face is the spatial polar of this face (using the polarity described above for the dual bowl), so the plane $Ax + By - z - C = 0$ meets the paraboloid at the point $(A,B,z) = (A,B,1/2(A^2 + B^2))$. Therefore $C = 1/2(A^2 + B^2)$. Thus in the construction used in Theorem 17.6, the height $h = \sqrt{2C - A^2 + B^2} = 0$, and we have a Dirichlet tessellation (Fig. 17-17). This argument and its converse prove:[28]

Theorem 17.10. A proper cell decomposition is a Dirichlet tessellation if and only if it is the vertical projection of a Maxwell bowl.

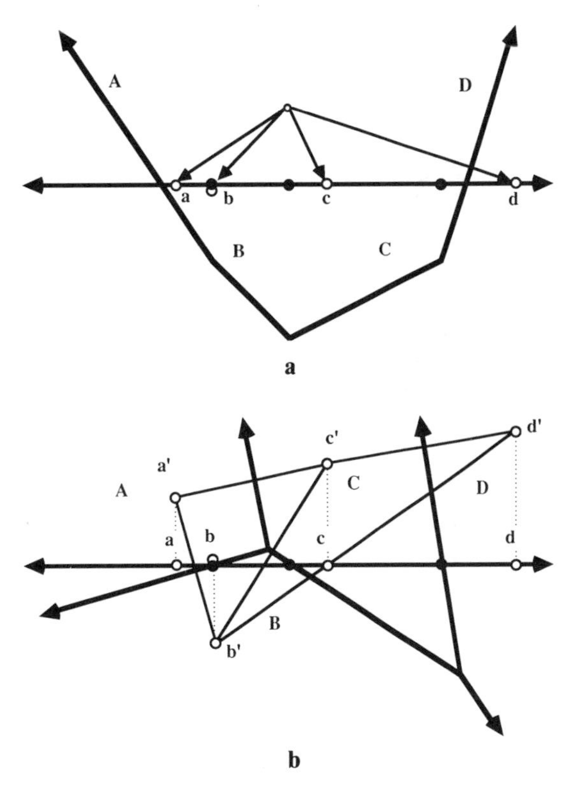

Fig. 17-15. A polygonal bowl projects to a cell decomposition with its reciprocal on the line (a) and the corresponding plane Dirichlet tessellation sections to the same cell decomposition with the same reciprocal (b).

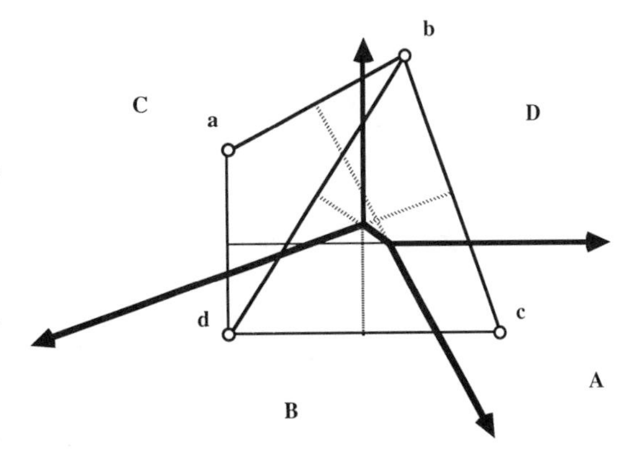

Fig. 17-16. A set of centers (*the circles*) produces a furthest-point Dirichlet tessellation (*heavy lines*) with a reciprocal diagram of centers (*medium lines*).

As we noted before, the furthest-point tessellation corresponds to taking the smallest z value among the planes over the point (x,y), or equivalently, the intersections of the lower half spaces of the planes. Thus a furthest-point tessellation is the projection of an inverse

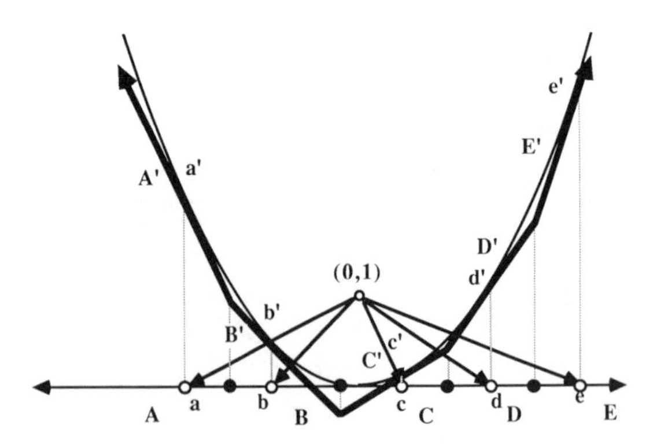

Fig. 17-17. Each Dirichlet tessellation on the line is the projection of a polygonal bowl (*heavy lines*) circumscribed about the Maxwell parabola. The centers are the projections of the points of contact of the edges.

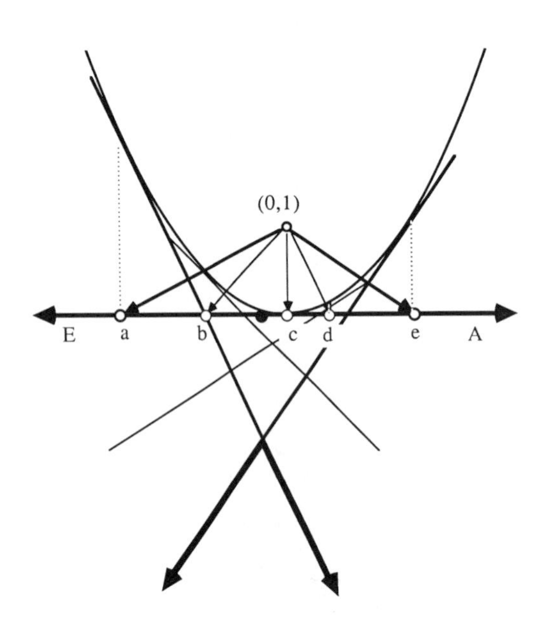

Fig. 17-18. Each furthest-point tessellation on the line is the projection of an inverse polygonal bowl (*the heavy lines at the bottom*) with edges tangent to the Maxwell parabola. The centers are the projections of the points of contact.

bowl all of whose faces are tangent to the Maxwell paraboloid: a *Maxwell inverse bowl* (Fig. 17-18). (It is not the projection of the inverse of a Maxwell bowl, and is not a Dirichlet tessellation.)

Theorem 17.11. A proper cell decomposition is a furthest-point Dirichlet tessellation if and only if it is the vertical projection of a Maxwell inverse bowl.

The plane at infinity is also tangent to the Maxwell paraboloid at the infinite point of projection. In the construction of Corollary 17.5 the projective transformation converts this paraboloid into the sphere $x^2 + y^2 + (z - 1)^2 = 1$. After a suitable translation of the Dirichlet tessellation and its centers, this creates a convex polyhedron with all faces tangent to the sphere that projects onto the Dirichlet tessellation from the point of contact (0,0,2) of a horizontal face (Fig. 17-19). This gives[29]

Corollary 17.12. A nontrivial proper cell decomposition is a Dirichlet tessellation of the plane if and only if it is the central projection of a convex polyhedron circumscribed about a sphere, from the point of contact of one face, onto a plane parallel to this face.

We note a curious consequence of this vision of Dirichlet tessellations. Given a convex polyhedron with all faces tangent to a sphere, we can turn the sphere so that any one face is parallel to the xy-plane, and project from the point of contact. This defines a new and unusual equivalence relation among Dirichlet tessellations: a tessellation with n cells is equivalent to n other tessellations.

It is easy to construct examples of proper cell decompositions of the plane which are not spider webs, or, equivalently, are not projections of convex polyhedral bowls. Consider the cell decomposition of Fig. 17-20. If this were the projection of a convex polyhedral bowl, the three planes over the cells A, B, and C would meet in a point. This point would be at the intersection of the three lines separating these cells; this intersection does not exist. Equivalently, if this were a spider web, the three forces in these three separating edges would be in equilibrium (since the forces on any cut set in a spider web will be in equilibrium) but three forces in the plane can reach equilibrium only if they are on concurrent lines. Finally, if this were a sectional Dirichlet tessellation, then the line of points equidistant from the spatial centers of the three exterior cells would lie in the three planes separating these cells and the intersection with the plane would be three concurrent lines.

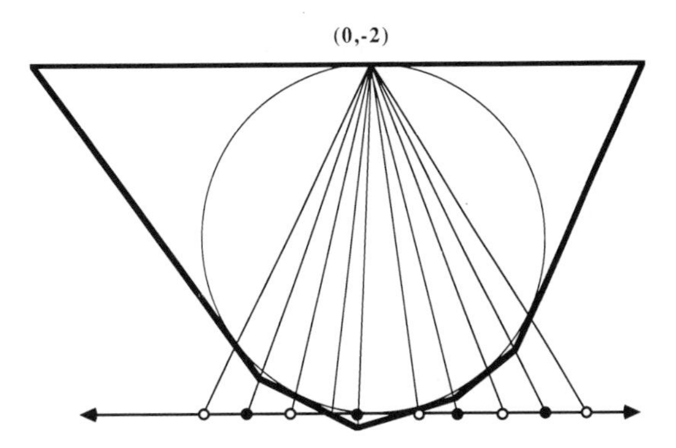

Fig. 17-19. Each Dirichlet tessellation on the line is the projection of a convex polygon with an inscribed circle, from the point of contact of one edge. The centers are the projections of the other points of contact.

Realizations of Abstract Graphs

Given a proper cell decomposition of the plane, we can consider the vertices, edges, and cells as a combinatorial structure G (whose precise definition will be given soon) and ask: "Which realizations of G as proper cell decompositions have convex reciprocals?" Work on plane stresses shows that answering this question is equivalent to finding the convex cone of entirely positive solutions to a homogeneous system of linear equations whose unknowns represent the positions of the vertices and the directions of the infinite edges.[30] To be specific, we have a set of finite vertices, V, and infinite vertices V^o, one for each positive direction of an unbounded edge, as well as a cyclic order for these infinite vertices. The finite and unbounded edges are given in the obvious way, with two unbounded edges sharing an infinite vertex if they go in the same direction. We take all realizations as proper cell decompositions which respect the order of the cycle of infinite vertices (or its reverse). From general results on planar graphs, this graph structure uniquely determines the possible cells. For such an abstract structure G there are five possibilities:

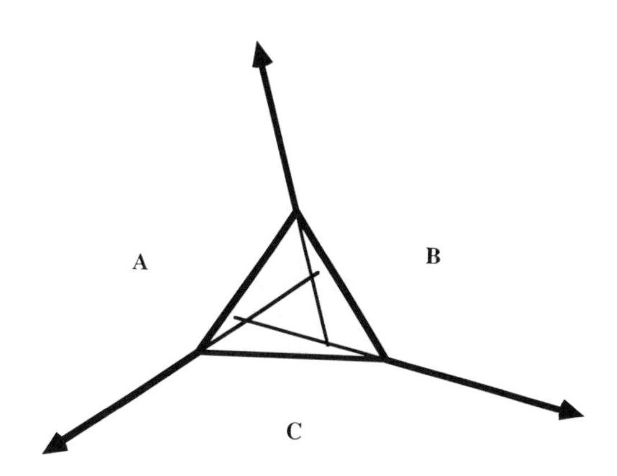

Fig. 17-20. This cell decomposition has no stress and no reciprocal figure, because the three extended lines are not concurrent.

(i) No realization of G as a proper cell decomposition has a convex reciprocal.

(ii) Some realizations of G as proper cell decompositions have a convex reciprocal; almost all small changes in the position of at least one vertex destroy the spider web stress. These special realizations must satisfy a *geometric condition* expressed by nontrivial polynomial equations in the coordinates of the vertices.

(iii) Many realizations of G as proper cell decompositions have a convex reciprocal; all realizations near a given spider web will also be a spider web. These correct realizations meet a *qualitative condition* expressed by nontrivial polynomial inequalities in the coordinates of the vertices.

(iv) Almost every realization of G as a proper cell decomposition is a spider web; certain special positions are improper. These improper positions, expressed by polynomial equations in the coordinates of the vertices, have zero tension in some edges.

(v) Every realization of G as a proper cell decomposition is a spider web.

Case (i) cannot happen. It is easy to check that all graphs of nontrivial proper cell decompositions are also triply connected planar graphs, if we add the polygon at infinity. Theorem 17.9 tells us that graphs of nontrivial spider webs are all constructed from convex polyhedra, by projecting from a point inside

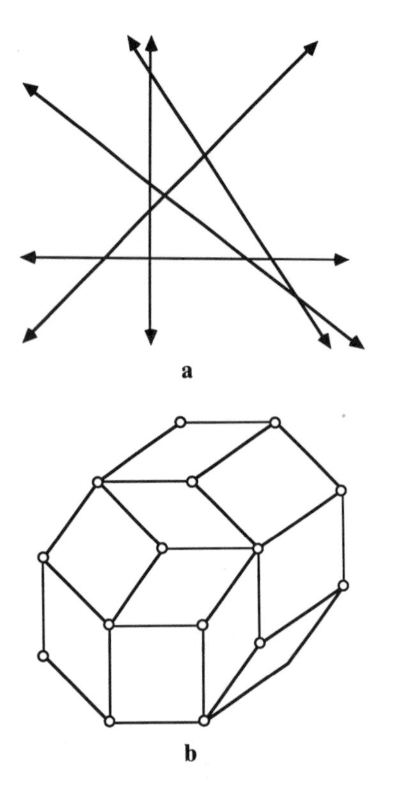

Fig. 17-21. Any arrangement of lines in the plane gives a cell decomposition (a) with a zonohedral cap as a convex reciprocal figure (b).

one face onto a plane parallel to this face. A classical theorem of Steinitz shows that any triply connected planar graph can be realized as a convex polyhedron.[31]

Figure 17-20 illustrates case (ii). If a graph has |V| finite vertices, and |E| edges, then the general theory of plane frameworks guarantees that when $|E| \leq 2|V|$ there are geometric conditions that must be satisfied if there is to be a stress on all members.[32] If the graph has all vertices 3-valent, the conditions for a spider web can be expressed as a set of equations, one for each finite cell in the graph.[33] Figure 17-6 illustrates case (iii). All realizations near Fig. 17-6a will be spider webs, since the convexity of the reciprocal is preserved by small changes. Similarly, all realizations near Fig. 17-6b will have only nonconvex reciprocals and will not be spider webs.

We conjecture that case (iv) cannot occur. As evidence, we cite the study of White and Whiteley[34] in which the authors conjecture that such boundary events, where one tension must be zero, lead to nearby points where the sign of the stress in the edge switches. The set of realizations as proper cell decompositions forms an open convex cone, so both signs of stress must appear in proper cell decompositions near the boundary event, putting us back in case (iii).

Cell decompositions with no polygons illustrate case (v). Examples are the trivial decompositions and trees with all vertices at least 3-valent. The interested reader can check directly that such cell decompositions always have a convex reciprocal and thus are always the projection of convex polyhedral bowls. Some other graphs of cell decompositions share this property, but we have not been able to characterize them. A *line arrangement* yields a cell decomposition that is always a spider web. A finite set of lines in the plane creates a cell decomposition (Fig. 17-21a). If for each line we choose a nonzero tension and assign this tension to all segments of the line, we have created a spider web stress. (At each vertex on the line, the two tensions in the line cancel; this local cancellations gives the equilibrium.) Therefore this cell decomposition has a convex reciprocal. The reader can check that this reciprocal will be a drawing of the zonohedral cap corresponding to the line arrangement (Fig. 17-21b), and the dual bowl over this reciprocal will be a zonohedral cap as found in Coxeter.[35]

What graphs have realizations that are nontrivial Dirichlet tessellations? By the theorem of Brown, such a realization, with the polygon at infinity added, must be the projection of a convex polyhedron with an insphere. There are theorems, also originating with Steinitz, which provide examples of graphs which can, and cannot, be the edge graphs of convex polyhedra with inspheres.[36] From these we can conclude that while the graph of Fig. 17-22 can be realized as a sectional Dirichlet tessellation, it cannot be realized as a Dirichlet tessellation since it is the projection of a polyhedron which cannot have an insphere.[37]

If a nontrivial graph can be realized as a Dirichlet tessellation, the realization must satisfy geometric conditions to be a Dirichlet tessellation; it must have a convex reciprocal with the dual edges bisected by the original edges. Ash and Bolker provide[38] a geometric characterization of proper cell decompositions that are Dirichlet tessellations. We will not attempt here to connect their characterization with that of Theorem 17.10 or Corollary 17.12. There may be some nice geometry waiting for someone who wishes to explore the connection.

We can characterize those graphs which can appear as nontrivial furthest-point Dirichlet tessellations: they are the trees with all vertices at least 3-valent. To prove this, consider such a tessellation, its reciprocal figure, and the corresponding Maxwell inverse bowl. Look at the convex hull of the points of contact with the Maxwell paraboloid. The upper surface of this hull projects to the reciprocal figure. Since the points are on the paraboloid, this figure is a convex polygon with some interior edges forming a "split polygon" (Fig. 17-23a) or it is a line segment. A split polygon is the dual of a tree (Fig. 17-23b) and a line segment is the dual of the 2-cell trivial decomposition. Conversely, every embedded tree has a split polygon as its dual, and we can arrange the points on the paraboloid to realize any combinatorial split polygon as the projection of an upper convex hull. Of course, the realizations of a tree which form furthest-point Dirichlet tessellations must satisfy additional geometric conditions, but we have not seen these explicitly worked out.

We close this section by remarking that all the results in it are special, convex cases of more general theorems. A planar graph that has a (possibly nonconvex) reciprocal is the projection of a general spherical polyhedron in 3-space (possibly self-intersecting). The reciprocal corresponds to a set of nonzero tensions and compressions in the edges of the graph, in a static equilibrium at each vertex. This general case was the one first studied in graphical statics.[39] These closing observations emphasize the important but often forgotten fact that statics and the equivalent theory of infinitesimal mechanics both truly belong to projective geometry.[40] So too does the study of projected polyhedra and general reciprocal diagrams.

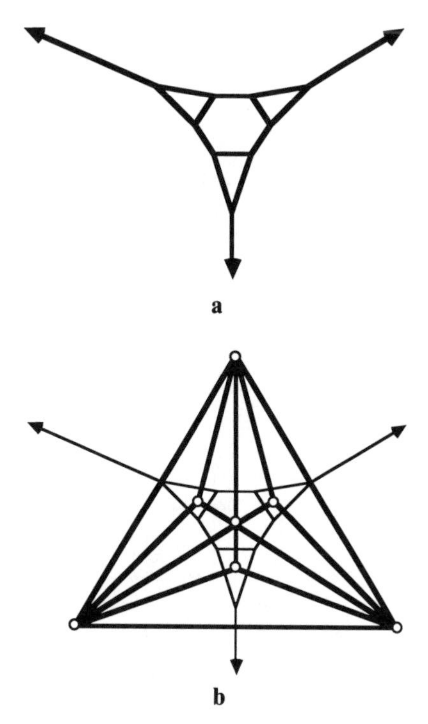

Fig. 17-22. The cell decomposition in (a) cannot be a Dirichlet tessellation, by its graph, but it is a sectional Dirichlet tessellation, by the convex reciprocal in (b).

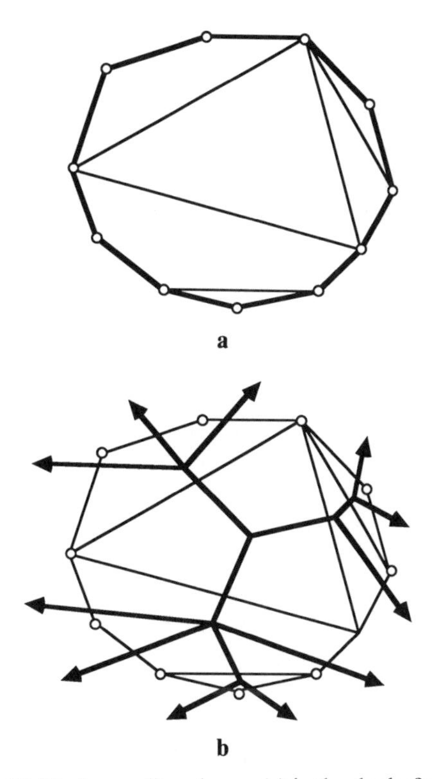

Fig. 17-23. Any split polygon (a) is the dual of a tree (b) which is the graph of some furthest-point Dirichlet tessellation.

Infinite Plane Examples

In order to study packings of the plane by identical circles, a number of mathematicians have considered the locally finite graph in the

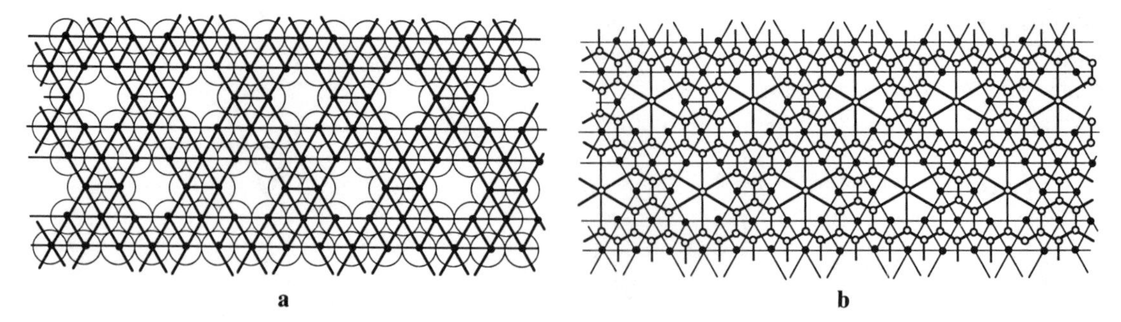

Fig. 17-24. The graph of centers of this circle packing gives an infinite cell decomposition of the plane (a) which has a convex reciprocal figure (b) which is also an infinite cell decomposition of the plane.

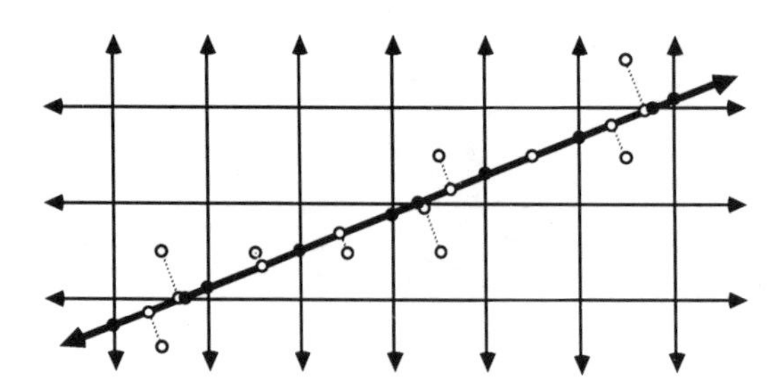

Fig. 17-25. A periodic Dirichlet tessellation of the plane (*light lines*) can section to a nonperiodic cell decomposition of the line (*the black dots on the heavy line*).

plane which has the centers of the circles for vertices and an edge joining two vertices whenever the two circles touch (Fig. 17-24a). Connelly has discovered[41] that there is an intimate connection between the problems: "Is this diagram a spider web?" and "Is this packing locally maximally dense?" He even uses convex reciprocal diagrams to study these examples (Fig. 17-24b).

Another class of infinite but locally finite examples comes from the study of Dirichlet tessellations arising from periodic lattices in the plane. Surprisingly, there are also aperiodic sections of periodic tessellations of 3-space. (Figure 17-25 shows an example on the line.) To study these and other similar examples, we change the definition of a proper cell decomposition by allowing infinitely many cells, but we require local finiteness: no point belongs to infinitely many cells.

We then refer to the cell decompositions of the previous section as *finite decompositions*. Our new infinite proper cell decompositions still have abstract dual graphs. The local finiteness implies that the dual polygon for each original vertex is finite, so the definition of a *convex reciprocal figure* is unchanged. However, since an unbounded cell of the decomposition may have infinitely many edges, the dual and the reciprocal may not be locally finite. We still include the trivial examples formed by an infinite number of parallel strips and their reciprocals.

An infinite set of centers, with only a finite number of centers in any bounded set, defines an *infinite Dirichlet tessellation*. More generally, an infinite sectional Dirichlet tessellation is a plane section of an infinite Dirichlet tessellation of space. The argument used for finite tessellations still works to show the following.

Theorem 17.13. An infinite sectional Dirichlet tessellation has a convex reciprocal figure.

Since the definition of a spider web refers to tensions in equilibrium at each vertex, local finiteness guarantees that we have an immediate extension of the definition and of Theorem 17.3:

Theorem 17.14. An infinite proper cell decomposition is a spider web if and only if it has a convex reciprocal figure.

An *infinite convex polyhedral bowl* is the intersection of the upper half spaces of an infinite set of nonvertical planes such that

(i) no finite region of space intersects an infinite number of the planes;
(ii) this intersection includes points over all points (x,y).

With this definition, which ensures that the projection of such a bowl is a proper infinite cell decomposition, Theorem 17.4 remains true.[42]

Theorem 17.15. An infinite proper cell decomposition has a convex reciprocal if and only if it is the vertical projection of an infinite convex polyhedral bowl.

However, if we add the plane at infinity to an infinite Maxwell bowl, we do not create a closed polyhedron. There are no analogues of Corollary 17.5. Note that each element of an expanding sequence of finite subpieces of an infinite cell decomposition may have a convex reciprocal, while the entire structure has no convex reciprocal. (An example is shown in Fig. 17-26.) Thus these finite pieces of an infinite cell decomposition may each be the projection of a convex polyhedral bowl, while the entire structure is not.

If our cell decomposition has only bounded polygons, then it must have infinitely many of them. If a convex reciprocal covers the plane, it is also an infinite cell decomposition; this reciprocal and the original decomposition form a reciprocal pair (Fig. 17-24). Such a reciprocal pair corresponds to a full bowl for which the total curvature is 2π, since the normals to the faces cover a hemisphere. For example Robert Connelly has observed that any infinite cell decomposition of the plane using only regular convex polygons as cells corresponds to such a bowl. In fact, the centers of the polygons form a reciprocal: the Dirichlet tessellation for the vertices of the regular polygons, as illustrated in Fig. 17-24b. How about the analogue of Theorem 17.6? For the spatial center corresponding to the plane $Ax + By - z - C = 0$ we took the point $(A,B,\sqrt{2C - A^2 - B^2})$. To make all these z-coordinates well defined, we translated the bowl by adding a sufficiently large constant d simultaneously to all values of C. With an infinite set of planes, this may be impossible, and there would be no Dirichlet tessellation to section. Consider the trivial example on the line formed by the plane polygon with vertices: $\ldots, (0, 0), (1, -2), (2, -6), (3, -12), \ldots, (n, -n(n + 1)), \ldots$. The face gradients have A: $\ldots, 2, 4, 6, 8, \ldots, 2n, \ldots$, and intercepts C: $0, -2, -6, \ldots, -n(n - 1), \ldots$. No constant d can make $2C + d - A^2 > 0$ for all n. This example with this reciprocal cannot correspond to any sectional Dirichlet tessellation. Each finite segment of the cell decomposition is a sectional Dirichlet tessellation with centers in the plane over the reciprocal vertices, but the entire object is not. (However, it is a Voronoi diagram in the Laguerre geometry, since this algebraic definition allows h^2 to be negative.[43])

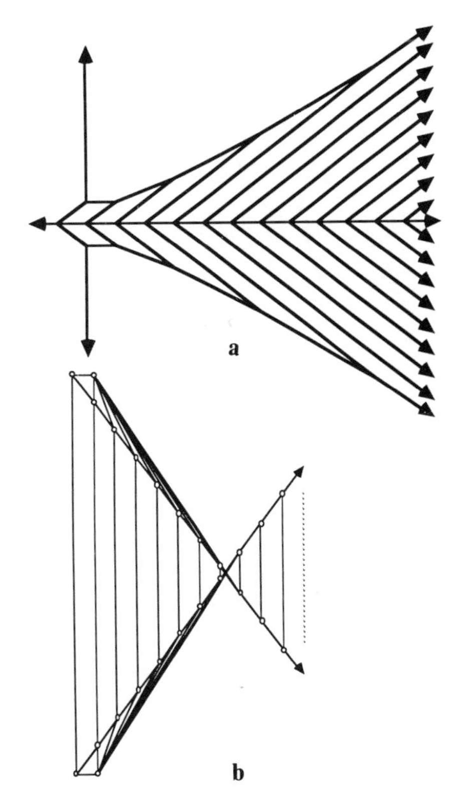

Fig. 17-26. The cell decomposition in (a) has a local convex reciprocal for arbitrarily large finite sets of vertices, but all global reciprocals are nonconvex (b).

The transformation from a sectional Dirichlet tessellation to a convex polyhedral bowl still applies.

Theorem 17.16. Each infinite sectional Dirichlet tessellation corresponds to an infinite convex polyhedral bowl, determined up to vertical translation.

Since it is not meaningful to talk about furthest-point Dirichlet tessellations for infinite

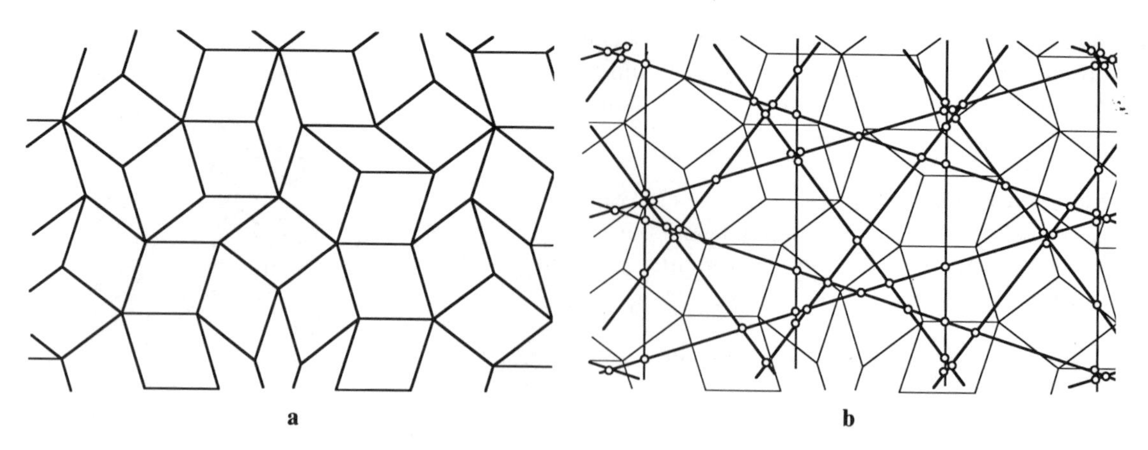

Fig. 17-27. The aperiodic rhombic tiling (a) has a pentagrid of five families of parallel lines as a convex reciprocal (b).

sets of centers, we have completed our chain of analogues. The following summarizes the limited analogue of Theorem 17.8.

Theorem 17.17. Given an infinite proper cell decomposition D in the plane, the following are equivalent: (i) D has a convex reciprocal figure; (ii) D is a spider web; (iii) D is the vertical projection of an infinite convex polyhedral bowl in 3-space.

For infinite Dirichlet tessellations, our construction did not require a vertical translation of the bowl and we have a complete analogue for Theorem 17.9.

Theorem 17.18. An infinite proper cell decomposition is a Dirichlet tessellation if and only if it is the vertical projection of an infinite Maxwell bowl.

We note that a periodic cell decomposition may have only nonperiodic reciprocals, and that even a periodic cell decomposition with a periodic reciprocal may induce no periodicity in the spatial polyhedron or in the spatial Dirichlet tessellation it sections.[44] Conversely, some of the interesting aperiodic tessellations of the plane are sections of periodic Dirichlet tessellations of some higher space, but neither the tessellation nor the reciprocal is periodic. For example, it has been observed[45] that a version of the nonperiodic Penrose tiling of the plane, drawn with rhombi (Fig. 17-27a), is a section of the regular cubic tessellation in R^5. The corresponding reciprocal figure for this tiling consists of a line arrangement of five families of parallel lines (Fig. 17-27b), called a pentagrid[46]; this pentagrid and the aperiodic rhombic tiling form a reciprocal pair.

Finally, we do not know an infinite analogue of Steinitz's theorem, so we cannot answer the question: "What infinite planar graphs can be realized as the edge skeleton of an infinite convex polyhedral bowl?" nor the equivalent question: "What infinite graphs can be represented as spider webs?" *We conjecture that any graph that can be realized as a proper infinite cell decomposition can also be realized as the edge skeleton of an infinite convex polyhedral bowl.*

Notes

Acknowledgments: Research supported in part by a grant to Henry Crapo from National Science and Engineering Research Council of Canada and in part by grants to Walter Whiteley from Fonds pour la Formation de Chercheurs et l'Aide à la Recherche Quebec and NSERC, Canada. This chapter grows out of previous joint work by Crapo and Whiteley, Ash and Bolker, and Bolker and Whiteley. The unification which is our theme developed while we were writing it. Because time was short, the pleasure of shaping the chapter and working out the details fell to the final author. We all share the responsibility for any errors in the text.

[1] The technical term is *infinitesimally rigid.* A bar-and-joint framework is infinitesimally rigid if

every set of velocity vectors (which preserves the lengths of the bars) at the joints represents a Euclidean motion of the whole space.

[2] J. C. Maxwell, "On Reciprocal Diagrams and Diagrams of Forces," *Philosophical Magazine*, ser. 4, *27* (1864):250–61. J. C. Maxwell, "On Reciprocal Diagrams, Frames and Diagrams of Forces," *Transactions of the Royal Society of Edinburgh 26* (1869–72):1–40.

[3] L. Cremona, *Graphical Statics*, English trans. (London: Oxford University Press, 1890).

[4] H. Crapo and W. Whiteley, "Statics of Frameworks and Motions of Panel Structures: A Projective Geometric Introduction," *Structural Topology 6*, (1982):43–82. W. Whiteley, "Motions and Stresses of Projected Polyhedra," *Structural Topology 7* (1982):13–38. H. Crapo and W. Whiteley, "Plane Stresses and Projected Polyhedra," preprint, Champlain Regional College, St. Lambert, Quebec, J4P-3P2 Canada, 1986.

[5] D. Huffman, "A Duality Concept for the Analysis of Polyhedral Scenes," in E. W. Elcock and D. Michie (eds.), *Machine Intelligence 8* [Ellis Horwood, England] (1977):475–92. A. K. Mackworth, "Interpreting Pictures of Polyhedral Scenes," *Artificial Intelligence 4* (1973):121–37.

[6] Crapo and Whiteley, "Plane Stresses and Projected Polyhedra."

[7] K. Q. Brown, "Voronoi Diagrams from Convex Hulls," *Information Processing Letters 9* (1979):223–28.

[8] P. Ash and E. Bolker, "Recognizing Dirichlet Tessellations," *Geometriae Dedicata 19* (1985):175–206.

[9] P. Ash and E. Bolker, "Generalized Dirichlet Tessellations," *Geometriae Dedicata 20* (1986):209–43.

[10] H. Edelsbrunner and R. Seidel, "Voronoi Diagrams and Arrangements," *Discrete and Computational Geometry 1* (1986):25–44.

[11] In a Dirichlet tesselation the centers of the cells which share a vertex are equidistant from that vertex. If the centers are chosen at random, then (with probability 1) no four lie on a circle. So the resulting Dirichlet tesselation has only 3-valent vertices.

[12] See note 8 supra.

[13] See note 9 supra.

[14] Edelsbrunner and Seidel, "Voronoi Diagrams and Arrangements."

[15] Ash and Bolker, "Recognizing Dirichlet Tessellations."

[16] R. Connelly, "Rigidity and Energy," *Inventiones Mathematicae 66* (1982):11–33.

[17] Crapo and Whiteley, "Statics of Frameworks and Motions of Panel Structures." Whiteley, "Motions and Stresses of Projected Polyhedra."

[18] R. Connelly, "Rigid Circle and Sphere Packings I. Finite Packings," *Structural Topology*, to appear. R. Connelly, "Rigid Circle and Sphere Packings II. Infinite Packings in Finite Motion," *Structural Topology*, to appear.

[19] Maxwell, "On Reciprocal Diagrams and Diagrams of Forces." Crapo and Whiteley, "Plane Stresses and Projected Polyhedra."

[20] Crapo and Whiteley, "Plane Stresses and Projected Polyhedra."

[21] Maxwell, "On Reciprocal Diagrams and Diagrams of Forces." Crapo and Whiteley, "Plane Stresses and Projected Polyhedra."

[22] Crapo and Whiteley, "Plane Stresses and Projected Polyhedra," sec. 4.

[23] Edelsbrunner and Seidel, "Voronoi Diagrams and Arrangements."

[24] Crapo and Whiteley, "Plane Stresses and Projected Polyhedra."

[25] Ash and Bolker, "Generalized Dirichlet Tessellations,"

[26] Edelsbrunner and Seidel, "Voronoi Diagrams and Arrangements."

[27] B. Roth and W. Whiteley, "Tensegrity Frameworks," *Transactions of the American Mathematical Society 265* (1981):419–45. Whiteley, "Motions and Stresses of Projected Polyhedra." Connelly, "Rigidity and Energy."

[28] Connelly, "Rigidity and Energy."

[29] Brown, "Voronoi Diagrams from Convex Hulls."

[30] Roth and Whiteley, "Tensegrity Frameworks."

[31] B. Grünbaum, *Convex Polytopes* (New York: Interscience, 1967).

[32] N. White and W. Whiteley, "The Algebraic Geometry of Stresses in Frameworks," *S.I.A.M. Journal of Algebraic and Discrete Methods 4* (1983):481–511.

[33] Davis, "The Set of Non-Linearity of a Convex Piece-wise Linear Function," *Scripta Mathematica 24* (1959):219–28.

[34] White and Whiteley, "The Algebraic Geometry of Stresses in Frameworks."

[35] H. S. M. Coxeter, "The Classification of Zonohedra by Means of Projective Diagrams," *Journal de Mathématiques Pures et Appliquées*, ser. 9, *42* (1962):137–56.

[36] E. Schulte, "Analogues of Steinitz's Theorem about Non-Inscribable Polytopes," preprint, Mathematisches Institut, Universität Dortmund, Dortmund, Federal Republic of Germany, 1985.

[37] Grünbaum, *Convex Polytopes*.

[38] Ash and Bolker, "Recognizing Dirichlet Tessellations."

[39] Maxwell, "On Reciprocal Diagrams and Diagrams of Force." Cremona, *Graphical Statics*.

Crapo and Whiteley, "Plane Stresses and Projected Polyhedra."

[40] Maxwell, "On Reciprocal Diagrams, Frames and Diagrams of Forces." W. Whiteley, "The Projective Geometry of Rigid Frameworks," in C. Baker and L. Batten, eds, *Proceedings of the Winnipeg Conference on Finite Geometries* (New York: Marcel Dekker, 1985), pp. 353–70.

[41] Connelly, "Rigid Circle and Sphere Packings: I and II."

[42] Edelsbrunner and Seidel, "Voronoi Diagrams and Arrangements."

[43] Ibid.

[44] Crapo and Whiteley, "Plane Stresses and Projected Polyhedra."

[45] N. G. de Bruijn, "Algebraic Theory of Penrose's Non-periodic Tilings I, II," *Akademie van Wetenschappen* [Amsterdam], *Proceedings*, ser. A, *43* (1981):32–52, 53–66.

[46] Ibid.

18

Unsolved Problems

The Shaping Space Conference concluded with a session on unsolved problems, organized by Douglas Dunham. We present a selection of them here.

A. Can Neighborly Polyhedra Be Realized Geometrically?

JOHN REAY

In a 1940 Hungarian competition, the following question was asked of high-school age students:

Prove that the tetrahedron is the only polyhedron with each pair of vertices joined by an edge.

As stated, the problem is false. The solution desired by the problem judges was to combine Euler's formula $v - e + f = 2$ with $3f = 2e = v(v - 1)$ and eliminate variables to show that $v = 4$. This solution is valid if the polyhedron is homeomorphic to a sphere (for example, when it is convex). However, if we use the more general theorem of Euler–Poincaré, $v - e + f = 2(1 - g)$, where g is the genus of an orientable 2-manifold M, and assume that a neighborly simplicial cell complex is embedded on M, then we obtain

$$(v - 3)(v - 4) = 12g. \qquad (18.A1)$$

We see that this equation has integer solutions when $v \equiv 0, 3, 4$, or $7 \pmod{12}$ which raises the question of the possible existence of an infinite family of simplicial neighborly polyhedra in E^3.

Historical Background

Polyhedron in these problems means a cell complex whose point set is a closed orientable 2-manifold, and each of whose 2-cells is an affine polygon which is not coplanar with any adjacent 2-cell. We will be primarily interested in polyhedra which are *simplicial* (each 2-cell is a triangle) and *neighborly* (each pair of vertices determines an edge). An arbitrary cell complex on a 2-manifold is *realized geometrically* if each of its 2-cells is an affine polygon, not coplanar with any adjacent 2-cell; that is to say, it is a polyhedron. The question raised in the title may now be stated precisely in the following problems:

Problem 18.A1. If $v \equiv 0, 3, 4$, or $7 \pmod{12}$ and $v \geq 4$, does there exist a simplicial neighborly cell complex with v vertices on the orientable 2-manifold of genus $g = (v^2 - 7v + 12)/12$?

Problem 18.A2. If the answer to Problem 18.A1 is yes, then can such a topological cell complex be realized geometrically? In other words, is the complex a polyhedron?

Each of these problems has an interesting historical background. The first is a special case of the famous Heawood Map-Coloring Conjecture. Heawood[1] in 1890 gave an upper bound, the number

$$v = \lfloor (7 + \sqrt{1 + 48g})/2 \rfloor$$

obtained from Eq. (18.A1), to the number of colors necessary to "color" any map on a closed orientable 2-manifold of genus $g \geq 1$, and perhaps believed that this was also the lower bound. Thus the chromatic number of each such surface would be determined by producing suitable examples, for example showing in part that the answer to Problem 18.A1 was yes. This famous conjecture stood open from 1890 to 1968. The part that answers the Problem 18.A1 affirmatively was done by Ringle, Youngs, and others between 1954 and 1963. See Harary, Ringel and Youngs for historical details, references, and interesting expository descriptions of this problem.[2]

Many people have conjectured that the answer to Problem 18.A2 is also yes, but it is unknown unless $g = 0$ or $g = 1$. Obviously the tetrahedron has $v = 4$ and $g = 0$. For the case $g = 1$, Möbius stated as early as 1865 that T_7, the neighborly triangulation of the torus using seven vertices, could be realized geometrically as a polyhedron by fitting together seven tetrahedra.[3] Actual models were apparently constructed before 1885 by C. Reinhardt.[4] Akos Császár, probably unaware of this earlier work, gave a constructive proof in 1949 that T_7 could be realized geometrically, and the polyhedron he described has been called the Császár polyhedron since that time.[5] In a fine survey article in 1970, Richard Duke[6] describes how T_7 can also be constructed geometrically by a suitable projection of a cyclic 4-polytope with seven vertices into one of its tetrahedral facets, since its 2-skeleton contains a subcomplex isomorphic to T_7. Branko Grünbaum has constructed such a model of T_7. (Also see the papers of Altschuler.[7]) A picture of the Császár polyhedron and a pattern with directions for its construction appeared in the widely read Mathematical Games section of the *Scientific American* by Martin Gardner.[8] Gardner, describing work of Donald Crowe,[9] showed how the Császár polyhedron is related to Steiner triple systems, bridge tournaments, certain magic squares (Hadamard matrices), and other problems.

Since the simplicial cell complex T_7 on the torus has seven vertices (each of valence six) and 14 faces, its topological dual on the torus will have 14 simple vertices, 21 edges, and 7 hexagonal faces. In 1977 Lajos Szilassi[10] described a computer-generated geometric realization of this dual of T_7. It may be shown that no such geometric realization exists with *convex* hexagons as faces. This lack of convexity is clear in the picture of this model and the pattern for its construction which was shown by Martin Gardner in a later *Scientific American* article.[11]

Unsolved Problems

Our first problem is the unsolved part of Problem 18.A2. $S_g(v)$ denotes the closed orientable 2-manifold of genus $g(v) = (v - 3)(v - 4)/12$.

Open Problem 18.A1. If $12 \leq v \equiv 0, 3, 4$, or $7 \pmod{12}$, can the neighborly (simplicial) 2-complex with v vertices on $S_{g(v)}$, or its dual, be realized geometrically?

Crowe[12] suggests that since $v > 16$ implies that the number of holes in such a polyhedron would exceed the number of vertices, its existence is unlikely. The first open case using 12 vertices on the sphere with six handles is particularly tempting. The problem might be more tractable if we weaken the neighborly property. Altschuler and Brehm[13] have recently defined a cell complex on a 2-manifold to be *weakly neighborly* if each pair of vertices is contained in some 2-cell of the complex. If we restrict the point set of the complex to be the torus, then there are exactly five weakly neighborly complexes, three of which (including T_7) can be realized geometrically, and two of which cannot be realized geometrically in any Euclidean space.

Open Problem 18.A2. What weakly neighborly cell complexes on S_g, with $g > 1$, are geometrically realizable?

In another recent paper Altschuler and Brehm[14] have produced a cell complex on the torus which may be realized geometrically, but not as the Schlegel diagram of any convex 4-polytope. In fact, this map on the torus is not isomorphic to a subcomplex of the boundary complex of any convex polytope. This shows the limitations of the technique mentioned in the previous section, where T_7 is realized geometrically by projecting an appropriate subcomplex of the cyclic 4-polytope with seven vertices into one of its tetrahedral facets.

Let us return to this geometric realization of T_7. When a three-dimensional model is obtained as a projection into a tetrahedron, its convex hull has four vertices, which we call the *exposed vertices*. Császár's model of the same polyhedron has five exposed vertices; the remaining two vertices lie interior to its convex hull. A suitable modification of Császár's model exists with six exposed vertices.

Open Problem 18.A3. Does there exist a geometric model of the Császár polyhedron with seven exposed vertices? That is, can each vertex of the polyhedron also be a vertex of the convex hull of the polyhedron?

It seems likely that this problem has a negative answer, but apparently no proof has been given. Each of the seven vertices may lie arbitrarily close to the boundary of the convex hull, but apparently only at the expense of "making the hole small." We can ask how large the hole can be. Let r be the radius of an infinite cylinder which may be placed "through the hole" of a geometric realization, and let R be the radius of a sphere containing the polyhedron.

Open Problem 18.A4. What polyhedron maximizes the ratio r/R for a Császár polyhedron? That is, what polyhedron maximizes the "size of the hole"?

Problem 18.A4 appears to be another problem that could be analyzed with a computer by considering small changes in the positions of the vertices of a known geometric realization. With this in mind, let us define two geometric realizations A and B of T_7 to be *equivalent* if there exists a continuous transformation of the seven vertices of A into the vertices of B so that each intermediate position of the corresponding cell complex remains a geometric realization of T_7.

Open Problem 18.A5. Are all geometric realizations of T_7 equivalent?

The conjecture is that the answer to Open Problem 18.A5 is no. In particular, the conjecture is that A and B are not equivalent if they have a different number of exposed vertices. For example, the polyhedral model of T_7 with six exposed vertices, obtained by "pulling" one of the interior vertices of the Császár model through a triangular face of the convex hull, does not appear to be equivalent to the Császár model itself.

B. How Many Faces Does a Polytope Have?

Margaret Bayer

Counting problems in polyhedral theory have intrigued mathematicians at least since Euler. Today they have taken on a new importance with the development of linear programming and related applications. Much of their appeal lies in the simplicity of the questions and in the wide range of techniques used to answer them.

The *f-vector* of a convex d-dimensional polytope P is the vector $f(P) = (f_0, f_1, \ldots, f_{d-1})$, where f_i is the number of i-dimensional faces of P. We can define a finer combinatorial measure by counting chains of faces of P. For $S \subseteq \{i_1, i_2, \ldots, i_s\}$, a subset of $\{0, 1, \ldots, d-1\}$, let $f_s(P)$ be the number of chains of faces $\phi \subset F_1 \subset F_2 \subset \ldots \subset F_s \subset P$ with dim $F_j = i_j$. (We shall drop the set brackets in subscripts.) Call the vector $(f_s(P))_{s\subseteq\{0,1,\ldots,d-1\}}$ the *extended f-vector* of P. The overall problem is to characterize the {f-vectors} and extended f-vectors of arbitrary convex polytopes.

The affine span of the f-vectors of all polytopes is the hyperplane defined by Euler's relation,

$$f_0 - f_1 + \cdots + (-1)^{d-1}f_{d-1} = 1 - (-1)^d.$$

For extended f-vectors the affine span is determined by the generalized Dehn–Sommerville equations, given by Bayer and Billera.[15]

Theorem. Let P be a d-dimensional polytope, and let S be a subset of $\{0, 1, \ldots, d-1\}$. If $\{i, k\} \subset S \cup \{-1, d\}$ and S contains no j such that $i < j < k$, then

$$\sum_{j=i+1}^{k-1} (-1)^{j-i-1} f_{S\cup j}(P) = f_S(P) \cdot (1 - (-1)^{k-i-1}).$$

Furthermore, these are all the linear relations holding on the extended f-vectors of all d-dimensional polytopes, and the dimension of the affine span of these extended f-vectors is $c_d - 1$, where c_d is the dth Fibonacci number ($c_d = c_{d-1} + c_{d-2}$; $c_0 = c_1 = 1$).

Little else is known for arbitrary polytopes, and it would be particularly interesting to find inequalities that f-vectors and extended f-vectors must satisfy.

In dimension three the vectors have easy characterizations. A vector (f_0,f_1,f_2) is the f-vector of a three-dimensional polytope if and only if $f_0 - f_1 + f_2 = 2$, $4 \le f_0 \le 2f_2 - 4$ and $4 \le f_2 \le 2f_0 - 4$. The extended f-vector is determined by the f-vector as follows: $f_{01} = f_{02} = f_{12} = 2f_1$ and $f_{012} = 4f_1$. But already in four dimensions we do not have a characterization. (The projections onto two components of the f-vectors are known, and various necessary inequalities are known.[16])

A polytope is simplicial if every i-dimensional face is the convex hull of $i + 1$ points. In this case the extended f-vector depends linearly on the f-vector. For f-vectors simplicial polytopes are extremal, that is, the f-vector of any d-polytope with n vertices is bounded above by the f-vector of some simplicial d-polytope with n vertices. The most important recent result in the combinatorial theory of polytopes is the characterization of the f-vectors of simplicial polytopes. The theorem was conjectured by McMullen[17] in 1970, and then proved ten years later by Stanley (necessity) and Billera and Lee (sufficiency).[18] The theorem is most easily stated in terms of the h-vector, obtained from the f-vector by a nonsingular linear transformation. For $0 \le i \le d$,

$$h_i = \sum_{j=0}^{i} (-1)^{i-j} \binom{d-j}{d-i} f_{j-1}.$$

The h-vector of a d-dimensional polytope P is then $h(P) = (h_0, h_1, \ldots, h_d)$. The characterization of the h-vectors of simplicial polytopes requires the definition of a certain nonlinear operator as follows. For positive integers h and i, we note that h can be written uniquely in the form

$$h = \binom{n_i}{i} + \binom{n_{i-1}}{i-1} + \cdots + \binom{n_j}{j}$$

where $n_i > n_{i-1} > \ldots > n_j \ge j \ge 1$. Define the ith pseudopower of h to be

$$h^{<i>} = \binom{n_i + 1}{i + 1} + \binom{n_{i-1} + 1}{i} + \ldots + \binom{n_j + 1}{j + 1}.$$

Put $0^{<i>} = 0$ for all i. Then the characterization is given by the following theorem.

Theorem. An integer vector $(h_0, h_1, \ldots, h_d)$ is the h-vector of a simplicial d-dimensional polytope if and only if

(i) $h_i = h_{d-i}$, $0 \leq i \leq [d/2]$;
(ii) $h_{i+1} \geq h_i$, $0 \leq i \leq [d/2] - 1$; and
(iii) $h_0 = 1$ and $h_{i+1} - h_i \leq (h_i - h_{i-1})^{<i>}$, $1 \leq i \leq [d/2] - 1$.

The linear equalities, known as the Dehn–Sommerville equations, and the linear inequalities, the generalized lower bound theorem, have combinatorial-geometric proofs. The necessity of the nonlinear inequalities has been proved (by Stanley) only using a very difficult theorem of algebraic geometry. A combinatorial proof of this result would be of great interest.

The classification of f-vectors and extended f-vectors would be greatly advanced by the solution of these problems:

1. Find a combinatorial proof of the nonlinear inequalities of the characterization of f-vectors of simplicial polytopes.
2. Characterize the f-vectors and extended f-vectors of four-dimensional polytopes.
3. Extend the generalized lower bound inequalities to arbitrary polytopes.

C. Problems on the Realizability and Rigidity of Polyhedra

WALTER WHITELEY

Realizability

A classical theorem of Steinitz characterizes the graphs of convex spherical polyhedra as the three-connected planar graphs.[19] However, many built structures take the form of non-convex spherical polyhedral with two-connected planar graphs.[20]

Problem 18.C1 Which planar graphs can be realized as *nonconvex* spherical polyhedra with

(a) convex plane faces and no self-intersection?
(b) nonconvex plane faces but no self-intersection?
(c) plane polygonal faces, with self-intersection allowed?

With the advent of space stations and other experimental designs, it is also interesting to ask about the graphs of other types of oriented polyhedra.

Problem 18.C2. Which abstract oriented polyhedra (graphs of the edges and vertices with an appropriate cycle structure of faces) can be realized as spatial polyhedra with plane faces (convex faces, or non-self-intersecting, etc.)?

Infinitesimal Rigidity

A triangulated polyhedron in space can be constructed with bars along the edges and universal joints at the vertices. Such a *bar-and-joint framework* is *infinitesimally rigid* if every set of vectors at the joints (velocities) which preserves the lengths of the bars represents a Euclidean motion of the whole space.[21] A theorem of Cauchy and Dehn states that any convex triangulated sphere is infinitesimally rigid when built as a bar-and-joint framework.[22] An extension of Alexandrov shows that any convex spherical polyhedron is infinitesimally rigid when the faces are triangulated and the structure is built as a bar-and-joint framework.[23] As a consequence almost all spherical polyhedra are infinitesimally rigid when built this way even if they are nonconvex.[24] (A configuration *almost always* works if the set of realizations of the graph which work contains a dense open set in the set of all realizations.) Recently purely combinatorial proofs have been given for this general result.[25]

Conjecture 18.C1. The graph of a triangulated polyhedron is almost always infinitesimally rigid as a bar-and-joint framework in 3-space.

We have a combinatorial proof of this conjecture for toroidal polyhedra,[26] but its complexity is awkward. These structures are going to be built, so it would be nice to understand them!

Unique Realizability

These rigidity results can also be interpreted as a local form of unique realizability—there is no second noncongruent realization nearby which has the same edge lengths. If we describe a convex triangulated spherical polyhedron by giving all the edge-lengths, then by the general theorem of Cauchy, this will uniquely determine the convex polyhedron. However there will often be a nonconvex realization with the same edge lengths.

Conjecture 18.C2. If all the edge-lengths of a nonspherical triangulated polyhedron realized in 3-space are given, then the polyhedron is uniquely realized, almost always.

If we describe a nontriangulated convex polyhedron it may not be enough to just give the face structure and the edge-lengths, as indicated by the example of a cube moving into a parallelepiped. However, this failure is probably rare.

Conjecture 18.C3. Given a planar graph that can be realized as a spherical polyhedron with proper plane polygons as faces, almost all such realizations are locally uniquely determined by the set of all edge-lengths.

Dihedral Angles and Locally Unique Realizability

If we take a triangulated convex sphere, built as bar-and-joint framework, and drop one bar, the object has finite motion and is not uniquely determined, even locally. Can we return to a local unique description by specifying one of the dihedral angles between adjacent faces?

Theorem (Whiteley). If all but one edge-length and any one dihedral angle are fixed in a 4-connected triangulated sphere then the polyhedron is locally uniquely determined, almost always.

In fact the result is more specific: the polyhedron is almost always locally uniquely determined if the quadrilateral of the omitted edge and the quadrilateral of the dihedral angle are connected by four vertex disjoint paths.

We can go further and ask which combinations of a set of dihedral angles and a set of edge-lengths will determine a locally unique polyhedron.

Problem 18.C3. Characterize the patterns of minimal sets of edge-lengths and dihedral angles which will almost always determine a locally unique spatial polyhedron with plane faces.

The following is an example of such a pattern.

Theorem (Whiteley.[27]). If in any spherical polyhedron we fix the lengths of each edge in a set which forms a spanning tree in the graph of the polyhedron, and we fix the size of each dihedral angle in a set that forms a spanning tree in the dual of the polyhedron, then the polyhedron is locally uniquely determined, almost always.

Some work has been done on the maximum number of dihedral angles which can appear in such a minimal set. This requires, at least, an understanding of the isogonalities (maps of the polyhedron which preserve all dihedral and facial angles of the polyhedron).[28] We have recently found that there is a strong connection between the isogonalities of a spherical polyhedron and an old static construction of "reciprocal diagrams" due to Maxwell,[29] and this is now being exploited. In addition to such information we must know if maps preserving the dihedral angles must preserve the facial angles and be isogonalities.

Conjecture 18.C4 (Stoker[30]). There does not exist a pair of convex polyhedra with the same combinatorial structure, the same dihedral angles, and different facial angles.

We have found a pair of nonconvex spherical polyhedra with the same dihedral angles and different facial angles, and we actually disbelieve Stoker's conjecture. However, we believe that this is locally rare.

Conjecture 18.C5. If all dihedral angles of a polyhedron are fixed then all facial angles are determined, both locally and infinitesimally, almost always.

Concluding Remarks on the Design of Spatial Polyhedra

The search for minimal sets of angles and edge-lengths also has applications to the design of polyhedra. Such sets are independent in the sense that we can make new choices for the values and still realize the polyhedron, at least if the total size of the changes is small. Certainly if the set is almost always dependent it is unreasonable to choose independent values for these measurements and expect to build the object.

To design a general polyhedron in space we will first choose a combinatorial structure (faces, vertices, edges) and then choose some edge-lengths, some dihedral angles and perhaps some other values (like facial angles, or nonadjacent faces to be parallel etc.). To make these choices intelligently we must know which are free choices and which will prevent any realization because the value is already determined by the previous choices. Some understanding of the problems presented here is indispensable to this process. This problem also provides a coherent geometric program which includes, as special cases, the classical geometric theorems of Cauchy, Steinitz, and Maxwell.

D. Problems Concerning Polyhedral 2-Manifolds

PETER GRITZMANN

A polyhedral 2-manifold P is a geometric cell complex (the 2-cells are planar convex polygons) in some Euclidean space, whose underlying point-set is a closed connected 2-manifold.

Let $v_k(P)$ denote the number of k-valent vertices and $p_k(P)$ the number of k-gonal facets of P. Furthermore let $v(P)$ be the valence-value of P, i.e.,

$$v(P) = \sum_{k \geq 3} (k - 3)v_k(P)$$

If P is different from the sphere we have

$$4 - \chi(P) \leq v(P) \qquad (18.D1)$$

where $\chi(P)$ is the Euler characteristic. In particular for polyhedral tori T in E^3,

$$6 \leq v(T). \qquad (18.D2)$$

Obviously, inequalities (18.D1) and (18.D2) are necessary conditions for the existence of polyhedral 2-manifolds. Inequality (18.D2) is best possible, but inequality (18.D1) is only "asymptotically" best possible.

Problem 18.D1. Find sharp lower bounds for the valence values of polyhedral 2-manifolds different from the sphere and the torus.

Once a sharp lower bound for $v(P)$ is attained, one may hope to solve several problems. In fact, inequality (18.D2) leads to a characterization of f-vectors of polyhedral tori in E^3.

Problem 18.D2. Characterize the f-vectors of polyhedral 2-manifolds different from the sphere and the torus.

Inequality (18.D2) even gives the main condition for a toroidal analogue to Eberhard's theorem.[31] In fact, let s, p_k $(k \leq 3, k \neq 6)$ be nonnegative integers; then there exists a polyhedral torus T in E^3 with $p_k(T) = p_k$ $(k \neq 6)$ and $v(T) = s$ if and only if

$$\sum_{k \geq 3} (6 - k)\, p_k = 2s \quad \text{and} \quad s \geq 6.$$

Problem 18.D3. For which sequences $(v_3, v_4, \ldots)$ does there exist a polyhedral torus T in E^3 with $v_k(T) = v_k$?

For arbitrary polyhedral 2-manifolds, inequality (18.D1) clearly gives a necessary Eberhard-type condition and there exist some sufficient conditions, too. But in general the problem of characterizing possible numbers $p_k(P)$, $v_k(P)$ remains open.

Problem 18.D4. Give analogues of Eberhard's theorem for polyhedral 2-manifolds different from the sphere and the torus.

E. Extending the Conway Criterion

Doris Schattschneider

A sufficient criterion[32] for a tile (a simple closed curve and its interior) to fill the plane using only half-turns was formulated by John H. Conway:

The Conway Criterion. A tile T can pave the plane by half-turns if there are six consecutive points $\nu_1, \ldots, \nu_6$, at least three distinct, on the boundary of T (consecutive in the sense of traveling a cycle around the boundary) which satisfy the following conditions:

(i) $\widehat{\nu_1\nu_2}$ is congruent to $\widehat{\nu_5\nu_4}$ by a translation τ in which $\tau(\nu_1) = \nu_5$ and $\tau(\nu_2) = \nu_4$.

(ii) $\widehat{\nu_2\nu_3}$, $\widehat{\nu_3\nu_4}$, $\widehat{\nu_5\nu_6}$, $\widehat{\nu_6\nu_1}$ are centrosymmetric.

The Conway criterion characterizes those tiles that pave "nicely" by half-turns: If a tile T paves the plane by half-turns that leave the entire tiling invariant (this implies that the tiling is periodic), then T satisfies the Conway criterion.

Problem 18.E1. If a tile T can pave the plane using half-turns only (the paving need not be periodic), must such a tile satisfy the Conway criterion?

Problem 18.E2. Is there a three-dimensional version of the Conway criterion for space-fillers? (Parallelopipeds, prisms, and suitable "halves" of these fill space using only half-turns—but what others do?)

Notes (for Chapter 18)

[1] P. J. Heawood, "Map Colour Theorems," *Quarterly Journal of Mathematics* [Oxford], ser. 2, *24* (1890):332–38.

[2] F. Harary, *Graph Theory* (Reading, Mass.: Addison-Wesley, 1969). G. Ringel, *Map Color Theorems* (New York: Springer-Verlag, 1974). J. W. T. Youngs, "The Mystery of the Heawood Conjecture," from B. Harris, ed., *Graph Theory and Its Applications* (New York: Academic Press, 1970).

[3] A. F. Möbius, "Über die Bestimmung des Inhalts eines Polyeders," *Berichte der K. Sachs. Ges. Wiss., Math. Phys. Klasse 17* (1865):31–68.

[4] C. Reinhardt, "Zu Möbius' Polyedertheorie," *Berichte der K. Sachs. Ges. Wiss., Math. Phys. Klasse* (March 1885).

[5] A. Császár, "A Polyhedron without Diagonals," *Acta Scientiarum Mathematicarum* [Szeged] 13 (1949):140–42.

[6] R. A. Duke, "Geometric Embedding of Complexes," *Mathematics Monthly 77* (1970):597–603.

[7] A. Altschuler, "Polyhedral Realization in R³ of Triangulations of the Torus and 2-Manifolds in Cyclic 4-Polytopes," *Discrete Mathematics 1,* (1971):211–38; "Construction and Enumeration of Regular Maps on the Torus," *Discrete Mathematics 4* (1973):201–17.

[8] M. Gardner, "On the Remarkable Császár Polyhedron and Its Applications in Problem Solving," *Scientific American* (March 1975):102–7.

[9] D. Crowe, "Steiner Triple Systems, Heawood's Torus Coloring, Császár's Polyhedron, Room Designs, and Bridge Tournaments," *Delta: An Undergraduate Mathematics Journal* [Waukesha, Wis.] *3*, no. 1 (1972–73):27–32.

[10] L. Szilassi, "A Polyhedron in Which Any Two Faces Are Continuous," *A Juhasz Gyula Tanarkepzo Foiskola Tudomanyos Kozlemenyei* [Szeged] *2* (1977).

[11] M. Gardner, "Mathematical Games . . . Modern Minimal Art," *Scientific American* (November 1978):20–24.

[12] Crowe, "Steiner Triple Systems."

[13] A. Altschuler and U. Brehm, "The Weakly Neighborly Polyhedral Maps on the Torus," to appear.

[14] A. Altschuler and U. Brehm, "A Non-Schlegelian Map on the Torus," *Mathematika 31,* no. 1 (1984):83–88.

[15] M. M. Bayer and L. J. Billera, "Generalized Dehn-Sommerville Relations for Polytopes, Spheres and Eulerian Partially Ordered Sets," *Inventiones Mathematicae 79* (1985):143–57.

[16] D. W. Barnette, "The Projection of the f-Vectors of 4-Polytopes onto the (E,S)-Plane," *Discrete Mathematics 10* (1974):201–216. D. W. Barnette and J. R. Reay, "Projections of f-vectors of Four-Polytopes," *Journal of Combinatorial Theory,* ser. A, vol. *15* (1973):200–209. B. Grünbaum, *Convex Polytopes* (New York: Wiley-Interscience, 1967).

[17] P. McMullen and G. C. Shephard, "Convex Polytopes and the Upper Bound Conjecture," *London Mathematical Society Lecture Notes Series 3* (1971).

[18] R. Stanley, "The Number of Faces of a Simplicial Convex Polytope," *Advances in Mathematics 35* (1980):236–38. L. J. Billera and C. W. Lee, "A Proof of the Sufficiency of McMullen's Conditions for f-Vectors of Simplicial Convex Polytopes," *Journal of Combinatorial Theory,* ser. A, *31* (1981):237–55.

[19] D. Barnette and B. Grünbaum, "On Steinitz's Theorem Concerning 3-Polytopes and Some Properties of Planar Graphs," in *The Many Facets of Graph Theory,* Lecture Notes in Mathematics no. 100 (Heidelberg: Springer-Verlag, 1969).

[20] W. Whiteley, "Realizability of Polyhedra," *Structural Topology, 1* (1979):46–59.

[21] W. Whiteley, "Infinitesimally Rigid Polyhedra I. Statics of Bar and Joint Frameworks," *Transactions of the American Mathematical Society, 285* (1984):431–65.

[22] Ibid. and H. Gluck, "Almost All Simply Connected Surfaces Are Rigid," in *Geometric Topology,* Lecture Notes in Mathematics no. 438 (Berlin: Springer-Verlag, 1975), pp. 225–39.

[23] Whiteley, "Infinitesimally Rigid Polyhedra I."

[24] Barnette and Grünbaum, "On Steinitz's Theorem." Whiteley, "Infinitesimally Rigid Polyhedra I." W. Whiteley, "Infinitesimally Rigid Polyhedra II," *Transactions of the American Mathematical Society,* to appear.

[25] T. S. Tay and W. Whiteley, "Generating Isostatic Frameworks in Space," *Structural Topology 11* (1985):20–69.

[26] W. Whiteley, "Infinitesimally Rigid Polyhedra III," preprint, Champlain Regional College, St. Lambert, Quebec, 1985.

[27] W. Whiteley, "Determination of Spherical Polyhedra," to appear.

[28] H. Karcher, "Remarks on Polyhedra with Given Dihedral Angles," *Communications on Pure and Applied Mathematics 21* (1968):169–261. J. J. Stoker, "Geometric Problems Concerning Polyhedra in the Large," *Communications on Pure and Applied Mathematics 21* (1968):119–68. Whiteley, "Determination of Spherical Polyhedra."

[29] J. C. Maxwell, "On Reciprocal Figures and Diagrams of Forces," *Philosophical Magazine,* ser. 4, *27* (1864):250–61; "On Reciprocal Diagrams, Frames, and Diagrams of Forces, *Transactions of the Royal Society of Edinburgh 26* (1869–72):1–40. W. Whiteley, "Motions, Stresses, and Projected Polyhedra," *Structural Topology 7* (1982):13–38.

[30] Stoker, "Geometric Problems Concerning Polyhedra in the Large."

[31] Peter Gritzmann, "The Toroidal Analogue to Eberhard's Theorem," *Mathematika 30* (1983):274–90.

[32] D. Schattschneider, "Will It Tile? Try the Conway Criterion!" *Mathematics Magazine 53* (1980):224–33.

Part V
Further Steps

Fig. 19-1. How do I measure, cut and glue pieces of cardboard to make a polyhedron that will stay together? Photograph by Stan Sherer.

Fig. 19-2. Do all the faces of a polyhedron point either "up" or "down"? Photograph by Stan Sherer.

Fig. 19-3. Can the faces of a polyhedron be ordered in a natural way? Photograph by Stan Sherer.

Fig. 19-4. What interesting polyhedral shapes can we make if we allow holes and tunnels? Photograph by Stan Sherer.

19

Polyhedra in the Curriculum?

MARJORIE SENECHAL AND GEORGE FLECK

We hope you have enjoyed this book: the introductory visit to the Polyhedron Kingdom, the lectures exploring the subject (its theory, its history, its applications, and its pedagogy) and the chapters describing many aspects of current research. We hope that you have made some beautiful models (in addition to looking at beautiful pictures), and have begun to ask and try to answer questions that may never have occurred to you before.[1] If so, then you will share our astonishment at the fact that *virtually none of the topics in this book is included anywhere in our present school curricula!*

From molecules to galaxies, from viruses to giant domes, from crystal architecture to large-scale housing projects: our world is embodied in geometric forms, many of them polyhedral. A wealth of activities, experiments, and experiences that involve the varied forms of our world is accessible through the study of polyhedra. Construction with wooden blocks is recognized by educators and parents alike as a valuable part of the education of young children, but it is an activity usually restricted to the kindergarten classroom and the toddler's playroom. Few educators seem to realize that this is in fact an introduction to the process of model building, the way we explore our surroundings throughout our lives. Both the small child playing with wooden blocks and the chemist constructing molecular models in the laboratory are testing their constructions against reality, and refining them in accordance with experience. Why isn't model-building—both hands-on and conceptual—a central part of our education, from elementary school through college? (See, for example, Fig. 19-5.)

It is widely agreed that in the United States there is a crisis in the teaching of geometry. The post-Sputnik restructuring of mathematics curricula decreased its traditional role in American schools, and the more recent back-to-basics movement has coincided with further compression of geometry instruction. Throughout this period, the fraction of the K–12 curriculum devoted to "mathematics" has remained reasonably constant, but there has been a marked shift of emphasis. Concerns that "Johnny can't add" have brought pressure to spend more time on arithmetic in early grades, and the perception that our society demands greater computational skills has led to the introduction of statistics, computing, and calculus in the upper grades. This shift is sometimes justified by pointing out that one of the traditional tasks of geometry education—to provide rigorous exercises in formal logical thought—is perhaps handled better in courses in computer programming. As a result of such pressures, solid geometry has been omitted from the curriculum, and the time spent on plane geometry has been reduced. This is a cause for serious concern.

[1] The children of the Smith College Campus School, who made polyhedral models for their exhibit at the Shaping Space Conference, obviously asked themselves some very interesting questions: How do I measure, cut, and glue pieces of cardboard to make a polyhedron that will stay together? (Fig. 19-1) Do all the faces of a polyhedron point either "up" or "down"? (Fig. 19-2) Can the faces of a polyhedron be ordered in a natural way? (Fig. 19-3) What interesting polyhedral shapes can we make if we allow holes and tunnels? (Fig. 19-4)

Fig. 19-5. From the Shaping Space Conference: Smith College students Beth Thurberg and Libby Walker. Photograph by Stan Sherer.

Fig. 19-6. The panel for the plenary discussion "Geometry in the Curriculum" at the Shaping Space Conference. From *left* to *right*: mathematicians Joseph Malkevitch, York College of The City University of New York; George Martin, State University of New York at Albany; Seymour Schuster (moderator), Carleton College; and Gerry Segal, Brooklyn College of The City University of New York. Photograph by Stan Sherer.

What can we do about it? At the Shaping Space Conference, a panel discussion (Fig. 19-6) devoted to this question generated lively and interesting exchanges, but produced no conclusive answers. Nor did we expect any: the difficulties in effecting nationwide curricular change are great indeed. Problems cited include our decentralized educational system, the tendency of textbook publishers to continue printing familiar material for which a large market already exists, and the need for extensive in-service training for teachers. Another central problem, of course, is deciding what the content of a revitalized geometry course should be. While it will take time to solve any of these problems, we believe that the last one at least can and should be addressed now.

Our view, supported we believe by the material in this book, is that in a restructured geometry curriculum a central role should be given to the study of three-dimensional forms. We can envision, for example, a high-school course on the principles of three-dimensional structure. It would begin with a survey of significant forms, both from nature and those of human design: the polyhedral shapes such as pods and crystals, molecular and crystal structures, and trusses and other bridge constructions. The next step would be to build some of these forms, focusing on basic structural units and how they fit together. This would lead to important questions about the geometric principles on which these forms are based. Students would be led gradually to systematize and formalize their geometric knowledge. In turn, this would suggest further questions.

We also believe that because spatial geometry is one of the small group of subjects that truly comprise "the basics," it should be integrated with instruction in art, biology, chemistry, and physics. The study of three-dimensional shapes could provide students with profound understanding of topics as diverse as chemistry, the science of materials, physics, body movement, sculpture, and geography. The geometry of fundamental shapes and forms is a practical tool for the machinist; it is also an intellectual tool for anyone wanting to understand the arts and the sciences. Indeed, without fundamental spatial concepts, even the most academically successful students lack the tools for understanding the principles of structure presented in their high school science textbooks. The sense of bewilderment of students when confronting chemical theory should be dismaying to educators. Yet the molecular models which teachers and students alike find so abstract are based in large part, as we saw in Chapters 5, 9, and 10, on the geometry of tetrahedra, cubes, and octahedra. Students who have never before seen a tetrahedron or octahedron, and who have never examined a cube in detail, are quickly lost. Those who have built models and have studied them become absorbed in the problems of molecular structure.

Is there still a place for the traditional geometry curriculum? We would argue that some of the classical axioms, theorems, and constructions should be retained, not only because they are necessary for understanding (and constructing) geometric forms, but also because *they are part of our cultural heritage*. Seen in this light, however, we are led to ask: Why only classical material? Many scientists and mathematicians, including contemporary ones, have made important contributions to elementary geometry. (We point out that in many cases these contributions were not exercises isolated from other intellectual pursuits. It is neither surprising nor coincidental that polyhedra have been found to be profoundly relevant to concerns about natural philosophy, systematic logic, representational art, the design of machines, and the architecture of the heavens.)

This is a good time for the development of pilot projects at various levels, to serve as examples of what might be done. One such example is Project Synergy: An Interdisciplinary Mathematics Experience for the Middle School, developed almost a decade ago by Gerry Segal, then a mathematics teacher at Robert F. Wagner Junior High School in Manhattan. This successful curriculum, built around theories of Buckminster Fuller, takes an investigative approach to structures in space. Among the lessons:

- Building a Newspaper Tower with Triangles
- Stress and Triangulation in the Cube
- Construction of a Tetrahedron with Toothpicks and Marshmallows
- Bonding of Molecules
- The Platonic Solids and Crystallography
- Space between Atoms in a Molecular System
- Construction of a Geodesic Dome
- The Golden Rectangle, the Icosahedron, and Laban Dance Notation
- Euler's Inventory of Crossings, Areas, and Lines
- Construction of a Tetrakite

The interdisciplinary, hands-on nature of the synergy curriculum was one of the reasons for its success with children; it touched on many aspects of the lives of the people in the classroom. Surely another reason for its success is that it focuses on *space*, the very place where we live our lives.

Many other examples could of course be cited. More important, many new examples can be created. We hope that this book will inspire you to innovative curricular design, and that you will let us hear about it.

The contributors to this book have given extensive references for further study. The bibliography that follows is intended as an additional resource for students, teachers, and scholars. We regret that it is incomplete, but we hope that it will provide a helpful beginning.

Resources

A. Architecture

Abramowvitz, Anita. *People and Spaces*. New York: The Viking Press, 1979.

Albarn, Keith, Jenny Smith, Stanford Steele, and Dinah Walker. *The Language of Pattern*. London: Thames and Hudson, 1974.

Baglivo, Jenny A., and Jack E. Graver, eds. *Incidence and Symmetry in Design and Architecture*. Cambridge, England: Cambridge University Press, 1983.

Barratt, Krome. *Logic and Design, The Syntax of Art, Science and Mathematics*. Westfield, N.J.: Eastview Editions, 1980.

Blackwell, William. *Geometry in Architecture*. New York: John Wiley and Sons, 1984.

Bolker, Ethan, and Henry Crapo. "How to Brace a One Story Building." *Environment and Planning B4* (1977):125–52.

Bryan, James, and Rolf Sauer, eds. *Structures Implicit and Explicit*. Philadelphia: Graduate School of Fine Arts, University of Pennsylvania, 1973.

Burns, Jim. *Arthropods: New Design Futures*. New York: Praeger, 1971.

Collier, Graham. *Form, Space, and Vision*. Englewood Cliffs, N.J.: Prentice-Hall, 1963.

Cowan, Henry J. *Architectural Structures*. Amsterdam: Elsevier, 1971.

———. *The Master Builders*. New York: Wiley Interscience, 1977.

Coy, P. H. *Structural Analysis of Unistrut Space Frame Roofs*. Ann Arbor: University of Michigan Press, 1955.

Cullen, Gordon. *Townscape*. New York: Van Nostrand Reinhold, 1961.

Doo, Peter C., and Alice Gray Read. *Architecture and Visual Perception*. Philadelphia: University of Pennsylvania Press, and Cambridge, Mass.: MIT Press, 1983.

Doxiadis, C. A. *Architectural Space in Ancient Greece*. Cambridge, Mass.: MIT Press, 1972.

Drabkin, David L. *Fundamental Structure: Nature's Architecture*. Philadelphia: University of Pennsylvania Press, 1975.

Engel, Heinrich. *Structure Systems*. New York: Praeger, 1968.

Faber, Colin. *Candels: The Shell Builder*. New York: Von Nostrand Reinhold, 1963.

Gomez-Alberto, Perez. *Architecture and the Crisis of Modern Science*. Cambridge, Mass.: MIT Press, 1984.

James, Albert. *Structures and Forces Stages 1 and 2*. MacDonald Education, 1972.

———. "Polyhedric Architecture." *Architectural Association Quarterly 4,* no. 3 (July–September 1972).

———. "The Geometry of My Polyhedral Sculpture." *Leonardo: International Journal of the Contemporary Artist 10* (1977):183–87.

———. "The Cube and the Dodecahedron in My Polyhedron Architecture." *Leonardo: International Journal of the Contemporary Artist 13* (1980):272–75.

Kepes, Gyorgy. *The Module, Proportion, Symmetry and Rhythm*. New York: George Braziller, 1966.

Kessler, Christel. *The Carved Masonry Domes of Medieval Cairo*. Cairo: The American University in Cairo Press, 1976.

Lynch, Kevin. *The Image of the City*. Cambridge, Mass.: MIT Press, 1960.

Nervi, Pier Luigi. *Aesthetics and Technology in Building*. Cambridge, Mass.: Harvard University Press, 1965.

Safdie, Moshe. *For Everyone a Garden*. Cambridge, Mass.: MIT Press, 1974.

Tarnai, Tibor. "Spherical Grids of Triangular Network." *Acta Technica Hungarian Academy of Sciences* [Budapest] (1974).

———. *Spherical Circle-Packing in Nature: Practice and Theory. Structural Topology Bulletin* [Montreal] *9* (1984).

Tompkins, Peter. *Secrets of the Great Pyramid.* New York: Harper Colophon, 1971.

Torroja, Eduardo. *Philosophy of Structures.* Berkeley: University of California Press, 1958.

Wong, Wucius. *Principles of Three-Dimensional Design.* New York: Van Nostrand Reinhold, 1968.

B. Art

Arnheim, Rudolf. *Visual Thinking.* Berkeley: University of California Press, 1969.

Audsley, W., and C. Audsley. *Designs and Patterns from Historic Ornament.* New York: Dover Publications, 1968.

Bager, Bertel. *Nature as Designer.* New York: Reinhold Publishing, 1906.

Bourgoin, J. *Arabic Geometrical Patterns and Design.* New York: Dover Publications, 1973.

Brisson, David, ed. *Hypergraphics: Visualizing Complex Relationship in Art, Science and Technology.* American Association for the Advancement of Science Selected Symposia Series. Boulder, Colo.: Westview Press, 1978.

Edgerton, Samuel Y. *The Renaissance Rediscovery of Linear Perspective.* New York: Basic Books, 1975.

Ernst, Bruno. *The Magic Mirror of M. C. Escher.* New York: Ballantine Books, 1976.

Escher, Maurits C. *The Graphic Work of M. C. Escher.* New York: Meredith Press, 1967.

Gombrich, E. H. *A Sense of Order.* Ithaca N.Y.: Cornell University Press, 1979.

Gregory, R. L., and E. H. Gombrich. *Illusion in Art and Nature.* London: Charles Scribner's Sons, 1973.

Hill, Anthony, ed. *DATA: Directions in Art, Theory, and Aesthetics.* Boston: Faber and Faber, 1968.

Kim, Scott. *Inversions.* New York: Byte Books, 1981.

Locher, J. L., et al. *Escher.* London: Thames and Hudson, 1982.

Malina, Frank J., ed. *Visual Art, Mathematics and Computers.* Oxford: Pergamon Press, 1979.

Munari, Bruno. *Discovery of the Square.* New York: George Wittenborn, 1965.

Munsell, A. H. *A Color Notation.* 10th ed. Baltimore: Munsell Color Company, 1954.

Murchie, Guy. *Music of the Stars (Spheres).* New York: Dover, 1961.

Nervi, Pier Luigi. *The Works of Pier Luigi Nervi.* New York: Praeger, 1957.

New York Gallery of Modern Art. *Salvador Dali.* New York: Foundation of Modern Art, 1965.

Panofsky, Erwin, and Fritz Saxl. *Dürer's Melencolia I.* New York: B. G. Teubner, 1923.

Portmann, Adolf. *Animal Form and Pattern.* Boston: Faber, 1952.

Spies, Werner. *Albers.* New York: Abrams, 1970.

Stevens, Peter S. *Handbook of Regular Patterns: An Introduction to Symmetry in Two Dimensions.* Cambridge, Mass.: MIT Press, 1980.

———. *String Art Encyclopedia.* New York: Sterling, 1976.

C. Geometry

Abbott, Edwin. *Flatland.* New York: Dover, 1983.

Abelson, Harold, and Andrea di Sessa. *Turtle Geometry.* Cambridge, Mass.: MIT Press, 1980.

Banchoff, Thomas, and John Werner. *Linear Algebra through Geometry.* New York: Springer-Verlag, 1983.

Boltianskii, V. G. *Hilbert's Third Problem.* Washington, D.C.: V. H. Winston and Sons, 1978.

Burns, Gerald, and A. M. Glazer, *Space Groups for Solid State Scientists.* New York: Academic Press, 1978.

Burt, M. *Spatial Arrangement and Polyhedra with Curved Surfaces and Their Architectural Applications.* Haifa: Technion.

Caravelli, Vito. *Le traité des icosoèdres.* Paris: A. Blanchard, 1959.

Coxeter, H. S. M. *Twelve Geometric Essays.* Carbondale: Southern Illinois University Press, 1968.

———. *Introduction to Geometry.* New York: John Wiley and Sons, 1969.

———. *Regular Polytopes.* 3rd ed. New York: Dover, 1973.

———. *Regular Complex Polytopes.* New York: Cambridge University Press, 1974.

Coxeter, H. S. M., P. Du Val, H. T. Flather, and J. F. Petrie. *The Fifty-Nine Icosahedra.* New York: Springer-Verlag, 1982.

Coxeter, H. S. M., and S. L. Greitzer. *Geometry Revisited.* Washington, D.C.: Mathematical Association of America, 1967.

Critchlow, Keith. *Order in Space.* New York: The Viking Press, 1970.

Cundy, H. Martyn. "Deltahedra." *The Mathematical Gazette 36* (1952):263–66.

Cundy, H. M., and A. P. Rollet. *Mathematical Models.* 2nd ed. Oxford: Oxford University Press, 1961. Reprinted by Tarquin Publications, Stradbroke, Diss, Norfolk, England.

Davis, C., B. Grünbaum, and F. A. Sherk, eds. *The Coxeter Festschrift: The Geometric Vein.* New York: Springer-Verlag, 1981.

Edmondson, Amy C. *A Fuller Explanation: The Synergetic Geometry of R. Buckminister Fuller.* Design Science Collection. Boston: Birkhäuser, Pro Scientia Viva, 1987.

Ehrenfeucht, Aniela. *The Cube Made Interesting.* New York: Pergamon Press, 1964.

Fejes Toth, L. *Regular Figures.* New York: Macmillan, 1964.

———. "What the Bees Know and What They Do Not Know." *Bulletin of the American Mathematical Society 20* (1964):468–81.

Fuller, R. Buckminster. *Synergetics I.* New York: Macmillan, 1975.

———. *Synergetics II.* New York: Macmillan, 1979.

Graziotti, Ugo Adriano. *Polyhedra: The Realm of Geometric Beauty.* San Francisco, 1962.

Grünbaum, B., and G. C. Shephard. *Tilings and Patterns.* San Francisco: W. H. Freeman, 1986.

———. "Tilings with Congruent Tiles." *Bulletin of the American Mathematical Society 3* (1980):951–73.

Gyhka, Matila. *The Geometry of Art and Life.* New York: Dover, 1977.

Hilbert, D., and S. Cohn-Vossen. *Geometry and the Imagination.* New York: Chelsea, 1952.

Holden, Alan. *Space, Shapes, and Symmetry.* New York: Columbia University Press, 1971.

———. *Orderly Tangles.* New York: Columbia University Press, 1983.

Hope, C. "The Nets of the Regular Star-faced and Star-pointed Polyhedra." *The Mathematical Gazette 35* (1951):8–11.

Ivins, William. *Art and Geometry.* New York: Dover, 1964.

Kenner, Hugh. *Geometric Math and How to Use It.* Los Angeles: University of California Press, 1976.

Klarner, David A. *The Mathematical Gardener.* Boston: Prindle, Weber, and Schmidt, 1981.

Klein, Felix. *Lectures on the Icosahedron.* London: Kegan Paul, 1913.

———. *The Icosahedron and the Solution of Equations of the Fifth Degree.* New York: Dover, 1956.

Lalvani, Haresh. *Transpolyhedra: Dual Transformations by Explosion-Implosion.* New York: Haresh Lalvani, 1977.

Lawler, Robert. *Sacred Geometry, Philosophy and Practice.* London: Thames and Hudson, 1982.

Lewis, F. T. "A Geometric Accounting for Diverse Shapes of 14-hedral Cells: The Transition from Dodecahedra to Tetrakaidecahedra." *American Journal of Botany 30* (1943):74–81.

Lindgren, H. *Geometric Dissections.* New York: Dover, 1972.

Lines, L. *Solid Geometry.* London: Macmillan, 1935.

Loeb, Arthur L. *Space Structures: Their Harmony and Counterpoint.* Reading, Mass.: Addison-Wesley, Advanced Book Program, 1976.

Lyusternik, L. A. *Convex Figures and Polyhedra.* Boston: D. C. Heath, 1966.

Mandelbrot, B. *The Fractal Geometry of Nature.* San Francisco: W. H. Freeman, 1983.

March, Lionel, and Philip Steadman. *The Geometry of Environment.* Cambridge, Mass.: MIT Press, 1971.

Miyazaki, Koji. *Shapes and Space.* Nade, Kobe, Japan: Kobe University.

———. *An Adventure in Multidimensional Space: The Art and Geometry of Polygons, Polyhedra, and Polytopes.* New York: John Wiley and Sons, 1986.

Pearce, Peter. *Structure in Nature Is a Strategy for Design.* Cambridge, Mass.: MIT Press, 1978.

Pearce, Peter, and Susan Pearce. *Polyhedra Primer.* New York: Van Nostrand Reinhold, 1978.

Pedoe, Daniel. *Geometry and the Visual Arts.* New York: Dover, 1983.

Preparata, F. P., and M. I. Shamos. *Computational Geometry.* New York: Springer-Verlag, 1985.

Pugh, Anthony. *An Introduction to Tensegrity.* Berkeley: University of California Press, 1976.

———. *Polyhedra: A Visual Approach.* Berkeley: University of California Press, 1976.

Senechal, M. "Which Tetrahedra Fill Space?" *Mathematics Magazine 54* (1981):227–43.

Steinhaus, H. *Mathematical Snapshots.* New York: Oxford University Press, 1969.

Stewart, B. M. *Adventures among the Toroids: A Study of Orientable Polyhedra with Regular Faces.* 2nd ed. Okemos, Mich. B. M. Stewart, 1980.

Structural Topology, an interdisciplinary journal published in English and in French for mathematicians, architects, and engineers. The journal focuses on static rigidity of structures, polyhedra, and space-filling (juxtaposition). The journal began in 1979 and is published a few times each year. *La Revue Topologie Structurale,* UQAM, C.P. 888, Succ. A, Montreal, Quebec, Canada H3C 3P8.

Stuart, Duncan. *Polyhedral and Mosaic Transformations.* Chapel Hill: University of North Carolina Press, 1963.

Thomson, Sir William (Lord Kelvin). "On the Division of Space with Minimum Partitional Area." *London, Edinburgh, and Dublin Philosophical Magazine and Journal of Science,* ser. 5, *24* (1887):503–14.

———. "On the Homogeneous Divison of Space." *Proceedings of the Royal Society.* [London] *55* (1894):1–16.

Wachman, A. M., M. Burt, and Kleinmann. *Infinite Polyhedra.* Haifa: Technion, 1974.

Wells, A. F. *Three-Dimensional Nets and Polyhedra.* Wiley Monographs in Crystallography. New York: John Wiley, 1977.

Williams, Robert. *The Geometric Foundation of Natural Structure.* New York: Dover 1979.

D. Instructional and Recreational Materials

Allyn and Bacon. *Molecular Model Set for Organic Chemistry.*

Ball, W. W. R., and H. S. M. Coxeter, *Mathematical Recreations and Essays.* New York: Dover, 1986.

Brown, S., and M. Walter, *The Art of Problem Posing.* Philadelphia: Franklin Institute Press, 1983. Republished by Lawrence Erlbaum Associates, Inc., Hillsdale, N.J.

Coffin, Stewart T. "Wooden Puzzles Easy to Make, but Tough to Solve." *Fine Woodworking* (Nov.–Dec. 1984).

D-Stix plastic rods and multipronged connectors. Spokane, Wash.: Geodestix.

Fantastix. Integrity Designs, Inc., Route 123, Another Place, Greenville, N.H. 03098.

Goldberg, Steven A. *Pholdit.* Hayward, Calif.: Activity Resources, 1977.

Haughton, E., and A. L. Loeb. "Symmetry: A Case History of a Program." *Journal of Research in Science Teaching 2* (1964):132.

Hunt, Leslie. *25 Kites That Fly.* New York: Dover, 1971.

Jenkins, Gerald, and Anne Wild. *Make Shapes 1, 2, 3,* and *Mathematical Curiosities 1, 2, 3.* Norfolk, England: Tarquin Publications. Distributed by Parkwest Publications, P.O. Caller Box A-10, Cathedral Station, New York, N.Y. 10025.

Johnson, Donavan A. *Mathmagic with Flexagons.* Hayward, Calif.: Activity Resources, 1974.

Laycock, Mary. *Bucky for Beginners.* Hayward, Calif. Activity Resources, 1984.

———. *Straw Polyhedra.* Mountain View, Calif.: Creative Publications.

Loeb, A. L. "Remarks on Some Elementary Volume Relations between Familiar Solids." *Mathematics Teacher 50* (58) (1965):417.

Loeb, A. L., and E. Haughton. "The Programmed Use of Physical Models." *Journal of Programmed Instruction 3* (1965):9–18.

Loeb, A. L., and G. M. Pearsall. "Moduledra Crystal Models." *American Journal of Physics 31* (1963):190–96.

McGowan, William E. "A Recursive Approach to Construction of the Deltahedra." *Mathematics Teacher* (March 1978):204–10.

Negahban, Bahman, and Ezat O. Negahban. Polyhedral Lampshades. 105 Glen, Ardmore, Penn. 19003.

O'Daffer, P. G., and S. P. Clemens. *Geometry: An Investigative Approach.* Reading, Mass.: Addison-Wesley, 1976.

Pearce, Peter, and Susan Pearce. *Experiments in Form: A Foundation Course in Three-Dimensional Design.* New York: Van Nostrand Reinhold, 1980.

Pedersen, Jean J., and A. Kent. "Geometric Playthings." Price, Stern and Sloane, 410 North La Cienega Boulevard, Los Angeles, Calif. 90048.

Rhombics Rhomblocks. Rhombics, 36 Pleasant Street, Watertown, Mass. 12172.

Row, Sundara T. *Geometrical Exercises in Paper Folding.* New York: Dover Publications.

Sanderson, R. T. *Teaching Chemistry with Models.* Princeton, N.J.: Van Nostrand, 1962.

Saunders, Kenneth. *Hexagrams.* New York: Parkwest Publications.

Segal, Gerry. *Synergy Curriculum.* New York: Board of Education of the City of New York, 1979.

Smart, Margret A., and Mary Laycock. *Create a Cube.* Hayward, Calif.: Activity Resources, 1985.

Stonerod, David. *Puzzles in Space.* Palo Alto, Calif.: Stokes, 1982.

Swienciki, Lawrence W. *Quadraflex Model Book.* San Jose, Calif.: A. R. Davis, 1976.

Symmetrics polystyrene model kits. Symmetrics, Inc., 9 Maple Street, Atkinson, N.H. 03811.

Termes, Dick A. TotalPhotos and Termespheres. Spherical Art, Inc., 2650 Jackson Boulevard, Rapid City, South Dakota 57702.

Walter, Marion. *The Mirror Puzzle Book.* Norfolk, England: Tarquin Publications, 1985. Distributed by Parkwest Publications, P.O. Caller Box A-10, Cathedral Station, New York, N.Y. 10015.

Wenninger, M. J. *Polyhedron Models.* New York: Cambridge University Press, 1971.

———. *Polyhedron Models for the Classroom.* 2nd ed. Reston, Va: National Council of Teachers of Mathematics, 1975.

———. *Spherical Models.* New York: Cambridge University Press, 1979.

———. *Dual Models.* New York: Cambridge University Press, 1983.

———. "Some Interesting Octahedral Compounds." *Mathematical Gazette 52,* no. 357 (1968):10.

Winter, John. *String Sculpture*. Palo Alto, Calif.: Creative Publications, 1972.

Wilson, Forrest. *Architecture: A Book of Projects for Young Adults*. New York: Van Nostrand Reinhold, 1968.

Young, Grace Chisolm. *Beginners' Book of Geometry*. New York: Chelsea Publishing Co., 1970.

Zeier, Franz. *Paper Constructions*. New York: Charles Scribner's Sons, 1974.

E. Science

Bacon, G. E. *The Architecture of Solids*. London: Taylor and Francis, 1981.

Bok, S. T. "On the Shape of Froth Chambers." *Proceedings, Akademie van Wetenschappen* [Amsterdam] *43* (1940):1180–90.

Boys, C. V. *Soap Bubbles and the Forces Which Mould Them*. Garden City, N.Y.: Doubleday-Anchor Books, 1959.

Bragg, Sir Lawrence. *Crystal Structures of Minerals*. Ithaca, N.Y.: Cornell University Press, 1965.

Bragg, W. L. *The Crystalline State*. New York: Macmillan, 1934.

Buerger, M. J. *Introduction to Crystal Geometry*. New York: McGraw-Hill, 1971.

Coxeter, H. S. M. "Close Packing and Froth." *Illinois Journal of Mathematics 2* (1958):746–58.

———. "Virus Macromolecules and Geodesic Domes." In Butcher, J. C., ed. *A Spectrum of Mathematics Essays Presented to H. G. Forder*. Oxford: Oxford University Press, 1971.

Erickson, R. O. "Tubular Packing of Spheres in Biological Fine Structure." *Science 181* (1973):705–16.

———. "The Geometry of Phyllotaxis." In J. E. Dale and F. L. Milthorp, eds., *The Growth and Functioning of Leaves*. New York: Cambridge University Press, 1982.

Erickson, R. O., and W. F. Harris. "Tubular Packings of Subunits; Continuous Contraction and Contraction by Passage of Edge Dislocations." *Proceedings of the International Symposium on Mathematical Topics in Biology: Kyoto* (1978):155–63.

Feininger, Andreas. *The Anatomy of Nature*. New York: Crown, 1956.

Frank, F. C., and J. S. Kasper. "Complex Alloy Structures Regarded as Sphere Packings." *Acta Crystallographica 11*(1957):184–90; *12* (1958): 483–99.

Gillespie, Ronald J. *Molecular Geometry*. London: Van Nostrand Reinhold, 1972.

Henderson, L. J. *The Order of Nature*. Cambridge, Mass.: Harvard University Press, 1925.

Holden, Alan. *The Nature of Solids*. New York: Columbia University Press, 1965.

———. *Bonds between Atoms*. New York: Oxford University Press, 1971.

Holden, Alan, and Phylis Singer. *Crystals and Crystal Growing*. Garden City, N.Y.: Anchor Books, 1960.

Harlow, William. *Patterns of Life*. New York: Dover, 1974.

Klug, A. "Architectural Design of Spherical Viruses." *Nature 303* (1983):378–79.

Thomson, William (Lord Kelvin). "The Molecular Constitution of Matter." *Mathematical and Physical Papers 3* (1910):395–427.

Lewis, P. T. "The Analogous Shapes of Cells and Bubbles." *Proceedings of the American Academy of Arts and Sciences 77* (1949):147–86.

Loeb, A. L. "A Systematic Survey of Cubic Crystal Structures." *Journal of Solid State Chemistry 1* (1970):237–67.

Needham, Joseph. *Order and Life*. Cambridge, England: Cambridge Unviersity Press, 1936.

Parthe, Erwin. *Crystal Chemistry of Tetrahedral Structures*. New York: Gordon and Breach, 1964.

Reeks, M. *Hints for Crystal Drawing*. London: Longmans, 1908.

Rhodin, Johannes A. G. *An Atlas of Ultrastructure*. Philadelphia: W. B. Saunders, 1963.

Rivier, N. "On the Structure of Random Tissues or Froths, and Their Evolutions." *Philosophical Magazine,* ser. 8, *47* (1983):L45–L49.

Rose, Gilbert. *The Power of Form*. New York: International Universities Press, 1980.

Ryschkewitsch, G. *Chemical Bonding and the Geometry of Molecules*. New York: Van Nostrand Reinhold, 1963.

Smith, Cyril Stanley. "Grain Shapes and Other Metallurgical Applications of Topology." *Metal Interfaces*. Cleveland: American Society for Metals, 1952; Ann Arbor, Mich.: University Microfilms OP 13754, 65–113.

Stevens, Peter S. *Patterns in Nature*. Boston: Little, Brown, 1974.

Strache, Wolf. *Forms and Patterns in Nature*. New York: Pantheon Press, 1956.

Thompson, D'Arcy Wentworth. *On Growth and Form*. London: Cambridge University Press, 1942, 1961.

Wells, A. F. *The Third Dimension in Chemistry*. Oxford: The Clarendon Press, 1956.

———. *Models in Structural Inorganic Chemistry*. New York: Oxford University Press, 1970.

Whyte, L. L. *Aspects of Form*. London: Humphries, 1968.

F. Symmetry

Bernal, Ivan, Walter C. Hamilton, and John S. Ricci. *Symmetry*. San Francisco: W. H. Freeman, 1972.

Critchlow, Keith. *Islamic Patterns*. New York: Schocken, 1976.

Dye, Daniel Sheets. *Chinese Lattice Designs*. New York: Dover, 1974

Edwards, Edward B. *Pattern and Design with Dynamic Symmetry*. New York: Dover, 1967.

El-Said, Issam, and Ayse Parman. *Geometric Concepts in Islamic Art*. London: World of Islam, 1976.

Emmer, M., ed. *M. C. Escher: Art and Science*. Amsterdam: North Holland, 1986.

Hargittai, István, ed. *Symmetry: Unifying Human Understanding*. New York: Pergamon Press, 1986.

Hargittai, István, and Magdolna Hargittai. *Symmetry through the Eyes of a Chemist*. Weinheim: VCH Verlagsgesellschaft, 1986.

Jones, Owen. *The Grammar of Ornament*. New York: Van Nostrand Reinhold, 1968.

Loeb, A. L. *Color and Symmetry*. New York: Wiley, 1971; New York: Krieger, 1978.

MacGillavry, Caroline. *Fantasy and Symmetry*. New York: Abrams, 1976.

Rosen, Joe. *Symmetry Discovered*. New York: Cambridge University Press, 1975.

Senechal, Marjorie, and George Fleck, eds. *Patterns of Symmetry*. Amherst: University of Massachusetts Press, 1977.

Weyl, Hermann. *Symmetry*. Princeton, N. J.: Princeton University Press, 1952.

Wigner, E. P. *Symmetries and Reflections*. Woodbridge, Conn.: Oxbow Press, 1979.

Yale, Paul. *Geometry and Symmetry*. San Francisco: Holden-Day, 1968.

Contributors*

ASH, PETER. Department of Mathematics and Computer Science, Saint Joseph's University, Philadelphia, PA 19131.

BANCHOFF, THOMAS F. Department of Mathematics, Brown University, Providence, R.I. 02912.

BARACS, JANOS. 21 Springfield, Westmount, Quebec H3Y 2K9, Canada.

BAUERMEISTER, MARY. Hedwigshöhe 31, 5064 Forsbach, Federal Republic of Germany.

BAYER, MARGARET. Department of Mathematics, Northeastern University, Boston, MA 02115.

BOLKER, ETHAN. Department of Mathematics and Computer Science, University of Massachusetts at Boston, Boston, MA 02125.

BRADLEY, MORTON C., Jr. 20 Maple Street, Arlington, MA 02174.

BRISSON, HARRIET. Department of Art, Rhode Island College, Providence, R.I. 02908.

BURNS, LEE. Department of Art, Smith College, Northampton, MA 01063.

CASPAR, DONALD L. D. Rosenstiel Basic Medical Sciences Research Center, Brandeis University, Waltham, MA 02254.

CHIEH, CHUNG. Department of Chemistry, University of Waterloo, Waterloo, Ontario N2L 3G1, Canada.

COXETER, H. S. M. Department of Mathematics, University of Toronto, Toronto, Ontario M5S 1A1, Canada.

CRAPO, HENRY. Bâtiment 24, Institut National de Recherche en Informatique et en Automatique, BR 105, Le Chesnay, 78153 Cedex, France.

DUNHAM, DOUGLAS. Department of Computer Science, University of Minnesota, Duluth, MN 55812.

ERICKSON, RALPH O. Department of Biology, University of Pennsylvania, Philadelphia, PA 19104.

FLECK, GEORGE. Department of Chemistry, Smith College, Northampton, MA 01063.

FREDENTHAL, ROBINSON. 3434 Sansom Street, Philadelphia, PA 19104.

GRITZMANN, PETER. Universität Siegen, Hälderinstrasse 3, Postfach 10 12 40, D-5900 Siegen, Federal Republic of Germany.

GRÜNBAUM, BRANKO. Department of Mathematics, University of Washington, Seattle, WA 98195.

HARGITTAI, ISTVÁN. Hungarian Academy of Sciences, Budapest VIII, Pushkin U. 11–13, pf. 117, H-1431, Hungary

HARGITTAI, MAGDOLNA. Hungarian Academy of Sciences, Budapest VIII, Pushkin U. 11–13, pf. 117, H-1431, Hungary

HARKER, DAVID. Medical Foundation of Buffalo, 73 High Street, Buffalo, NY 14203.

HAUER, ERWIN. 180 York Street, New Haven, CT 06520.

HECKER, ZVI. 22 David Yellin Street, Tel Aviv, Israel.

LOEB, ARTHUR L. Department of Visual and Environmental Studies, Harvard University, Cambridge, MA 02138.

MALKEVITCH, JOSEPH. Department of Mathematics, York College, City University of New York, Jamaica, NY 11451.

MONSON, BARRY. University of New Brunswick, P. O. Box 4400, Fredericton, New Brunswick E3B 5A3, Canada.

NEGAHBAN, BAHMAN. 60 Sheldon Terrace #2, New Haven, CT, 06511.

PEDERSEN, JEAN. Department of Mathematics, University of Santa Clara, Santa Clara, CA 95053.

PERRY, CHARLES O. Shorehaven Road, Norwalk, CT 06850.

REAY, JOHN. Department of Mathematics, Western Washington University, Bellingham, WA 98225.

* Including artists who contributed art work to Shaping Space Conference.

SCHATTSCHNEIDER, DORIS. Department of Mathematics, Moravian College, Bethlehem, PA 18018.

SEGAL, GERRY. Center for Academic Computing, Brooklyn College, 1210 Plaza, Brooklyn, NY 11210.

SENECHAL, MARJORIE. Department of Mathematics, Smith College, Northampton, MA 01063.

SCHULTE, EGON. Mathematisches Institut, Universität Dortmund, D-4600 Dortmund, Federal Republic of Germany.

SHEPHARD, G. C. Department of Mathematics, University of East Anglia, Norwich NR4 7TJ, England.

SHERER, STAN. 349 South Street, Northampton, MA 01060.

TOUSSAINT, GODFRIED. School of Computer Science, 805 Sherbrooke Street West, McGill University, Montreal, Quebec H3A 2K6, Canada.

WALTER, MARION. Department of Mathematics, University of Oregon, Eugene, OR 97403.

WENNINGER, MAGNUS. St. John's Abbey, Collegeville, MN 56321.

WHITELEY, WALTER. Department of Mathematics, Champlain Regional College, 900 Riverside Drive, St. Lambert, Quebec J4P-3P2, Canada.

WILLS, J. M. Lehrstuhl für Mathematik II, Universität Siegen, Hölderlinstrasse 3, D-5900 Siegen, Federal Republic of Germany.

Index